T0223285

Lecture Notes in Computer Science 1464
Edited by G. Goos, J. Hartmanis and J. van Leeuwen

Springer

Berlin
Heidelberg
New York
Barcelona
Budapest
Hong Kong
London
Milan
Paris
Singapore
Tokyo

Horace H.S. Ip Arnold W.M. Smeulders (Eds.)

Multimedia Information Analysis and Retrieval

IAPR International Workshop, MINAR'98
Hong Kong, China, August 13-14, 1998
Proceedings

Springer

Series Editors

Gerhard Goos, Karlsruhe University, Germany
Juris Hartmanis, Cornell University, NY, USA
Jan van Leeuwen, Utrecht University, The Netherlands

Volume Editors

Horace H.S. Ip
City University of Hong Kong, Department of Computer Science
Tat Chee Avenue, Kowloon, Hong Kong, China
E-mail: cship@cityu.edu.hk

Arnold W.M. Smeulders
University of Amsterdam, Intelligent Sensory Information Systems
Kruislaan 403, 1098 SJ Amsterdam, The Netherlands
E-mail: smeulders@wins.uva.nl

Cataloging-in-Publication data applied for

Die Deutsche Bibliothek - CIP-Einheitsaufnahme

Multimedia information analysis and retrieval : IAPR international workshop ; proceedings / MINAR '98, Hong Kong, China, August 13 - 14, 1998. Horace H. S. Ip ; Arnold W. M. Smeulders (ed.). - Berlin ; Heidelberg ; New York ; Barcelona ; Budapest ; Hong Kong ; London ; Milan ; Paris ; Singapore ; Tokyo : Springer, 1998
(Lecture notes in computer science ; Vol. 1464)
ISBN 3-540-64826-7

CR Subject Classification (1991): I.4, I.5, H.5.1, H.4.3, H.5.4, H.3

ISSN 0302-9743
ISBN 3-540-64826-7 Springer-Verlag Berlin Heidelberg New York

Typesetting: Camera-ready by author
SPIN 10638431 06/3142 – 5 4 3 2 1 0 Printed on acid-free paper

Preface

The advent of Internet coinciding with the general use of digital sensors have given the field of computer vision and pattern recognition a boost in its application. Whilst the digital camera captures images for a variety of professional and consumer markets, the Internet spreads the use of pictures in digital form over the world. In short, the Internet has turned visual rather than textual. With the new tools come new usage, new audiences and, hence, new challenges for computer vision and recognition. Such challenges come in three different forms: the much enlarged dataset, new application domains, and the needs to cater to naive users.

Pictorial databases of 500 000 to over 1 000 000 images are quite common in Internet search. This imposes great demands on computational efficiency and issues relating to indexing compressed pictures and videos become relevant. Moreover, the requirement of robustness against varying circumstances in recording and interpretation in image browsing via the Internet is also larger than previously seen.

Before the Internet age, typical application domains of picture analysis were usually complicated tasks in narrow domains (e.g. medical image analysis, bank cheque verification, pharmaceutical product classification) or quick tasks in simplified environments (e.g. industrial inspection, printed character recognition). In the Internet age, the domain shifts to more general pictures from consumer video and large archives, where we have very little control over the imaging conditions, and where prior domain knowledge is not easily applied to simplify the search and retrieval process. This poses interesting new problems each time when conditions of illumination and query purpose alter. A general framework which addresses all these problems is not yet in sight.

Apart from new domains, the Internet and digital sensor revolution also generates new questions for vision research. Searching a database or the Internet requires similarity measures over very large numbers of pictures. Also, interactive browsing of an unknown query where the emphasis is on perceptual similarity is a relative unknown topic in computer vision.

Sensing the importance and the needs of new techniques for browsing large image and multimedia database, the International Association of Pattern Recognition (IAPR) established a new technical committee on Multimedia Systems (TC12) in 1995. This technical committee directs its attention to the vast number of scientific questions related to multimedia content analysis and retrieval.

In addition, TC12 also initiated the IAPR International Workshop on Multimedia Information Analysis and Retrieval (MINAR'98) which aims to bring together researchers who are working in the field of interests of the TC.

This volume is devoted to major research issues in content-based image and video search, and contains papers presented at MINAR'98 held in Hong Kong. Among these papers are topics on the exploitation of invariant properties, colour or geometric, of images for robust image and video retrieval, fusion of pictorial with other media such as text, image indexing and retrieval in compressed domain, pictorial query languages, video segmentation by content, as well as efficient storage organisation for multimedia data.

We would like to thank the following who served on the programme committee of MINAR'98: S.-K. Chang, A. del-Bimbo, W. Grosky, R. Kasturi, T. Kato, C. Leung, S.D. Ma, D. Petkovic, H. Samet, S. Smoliar, R. Srihara, H. Tagare and J.K. Wu. We also thank Ramesh Jain who gave the keynote talk on presence technology.

We are also grateful to Charmaine Yeung for taking care of the logistics of paper collection and the workshop organisation; T.C. Pong, Irwin King and Ken Law for publication, publicity and the local arrangements.

MINAR'98 was organised by IAPR TC12 and co-sponsored by the IEEE (Hong Kong Section) Computer Chapter.

Horace H.S. Ip
Department of Computer Science
City University of Hong Kong
Hong Kong

Arnold W.M. Smeulders
Intelligent Sensory Information Systems
University of Amsterdam
The Netherlands

June 1998

Contents

Invited Talk

Image Retrieval

Video Retrieval

Invited Talk

Image Analysis

Video Segmentation and Spatial Query

Indexing and Storage

Posters

Presence Technology: Accessing Live Multimedia Information Systems

Ramesh Jain

PRAJA inc. and University of California, San Diego

This paper introduces the *Presence Technology* (PT) - a technology targeted to the needs of people who want to be part of a remote, live environment. Presence Technology blends component technologies like computer vision, signal understanding, heterogeneous sensor fusion, live-media delivery, telepresence and multimedia information systems into a novel set of functionality that enables the user to perceive, move around, enquire about, and interact with the remote, live environment through her reception and control devices. PT creates the opportunity to perform different tasks: watch an event, tour and explore a location, meet and communicate with others, monitor the environment for a potential situation, perform a query on the perceived objects and events, and recreate past observations. Technically, the framework (Figure 1) offers computer-mediated access to multi-sensory information in an environment, integrates the sensory information into a situation model of the environment, and delivers, at the user's request, the relevant part of the assimilated information through a multimodal interface.

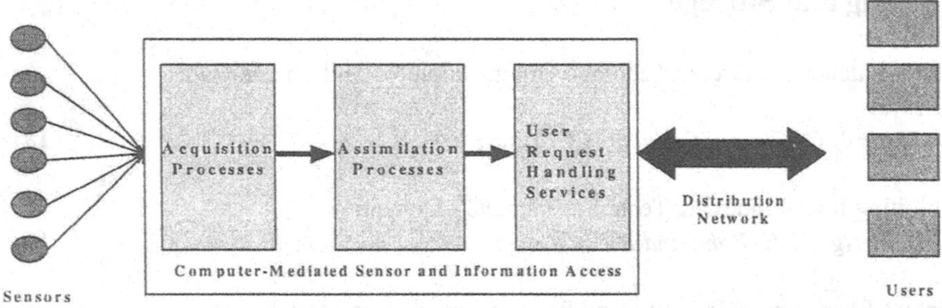

This framework departs from all previous architectures for multimedia content delivery or retrieval systems in two important ways. First, it does not just acquire and passively route a sensor content to a user (although it can be made to do so) like video streamers or web cameras. It integrates all sensor inputs into a composite model of the live environment. This model, called the *environment model* (EM), acts like a short-term database, and maintains the spatiotemporal state of the complete environment, as observed from all sensors taken together. By virtue of this integration, the EM holds a situationally complete view of the observed space. We refer to the EM' s view as the *Gestalt View* of the environment. As we will detail in a later section, maintenance of this Gestalt View provides an additional benefit that it can be a exported to a visualization tool at the client, producing a richer, 4-dimensional (space and time) "model view" of the environment for a user to move around it. The second point of departure is that in a Presence system, the user not only perceives sensory inputs, but actively interacts with it. This interaction, whether effected by an explicit query for more contextual information about an observed object, a request to track an object in the environment, or a notification request when any observed object enters a user-designated region, transcends any query operations found in current state of the art multimedia information retrieval systems. Indeed, we contend that *content-based interactivity* on live, dynamic objects is the next generation of capabilities beyond the content-based query operations on stored, mono-stream media objects offered by today's systems. Hence we believe that the ability to *perform an action on the remote environment* by a client-side action is a significant aspect of the Presence Technology. Albeit somewhat futuristic, an example of such remote action can be a remote distributed racing game, where the racing cars are controlled by individuals from their terminals or TV monitors. It is crucial to have the Gestalt View of the environment to know the positions of other vehicles and tracks at every instant.

We must emphasize that although throughout this paper we refer to live sensors as the primary source of dynamic information in the system, we treat the term "sensor" more generally. We include non-live information sources such as user annotations and database results to be special, controllable transducers that provide alphanumeric information to a Presence System.

PT is an extension of the Multiple Perspective Interactive Video project at the Visual Computing Laboratory, University of California, San Diego. In this paper we will present results from PRAJA Presence system implemented bring an early version of PT for different application. An interface for an art gallery application is shown below.

We will present a demo of this system to explain different technical components of the system.

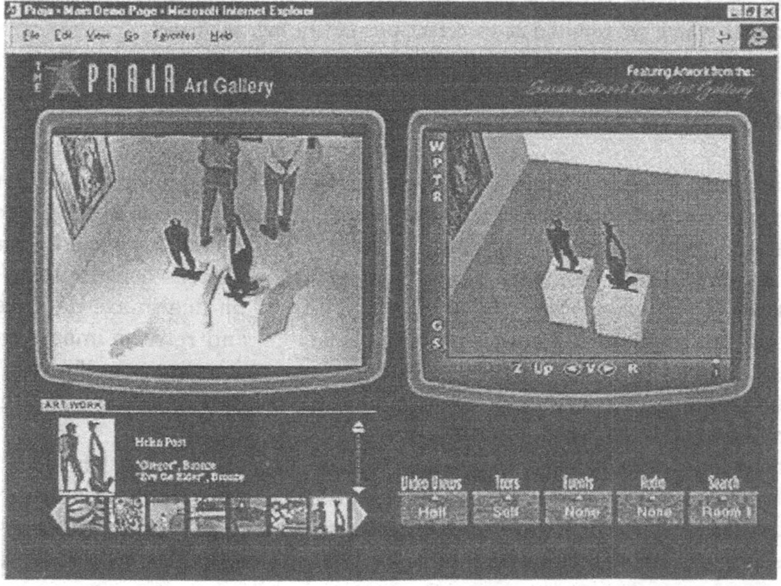

Content-Based Image Database Retrieval Using Variances of Gray Level Spatial Dependencies

Selim Aksoy and Robert M. Haralick

Intelligent Systems Laboratory
Department of Electrical Engineering
University of Washington
Seattle, WA 98195-2500
{aksoy,haralick}@isl.ee.washington.edu
http://isl.ee.washington.edu

Abstract. In this paper, we discuss how we use variances of gray level spatial dependencies as textural features to retrieve images having some section in them that is like the user input image. Gray level co-occurrence matrices at five distances and four orientations are computed to measure texture which is defined as being specified by the statistical distribution of the spatial relationships of gray level properties. A likelihood ratio classifier and a nearest neighbor classifier are used to assign two images to the relevance class if they are similar and to the irrelevance class if they are not. A protocol that involves translating a K × K frame throughout every image to automatically construct groundtruth image pairs is proposed and performance of the algorithm is evaluated accordingly. From experiments on a database of 300 512 × 512 grayscale images with 9,600 groundtruth image pairs, we were able to estimate a lower bound of 80% correct classification rate of assigning sub-image pairs we were sure were relevant, to the relevance class. We also argue that some of the assignments which we counted as incorrect are not in fact incorrect.

1 Introduction

Large amount of images that are generated by various applications and the advances in computation power, storage devices, scanning, networking, image compression, desktop publishing, and the World Wide Web have made image databases increasingly popular. The advances in these areas contribute to an increase in the number, size, use, and availability of on-line image databases. New tools are required to help users create, manage, and retrieve images from these databases. The value of these systems can greatly increase if they can provide the ability of searching directly on non-textual data, instead of searching only on the associated textual information.

In a typical content-based image database retrieval application, the user has an image he or she is interested in and wants to find similar images from the entire database. The image retrieval scenario we address here begins with a query expressed by an image. The user inputs an image or a section of an image and desires to retrieve images from the database having some section in them that is like the user input image.

In content-based retrieval, the problem is first to find efficient features for image representation, then to use an effective measure to establish similarity between two images. The features and the similarity measure should be efficient enough to match similar images as well as being able to discriminate dissimilar ones. In this paper, we discuss how we use variances of gray level spatial dependencies as textural features to retrieve images from a database of grayscale images. Then, we propose a protocol to automatically construct groundtruth image pairs to evaluate the performance of the algorithm accordingly. Given these groundtruths, we find the best case and worst case classification efficiencies of the algorithm.

The paper is organized as follows. First, some of the previous approaches to texture and its use in content-based retrieval are discussed in Section 2. Second, we discuss our textural features in Section 3. Section 4 describes the decision methods for similarity measurement. Next, we present our experiments and results in Section 5. Finally, we discuss the conclusions and suggestions for future work in Section 6.

2 Background and Motivation

Texture has been one of the most important characteristics which have been used to classify and recognize objects and scenes. Texture can be characterized by the spatial distribution of gray levels in a neighborhood. Numerous methods, that were designed for a particular application, have been proposed in the literature. However, there seems to be no general method or a formal approach which is useful in a broad range of images.

In his texture survey, Haralick [6] characterized texture as a concept of two dimensions, the tonal primitive properties and the spatial relationships between them. He pointed out that tone and texture are not independent concepts, but in some images tone is the dominating one and in others texture dominates. Then, he gave a review of two kinds of approaches to characterize and measure texture: *statistical* approaches like autocorrelation functions, optical transforms, digital transforms, textural edgeness, structuring elements, spatial gray level run lengths and autoregressive models, and *structural* approaches that use the idea that textures are made up of primitives appearing in a near regular repetitive arrangement.

Rosenfeld and Troy [13] also defined texture as a repetitive arrangement of a unit pattern over a given area and tried to measure coarseness of texture using amount of edge per unit area, gray level dependencies, autocorrelation, and number of relative extrema per unit area.

Many researchers used texture in finding similarities between images in a database. In the QBIC Project, Niblack *et al.* [4] used features like color, texture and shape that are computed for each object in an image as well as for each image. For texture, they extracted features based on coarseness, contrast, and directionality. In the Photobook Project, Pentland *et al.* [12] used features based on appearance, 2-D shape and textural properties. For texture, they used 2-D Wold-based decompositions. In the CANDID Project, Kelly *et al.* [9] used Laws' texture energy maps to extract textural features and introduced a global signature based on a sum of weighted Gaussians to describe the texture. Manjunath and Ma [11] used Gabor filter-based multiresolution representations to extract texture information. They used means and standard deviations of Gabor transform coefficients, computed at different scales and orientations, as features. Li *et al.* [10] used 21 different spatial features like gray level differences (mean, contrast, moments, directional derivatives, etc.), co-occurrence matrices, moments, autocorrelation functions, fractals and Robert's gradient on remote sensing images. Carson *et al.* [2] developed a region-based query system called "Blobworld" by first grouping pixels into regions based on color and texture using expectation-maximization and minimum description length principles, then by describing these regions using color, texture, location and shape properties. Texture features they used are anisotropy, orientation and contrast computed for each region.

We define texture as being specified by the statistical distribution of the spatial relationships of gray level properties. Julesz [8] was the first to conduct experiments to determine the effects of high-order spatial dependencies on the visual perception of synthetic textures. He showed that, although with few exceptions, textures with different first- and second-order probability distributions can be easily discriminated but differences in the third- or higher-order statistics are irrelevant.

One of the early approaches that use spatial relationships of gray levels in texture discrimination is [5], where Haralick used features like the angular second moment, angular second moment difference, angular second moment inverse difference, and correlation, computed from the co-occurrence matrices for automatic scene identification of remote sensing images and achieved 70% accuracy.

In [7], Haralick *et al.* again used features computed from co-occurrence matrices to classify sandstone photomicrographs, panchromatic aerial photographs, and ERTS multispectral satellite images. Although they used only some of the features they defined and did not use the same classification algorithm in their tests for different data sets, it can be concluded that features they compute from co-occurrence matrices performed well in distinguishing between different texture classes in many kinds of image data.

Weszka *et al.* [15] made a comparative study of four texture classification approaches; Fourier power spectrum, co-occurrence matrices, gray level difference statistics, and gray level run length statistics, to classify aerial photographic terrain samples and also LANDSAT images. They obtained results similar to Haralick's [7] and concluded that features computed from co-occurrence matrices perform as well as or better than other algorithms.

Another comparative study is done by Conners and Harlow [3]. They used Markov-generated images to evaluate the performances of different texture analysis algorithms for automatic texture discrimination and concluded that the spatial gray level dependencies method performed better than the gray level run length method, power spectrum method, and gray level difference method.

From the experiments on wide class of images, it can be concluded that spatial gray level dependencies carry much of the texture information [6] and they are more general and perform better than other methods [15, 3]. More information on this topic will be given in Section 3.1.

3 Feature Extraction

Structural approaches have been one of the major research directions for texture analysis. They use the idea that texture is composed of primitives with different properties appearing in particular arrangements. On the other hand, statistical approaches try to model texture using statistical distributions either in the spatial domain or in a transform domain. One way to combine these two approaches is to define texture as being specified by the statistical distribution of the properties of different textural primitives occurring at different spatial relationships.

A pixel, with its gray level as its property, is the simplest primitive that can be defined in a digital image. Consequently, distribution of pixel gray levels can be described by first-order statistics like mean, standard variation, skewness and kurtosis or second-order statistics like the probability of two pixels having particular gray levels occurring at particular spatial relationships. This information can be summarized in two-dimensional co-occurrence matrices computed for different distances and orientations. Coarse textures are ones for which the distribution changes slightly with distance, whereas for fine textures the distribution changes rapidly with distance.

In the following sections we describe the co-occurrence matrices and the features we compute from them.

3.1 Gray Level Co-Occurrence

Gray level co-occurrence can be specified in a matrix of relative frequencies $P(i,j;d,\theta)$ with which two neighboring pixels separated by distance d at orientation θ occur in the image, one with gray level i and the other with gray level j. For example, for a 0° angular relationship, $P(i,j;d,0°)$ averages the probability of a left-right transition of gray level i to gray level j at a distance d.

In our derivations, we define the origin of the image as the upper-left corner pixel. Let $L_r = \{0,1,\ldots,N_r-1\}$ and $L_c = \{0,1,\ldots,N_c-1\}$ be the spatial domains of row and column dimensions, and $G = \{0,1,\ldots,N_g-1\}$ be the domain of gray levels. The image I can be represented as a function which assigns a gray level to each pixel in the domain of the image; $I: L_r \times L_c \to G$. Then, for the orientations shown in Figure 1, co-occurrence matrices can be defined as

$$
\begin{aligned}
P(i,j;d,0^\circ) &= \#\{((r,c),(r',c')) \in (L_r \times L_c) \times (L_r \times L_c) | \\
&\quad r'-r=0, |c'-c|=d, I(r,c)=i, I(r',c')=j\} \\
P(i,j;d,45^\circ) &= \#\{((r,c),(r',c')) \in (L_r \times L_c) \times (L_r \times L_c) | \\
&\quad (r'-r=d, c'-c=d) \text{ or } (r'-r=-d, c'-c=-d), \\
&\quad I(r,c)=i, I(r',c')=j\} \\
P(i,j;d,90^\circ) &= \#\{((r,c),(r',c')) \in (L_r \times L_c) \times (L_r \times L_c) | \\
&\quad |r'-r|=d, c'-c=0, I(r,c)=i, I(r',c')=j\} \\
P(i,j;d,135^\circ) &= \#\{((r,c),(r',c')) \in (L_r \times L_c) \times (L_r \times L_c) | \\
&\quad (r'-r=d, c'-c=-d) \text{ or } (r'-r=-d, c'-c=d), \\
&\quad I(r,c)=i, I(r',c')=j\}.
\end{aligned}
\tag{1}
$$

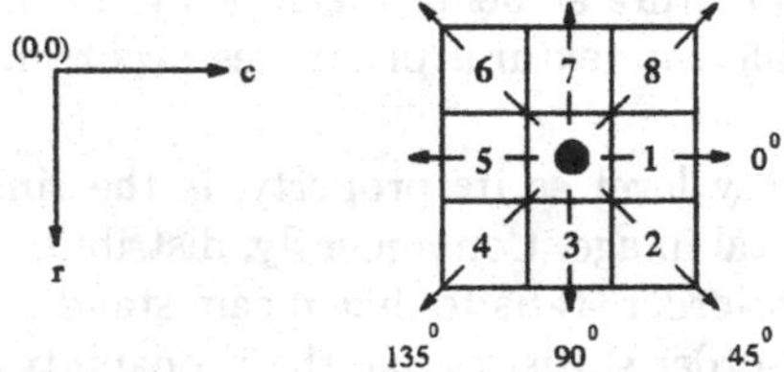

Fig. 1. Spatial arrangements of pixels

Resulting matrices are symmetric. The distance metric used in Equation (1) can be explicitly defined as

$$\rho((r,c),(r',c')) = \max\{|r-r'|, |c-c'|\}.$$

We can normalize these matrices by dividing each entry in a matrix by the number of neighboring pixels used in computing that matrix. Given distance d, this number is $2N_r(N_c - d)$ for 0° orientation, $2(N_r - d)(N_c - d)$ for 45° and 135° orientations, and $2(N_r - d)N_c$ for 90° orientation.

3.2 Textural Features

In order to use the information contained in the gray level co-occurrence matrices, Haralick [7] defined 14 statistical measures which measure textural characteristics like homogeneity, contrast, organized structure, complexity, and nature of gray level transitions. Since from many distances and orientations we obtain a very large number of values, computation of co-occurrence matrices and extraction of textural features from them become infeasible for an image retrieval application which requires fast computation of features. We decided to use only the variance

$$v(d,\theta) = \sum_{i=0}^{N_g-1} \sum_{j=0}^{N_g-1} (i-j)^2 P(i,j;d,\theta) \tag{2}$$

which is a difference moment of P that measures the contrast in the image. Rosenfeld and Troy [13] called this feature the moment of inertia. It will have a large value for images which have a large amount of local spatial variation in gray levels and a smaller value for images with spatially uniform gray level distributions.

Here a problem arises as deciding on which distances to use to compute the co-occurrence matrices. Researchers tried to develop methods to select the co-occurrence matrices that reflect the greatest amount of texture information from a set of candidate matrices obtained by using different spatial relationships. Zucker and Terzopoulos [16] interpreted intensity pairs in an image as samples obtained from a two-dimensional random process and defined a chi-square test to determine whether their observed frequencies of occurrences appear to have been drawn from a distribution where two intensities are independent of each other. In [14], Tou and Chang used an eigenvector-based approach and Karhunen-Loeve expansion to eliminate dependent features. Currently we are developing methods to select the distances that perform the best according to some statistical measures. In this work we compute the variance feature for 1 to 5 pixel distances and four orientations. This constitutes a 20-dimensional feature vector.

Note that angularly dependent features are not invariant to rotation. We can argue whether we want rotation invariance in a content-based retrieval system or not. One can say that a rotated image is not the same as the original image anymore. For example, people standing up and people lying down can be regarded as two different situations so these images can be perceived as quite different. On the other hand, in a military target database we do not want to miss a tank when it is in a different orientation in the image in our database than the orientation in our query image. This dilemma is also present in object-based queries. In this work, we will use the feature vector described above which is rotation variant. We are in the process of modifying our feature vector to include rotation invariance as discussed in [7] and are going to do experiments with the new feature vector on the same database.

Since our goal is to find a section in the database which is relevant to the input query, before retrieval, each image in the database is divided into overlapping sub-images using the protocol which will be discussed in Section 4.1. We then compute a 20-dimensional feature vector for each sub-image in the database.

4 Decision Methods

After computing the feature vectors for all images in the database, given a query, we have to decide which images in the database are relevant to it, and we have to retrieve the most relevant ones as the results of the query. In our experiments we use two different types of decision methods; a likelihood ratio approach which is a Gaussian classifier, and a nearest neighbor rule based approach. In the following sections we discuss these two approaches.

4.1 Likelihood Ratio

In the likelihood ratio approach, we define two classes, namely the relevance class A and the irrelevance class B. Given the feature vectors of a pair of images, if these images are similar, they should be assigned to the relevance class, if not, they should be assigned to the irrelevance class.

In the following two sections we describe first, how to determine the parameters of the two classes, and second, how to construct the likelihood ratio.

Determining the Parameters The protocol for constructing groundtruths to determine the parameters of the likelihood ratio classifier involves making up two different sets of sub-images for each image i, $i = 1 \ldots \mathrm{I}$, in the database. The first set of sub-images begins in row 0 column 0 and partitions each image i into M_i $\mathrm{K} \times \mathrm{K}$ sub-images. These sub-images are partitioned such that they overlap by half the area. We ignore the partial sub-images on the last group of columns and last group of rows which cannot make up the $\mathrm{K} \times \mathrm{K}$ sub-images. This set of sub-images will be referred as the *main database* in the rest of the paper.

The second set of sub-images are shifted versions of the ones in the *main database*. They begin in row $\mathrm{K}/4$ and column $\mathrm{K}/4$ and partition the image i into N_i $\mathrm{K} \times \mathrm{K}$ sub-images. We again ignore the partial sub-images on the last group of columns and last group of rows which cannot make $\mathrm{K} \times \mathrm{K}$ sub-images. This second set of sub-images will be referred as the *test database* in the rest of the paper.

To construct the groundtruth to determine the parameters, we record the relationships of the shifted sub-images in the *test database* with the sub-images in the *main database* that were computed from the same image. The feature vector for each sub-image in the *test database* is strongly related to four feature vectors in the *main database* in which the sub-image overlap is 9/16 of the sub-image area. From these relationships, we establish a ***strongly related sub-images*** set $R_s(n)$ for each sub-image n where $n = 1 \ldots \mathrm{N}_i$.

We assume that, in an image, two sub-images that do not overlap are usually not relevant. From this assumption, we randomly select four sub-images that have no overlap with the sub-image n. These four sub-images form the *other sub-images* set $R_o(n)$.

These groundtruth sub-image pairs constitute the relevance class A_i,

$$\mathrm{A}_i = \{(n,m) | \, m \in R_s(n) \, , \, n = 1 \ldots \mathrm{N}_i\},$$

and the irrelevance class B_i,

$$\mathrm{B}_i = \{(n,m) | \, m \in R_o(n) \, , \, n = 1 \ldots \mathrm{N}_i\}$$

for each image i. Then, the overall relevance class becomes $A = A_1 \cup A_2 \cup \cdots \cup A_I$ and the overall irrelevance class becomes $B = B_1 \cup B_2 \cup \cdots \cup B_I$.

An example for the overlapping concept is given in Figure 2 where the shaded region shows the 9/16 overlapping. For K = 128, sub-images with upper-left corners at (0,0), (0,64), (64,0), (64,64) and (192,256) are examples from the *main database.* The sub-image with upper-left corner at (32,32) is a sub-image in the *test database.* For this sub-image, R_s will consist of the sub-images at (0,0), (0,64), (64,0), and (64,64), because they overlap by the required amount. On the other hand, R_o will consist of four randomly selected sub-images, one being the sub-image at (192,256) for example, which are not in R_s and have no overlap with the test sub-image. The pairs formed by the test sub-image and the ones in R_s and R_o form the groundtruths for the relevance and irrelevance classes respectively. Note that for any sub-image which is not shifted by (K/4,K/4), there is a sub-image which it overlaps by more than half the area. We will use this property to evaluate the performance of the algorithm in Section 5.

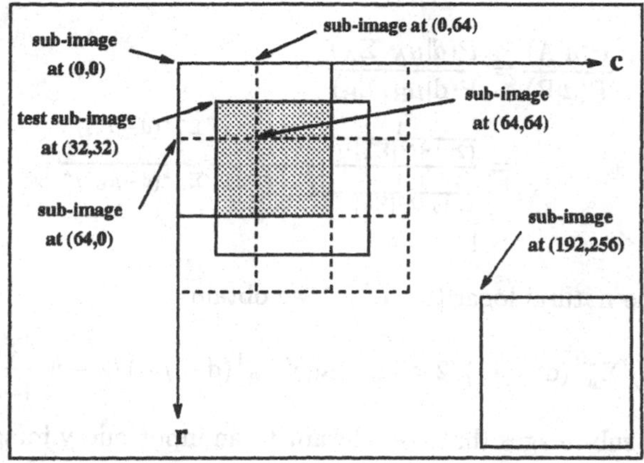

Fig. 2. The shaded region shows the 9/16 overlapping between two sub-images

As the database structure is concerned, our first sub-image database (*main database*) contains a unique sub-image i.d., bounding box, and the feature vector for each sub-image $m = 1 \ldots M_i$ and $i = 1 \ldots I$. The second sub-image database (*test database*) contains a unique sub-image i.d., bounding box, $R_s(n)$, $R_o(n)$, and the feature vector for each sub-image $n = 1 \ldots N_i$ and $i = 1 \ldots I$.

In order to estimate the distribution of the relevance class, we first compute the differences d, $d = x^{(n)} - y^{(m)}$, $(n, m) \in A$, $x^{(n)}, y^{(m)} \in \mathcal{R}^Q$ where Q is 20 for our features, and $x^{(n)}$ and $y^{(m)}$ are the feature vectors of sub-images n and m respectively. Then, we compute the sample mean, μ_A, and the sample covariance, Σ_A, of these differences. We assume that these differences for the relevance class have a normal distribution with mean μ_A, and covariance Σ_A. Similarly, we compute the differences d, $d = x^{(n)} - y^{(m)}$, $(n,m) \in \mathrm{B}$, $x^{(n)}, y^{(m)} \in \mathcal{R}^{\mathrm{Q}}$, then the sample mean, μ_{B}, and the sample covariance, Σ_{B}, for the irrelevance class.

Making the Decision Suppose for the moment that the user query is a $\mathrm{K} \times \mathrm{K}$ image. First, its feature vector x is determined. Then, the search goes through all the feature vectors $y^{(m)}$ in the *main database* where $m = 1 \ldots (\sum_{i=1}^{\mathrm{I}} \mathrm{M}_i)$, M_i being the number of sub-images in the i'th image. For each feature vector pair $(x, y^{(m)})$, the difference $d = x - y^{(m)}$ is computed.

The probability that the input query image with feature vector x, and a sub-image in the database with feature vector $y^{(m)}$ are relevant is $P(\mathrm{A}|d) = P(d|\mathrm{A})P(\mathrm{A})/P(d)$ and that they are irrelevant is $P(\mathrm{B}|d) = P(d|\mathrm{B})P(\mathrm{B})/P(d)$. We can define the likelihood ratio as

$$r(d) = \frac{P(\mathrm{A}|d)}{P(\mathrm{B}|d)}. \tag{3}$$

If this ratio is greater than 1, the sub-image m is considered to be relevant to the input query image. If we assume two classes are equally likely, equation (3) becomes

$$\begin{aligned} \frac{P(d|\mathrm{A})}{P(d|\mathrm{B})} &= \frac{P(d|\mu_{\mathrm{A}}, \Sigma_{\mathrm{A}})}{\mathrm{P}(\mathrm{d}|\mu_{\mathrm{B}}, \Sigma_{\mathrm{B}})} \\ &= \frac{\frac{1}{(2\pi)^{\mathrm{Q}/2}|\Sigma_{\mathrm{A}}|^{1/2}} e^{-(\mathrm{d}-\mu_{\mathrm{A}})'\Sigma_{\mathrm{A}}^{-1}(\mathrm{d}-\mu_{\mathrm{A}})/2}}{\frac{1}{(2\pi)^{\mathrm{Q}/2}|\Sigma_{\mathrm{B}}|^{1/2}} e^{-(\mathrm{d}-\mu_{\mathrm{B}})'\Sigma_{\mathrm{B}}^{-1}(\mathrm{d}-\mu_{\mathrm{B}})/2}} \\ &> 1. \end{aligned} \tag{4}$$

After taking the natural logarithm of (4) we obtain

$$(d - \mu_{\mathrm{A}})'\Sigma_{\mathrm{A}}^{-1}(\mathrm{d} - \mu_{\mathrm{A}})/2 < (\mathrm{d} - \mu_{\mathrm{B}})'\Sigma_{\mathrm{B}}^{-1}(\mathrm{d} - \mu_{\mathrm{B}})/2 + \ln \frac{|\Sigma_{\mathrm{B}}|^{1/2}}{|\Sigma_{\mathrm{A}}|^{1/2}}. \tag{5}$$

To find the sub-images that are relevant to an input query image, likelihood ratios for all sub-images in the database are computed as in (3) and the sub-images are ranked by these likelihood ratios. Among them, k sub-images having the highest r-values are retrieved as the most relevant ones.

4.2 Nearest Neighbor Rule

In the nearest neighbor approach we assume each sub-image m in the database is represented by its feature vector $y^{(m)}$ in the Q-dimensional feature space. Given the feature vector x for the input query, we want to find the y's which are the closest neighbors of x by a distance measure. Then, the k-nearest neighbors of x will be retrieved as the most relevant ones.

The problem of finding the k-nearest neighbors can be formulated as follows. Given the set $Y = \{y^{(m)} | y^{(m)} \in \mathcal{R}^{\mathrm{Q}}, m = 1, \ldots, \mathrm{M}\}$ and feature vector $x \in \mathcal{R}^{\mathrm{Q}}$, find the set of sub-images $R \subseteq \{1, \ldots, \mathrm{M}\}$ such that $\#R = k$ and

$$\rho(x, y^{(r)}) \leq \rho(x, y^{(p)}), \quad \forall r \in R, p \in \{1, \ldots, \mathrm{M}\} \backslash R$$

where $M = \sum_{i=1}^{I} M_i$, M_i being the number of sub-images in the i'th image.

For the distance metric ρ we use the Euclidean distance

$$\rho(x, y) = \|x - y\|$$

or the infinity norm

$$\rho(x, y) = \max_{i=1,\ldots,Q} |x_i - y_i|$$

where x_i and y_i are the i'th components of the corresponding feature vectors.

5 Experiments and Results

Testing content-based retrieval systems and comparing the performances of two different algorithms is an open question. Two traditional measures for retrieval performance are precision and recall. Precision is the percentage of retrieved images that are relevant and recall is the percentage of relevant images that are retrieved. Note that computation of these measures requires image-level groundtruthing of the database. We created two databases of sub-images according to the protocol in Section 4.1 but since these automatically generated sub-image-level groundtruths are not the ones required for precision and recall, we use modified versions of these measures to evaluate the performance of our algorithm. After manually grouping a smaller set of images in our database, we will evaluate the performance using precision and recall too.

In the following sections we describe the database population and two experimental procedures for our decision methods.

5.1 Database Population

To populate the database, we used the Fort Hood Data, supplied for the RADIUS program by the Digital Mapping Laboratory at Carnegie Mellon University. These images consist of visible light images of the Fort Hood area at Texas. We divided these aerial images into 300 512×512 images. After the database was constructed, we carried out the approach described in Section 4.1 which involved translating a 256×256 frame throughout every image and extracted the desired features for all sub-images.

5.2 Experimental Set-up

To test the classification effectiveness using the Gaussian classifier, we can apply the classification algorithm to each groundtruth pair (n,m) described in Section 4.1. Since we know which non-shifted sub-images and shifted sub-images overlap, we also know which sub-image pairs should be assigned to class A and which to class B. So, to test our approach, we then check whether each pair that should be classified into class A or B is classified into class A or B correctly.

To test the retrieval performance of the algorithm, we use the following procedure. Given an input query image of size $K \times K$, we create a list of retrieved images in descending order of likelihood ratio or ascending order of distance for nearest neighbor rule. If the correct image is retrieved as one of the k best matches, it is considered a success. This can also be stated as a nearest neighbor classification problem where the relevance class is defined to be the best k matches and the irrelevance class is the rest of the images. We also compute the average rank of the correct image among retrieved images. For this experiment, we use the non-shifted sub-images to compute the best case performance and the shifted sub-images to compute the worst case performance. We call this the worst case performance because the shifted sub-images overlap by approximately half the area of a sub-image in the database. All other possible sub-images have a sub-image in the database which they overlap by more than half the area. This experimental procedure is appropriate to our problem of retrieving images which have some section in them that is like the user input image.

5.3 Results

Classification Effectiveness In this experiment, the *main database* consists of 2,700 256 × 256 sub-images and the *test database* consists of 1,200 256 × 256 sub-images. There are 4 relevant and 4 irrelevant non-shifted sub-images for each of the 1,200 shifted sub-images, which make a total of 9,600 groundtruth sub-image pairs. As can be seen in Table 1, 79.75% of the groundtruth A pairs were assigned to A with an overall success of 62.96%.

Table 1. Confusion matrix for the classification effectiveness test.

	Assigned Relevant	Assigned Irrelevant	Success (%)
Relevant Pair G.truth	3,828	972	79.75
Irrelevant Pair G.truth	2,584	2,216	46.17
Overall	6,412	3,188	62.56

We can say that most of the groundtruth A pairs were assigned to A but the groundtruth B pairs seem to be split between being assigned to A or B. The cause of this problem can be explained as follows. Although the assumption that overlapping sub-images are relevant almost always holds, we can not always guarantee that non-overlapping sub-images are irrelevant. Obvious examples are images which have the same texture pattern at more than one location. Illustration of this fact can be found in [1] where we manually eliminated some images with large regions of constant gray values from the Fort Hood Dataset and obtained a 42% decrease in the false alarm rate. Hence, some of the assignments which we count as incorrect are not in fact incorrect. Thus the approximate 80% correct relevant pair rate is a lower bound.

Retrieval Performance Results for the retrieval performance experiments are summarized in Table 2 as the number of tests, number of successes, and average rank of the correct image. For the best case analysis 2,700 sub-image queries were used. As explained before, these are the non-shifted sub-images in the database. For the worst case analysis, 1,200 shifted sub-images in the *test database* are used. To illustrate the bounds found in these experiments, the database was queried with 500 randomly extracted 256×256 sections from images in the database. In all of these experiments a success means the correct image is retrieved as one of the best 20 matches.

Table 2. Results for the retrieval performance test.

	Original sub-images (no of tests = 2,700)		
	Likelihood Ratio	Euclidean Distance	Infinity Norm
# successes	2,536	2,683	2,684
% success	93.93	99.37	99.41
avg. rank	4.0430	2.0078	2.0138

	Shifted sub-images (no of tests = 1,200)		
	Likelihood Ratio	Euclidean Distance	Infinity Norm
# successes	683	701	681
% success	56.92	58.42	56.75
avg. rank	6.1830	5.5706	5.5727

	Random sub-images (no of tests = 500)		
	Likelihood Ratio	Euclidean Distance	Infinity Norm
# successes	326	330	325
% success	65.20	66.00	65.00
avg. rank	5.7301	4.2576	4.2523

As can be seen in Table 2, the algorithm successfully retrieved the correct image as one of the 20 best matches 56 percent of the time at the worst case. Euclidean distance performed slightly better than the infinity norm. Although the worst case and random query results of both likelihood ratio and nearest neighbor decision methods were almost equal, nearest neighbor method performed slightly better in the best case analysis. Also the nearest neighbor rule retrieved the correct image at a higher rank than the likelihood ratio which is 2 at the best case and 5 at the worst case on the average.

Experimenting on sub-image size showed that smaller sub-images give better results because co-occurrence features are measures of micro texture and texture tends to be more homogeneous as sub-image size gets smaller.

Some example queries are given in Figures 3(a)-3(f). In all of these figures the upper-left image is the 256×256 query image. First three rows show the best 12 matches among the 512×512 images in the database. Last row shows 4 images that are found to be the most irrelevant to the query image. Our system also displays the most irrelevant images to help the user understand how the system decides what is relevant and what is not. By looking at these irrelevant images and comparing them with the relevant ones, the user can refine his query in a more effective way. In each retrieved image, matched 256 × 256 sub-images are marked with a white border. More examples can be found at http://isl.ee.washington.edu/~aksoy/research/database.shtml.

6 Conclusions

In this paper, we discussed a system that allows a user to input an image or a section of an image and retrieves all images from a database having some section in them that is like the user input image.

To achieve this goal, texture was defined as being specified by the statistical distribution of the spatial relationships of gray level properties and variances computed from two-dimensional gray level co-occurrence matrices at 1 to 5 pixel distances and four orientations were used to extract this information.

A likelihood ratio classifier was defined to measure the relevancy of two images, one being the query image and one being a database image, so that image pairs which had a high likelihood ratio were classified as relevant and the ones which had a lower likelihood ratio were classified as irrelevant. Also k-nearest neighbor rule was used to retrieve k images which have the closest feature vector to the feature vector of the query image in the 20-dimensional feature space.

Testing content-based retrieval systems and comparing the performance of two different algorithms is an open question. A protocol which involved translating a K×K frame throughout every image to automatically construct groundtruth image pairs for the relevance and irrelevance classes was proposed and performance of the algorithm was evaluated accordingly.

Experiments were done on a database of 300 images to check the effectiveness of the features in representing images. Results of the classification effectiveness tests showed that the algorithm assigned 79.75% of the sub-image pairs we were sure were relevant, to the relevance class correctly when the database was partitioned into 9,600 256 × 256 sub-image pairs even with an offset of quarter of the image size which was 64 pixels in the tests. Results of the retrieval performance tests showed that all of the decision methods retrieved correct images successfully as one of the best 20 matches, which is less than 1 percent of the total, in more than 93 percent of the 2,700 experiments for the best case analysis and in more than 56 percent of the 1,200 experiments for the worst case analysis.

An interesting study will be to examine images that are successfully retrieved with the nearest neighbor rule but missed with the likelihood ratio, and vice versa. Although being a micro texture measure, our features showed significant performance on a database of complex aerial images. We are currently adding more features that will capture the texture information at higher scales [1]. This will result in a more compact representation that is needed for large databases containing different types of complex images.

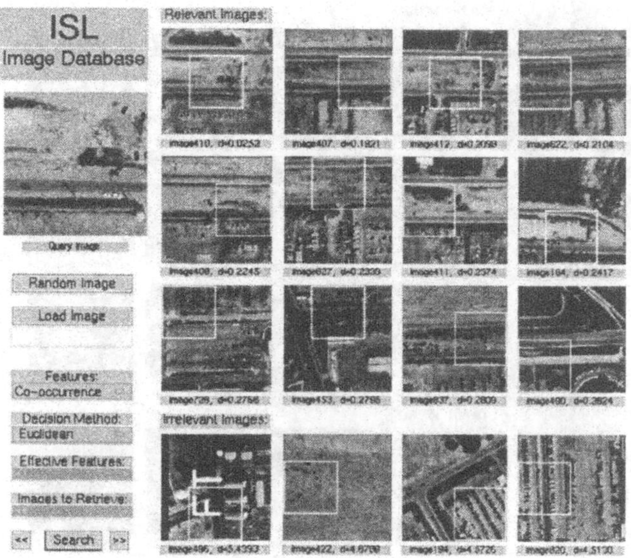

(a) Query by a sub-image from the main database using Euclidean distance.

(b) Query by a sub-image from the test database using Euclidean distance.

Fig. 3. Example queries using different distance methods

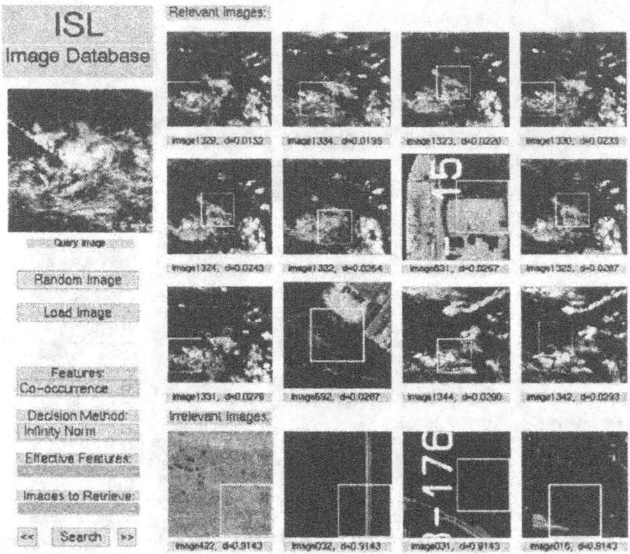

(c) Query by a sub-image from the test database using Infinity norm.

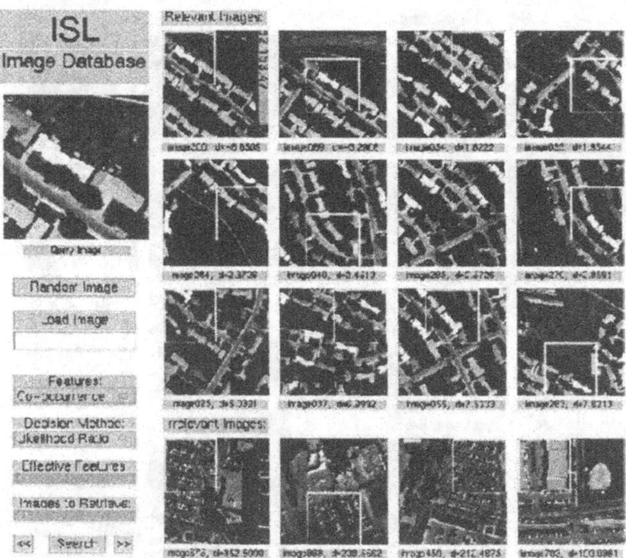

(d) Query by a sub-image from the main database using Likelihood ratio.

Fig. 3. Example queries using different distance methods (cont.)

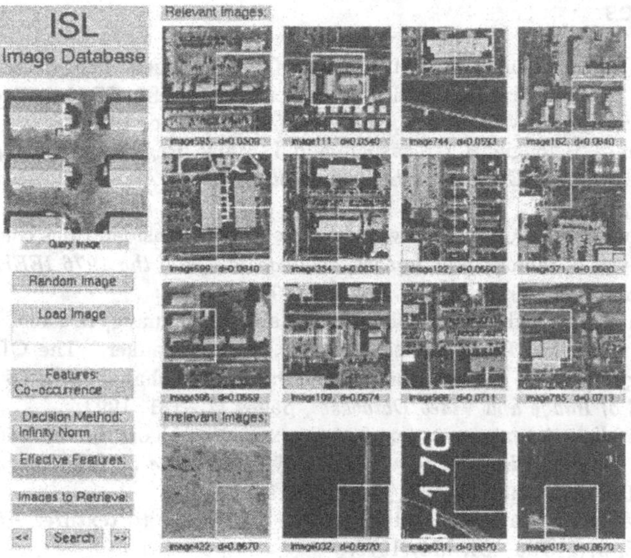

(e) **Query by a sub-image taken from another Ft.Hood set using Infinity norm.**

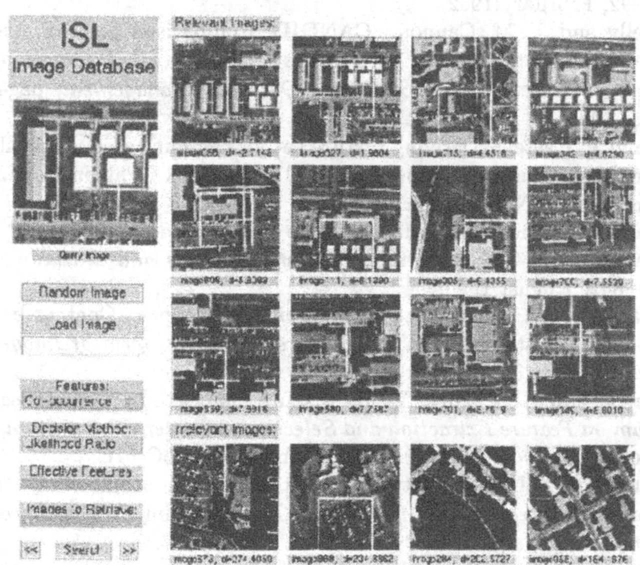

(f) **Query by a sub-image from the main database using Likelihood ratio.**

Fig. 3. Example queries using different distance methods (cont.)

References

1. S. Aksoy, M. L. Schauf, and R. M. Haralick. Content-based image database retrieval based on line-angle-ratio statistics. Technical Report ISL-TR, Intelligent Systems Lab., University of Washington, Seattle, WA, November 1997.
2. C. Carson, S. Belongie, H. Greenspan, and J. Malik. Region-based image querying. In *Proceedings of IEEE Workshop on Content-Based Access of Image and Video Libraries*, 1997.
3. R. W. Conners and C. A. Harlow. Some theoretical considerations concerning texture analysis of radiographic images. In *Proceedings of the 1976 IEEE Conference on Decision and Control*, pages 162–167, 1976.
4. M. Flickner, H. Sawhney, W. Niblack, J. Ashley, Q. Huang, B. Dom, M. Gorkani, J. Hafner, D. Lee, D. Petkovic, D. Steele, and P. Yanker. The QBIC project: Querying images by content using color, texture and shape. In *SPIE Storage and Retrieval of Image and Video Databases*, pages 173–181, 1993.
5. R. M. Haralick. A texture-context feature extraction algorithm for remotely sensed imagery. In *Proceedings of the 1971 IEEE Conference on Decision and Control*, pages 650–657, Gainesville, FL, December 1971.
6. R. M. Haralick. Statistical and structural approaches to texture. *Proceedings of the IEEE*, 67(5):786–804, May 1979.
7. R. M. Haralick, K. Shanmugam, and I. Dinstein. Textural features for image classification. *IEEE Transactions on Systems, Man, and Cybernetics*, SMC-3(6):610–621, November 1973.
8. B. Julesz. Visual pattern discrimination. *IRE Transactions on Information Theory*, pages 84–92, February 1962.
9. P. M. Kelly and T. M. Cannon. CANDID: Comparison algorithm for navigating digital image databases. In *Proceedings of the Seventh International Working Conference on Scientific and Statistical Database Management*, pages 252–258, September 1994.
10. C. S. Li and V. Castelli. Deriving texture set for content based retrieval of satellite image database. Technical Report RC20727, IBM T.J. Watson Research Center, Yorktown Heights, NY, February 1997.
11. B. S. Manjunath and W. Y. Ma. Texture features for browsing and retrieval of image data. *IEEE Transactions on Pattern Analysis and Machine Intelligence*, 18(8):837–842, August 1996.
12. A. Pentland, R. W. Picard, and S. Sclaroff. Photobook: Content-based manipulation of image databases. In *SPIE Storage and Retrieval of Image and Video Databases II*, pages 34–47, February 1994.
13. A. Rosenfeld and E. B. Troy. Visual texture analysis. In *Conference Record for Symposium on Feature Extraction and Selection in Pattern Recognition*, pages 115–124, Argonne, IL, October 1970. IEEE Publication: 70C-51C.
14. J. T. Tou and Y. S. Chang. Picture understanding by machine via textural feature extraction. In *Proceedings of 1977 IEEE Conference on Pattern Recognition and Image Processing*, pages 392–399, Troy, NY, June 1977.
15. J. S. Weszka, C. R. Dyer, and A. Rosenfeld. A comparative study of texture measures for terrain classification. *IEEE Transactions on Systems, Man, and Cybernetics*, SMC-6(4):269–285, April 1976.
16. S. W. Zucker and D. Terzopoulos. Finding structure in co-occurrence matrices for texture analysis. *Computer Graphics and Image Processing*, 12:286–308, March 1980.

Content-Based Access of VRML Libraries

Eric Paquet and Marc Rioux
Visual Information Technology
Institute for Information Technology
National Research Council
Building M-50, Montreal Road
Ottawa (Ontario) K1A 0R6 Canada
E-mail: eric.paquet@nrc.ca, rioux@iit.nrc.ca

Abstract

This paper presents a new approach for the classification and retrieval of three-dimensional images and models from databases. A set of retrieval algorithms is introduced. These algorithms are content-based, meaning that the input is not made out of keywords but of three-dimensional models. Tensor of inertia, distribution of normals, distribution of cords and multiresolution analysis are used to describe each model. The database can be searched by scale, shape or color or any combination of these parameters. A user friendly interface makes the retrieval operation simple and intuitive and make it possible to edit reference models according to the specifications of the user. Experimental results using a database of more than 1000 VRML models are presented

1. Introduction

The use of three-dimensional image and model databases throughout the world is growing both in number and in size. The diffusion of three-dimensional models is also spreading and their use is not limited to specialists anymore. With the advent of VRML [1] as a *de facto* standard for three-dimensional information exchange, three-dimensional databases may soon become as widespread as two-dimensional image databases. In the following, we shall focus on three-dimensional models [2-3] and ignore two-dimensional images [4]. There are many reasons for this. A bidimensional image does not have any scale unless a calibration is performed during its acquisition. The back part of the object is occluded by itself and consequently a significant fraction of the information is lost. The object can have any position and orientation. Also, it should not be forgotten that out of plane rotations are difficult to handle in two dimensions. Except for the contour and discontinuities, most of the geometrical information is lost. This is not the case for three-dimensional models, the scale is known and the geometrical information is available all over its surface. Consequently it is much easier to handle the rotation and translation problem.

2. General Considerations on the Algorithms

From the user point of view, a model can be analyzed in three different ways, based on the scale, shape or color. The scale is simply the size or dimension of the object. Even if it is a simple characteristic it is a powerful tool for discrimination. The shape is related to the geometrical appearance of the object. It can be analyzed at three levels: local, regional and global depending on the size of the region described by a given set of parameters. The shape can also be analyzed in two different ways: the model is viewed as a set of surfaces or as a volume. The color aspects include the color distribution, the texture and the materials description.

3. Analysis of the Geometry

The global analysis is performed in order to define a reference frame that shall be used by the other algorithms. The reference frame is defined as the principal axes of the tensor of inertia which is defined as

$$I = \left[I_{qr}\right] = \left[\frac{1}{n}\sum_{i=1}^{n}\left[S_i\left(q_i - q_{CM}\right)\left(r_i - r_{CM}\right)\right]\right] \tag{1}$$

where S_i is the surface of a triangular face (assuming a triangular decomposition of the object), CM is the center of mass of the object and q and r are equal to x, y and z. If the model is not made out of triangles, the triangulation is generated automatically by our software based on the Open Inventor library (SGI). The principal axes are obtained by computing the eigen vectors of the tensor

$$\left\lfloor I\mathbf{a}_i = \lambda_i \mathbf{a}_i \right\rfloor_{i=1,2,3} \tag{2}$$

Comparing the eigen values identifies the axes. The eigen vector with the highest eigen value is labeled one, the second highest is labeled two and the remaining axis is labeled three. The tensor of inertia has a mirror symmetry problem which can be handled by computing the statistical distribution of the mass in the positive and negative direction in order to identify the positive direction. For each axis, the points are divided between "North" and "South": a point belongs to the North group if the angle between the corresponding cord and a given axis is smaller than 90° and to the South group if it is greater than 90°. A cord is defined as a vector that goes from the center of mass of the model to the center of mass of the triangle. The standard deviation for the length of the cords is calculated for each group of each axis and it is defined as

$$s = \sqrt{\frac{n\sum_{i=1}^{n} d_i^2 - \left(\sum_{i=1}^{n} d_i^2\right)^2}{n(n-1)}} \tag{3}$$

where d is the length of a cord and n the number of points. If the standard deviation of the North group is higher than the standard deviation of the South group then the direction of the corresponding eigen vector is not changed while in the other case, the direction is flipped by 180°. This technique is applied to the first and second axes. Then the outer product between them is calculated. If the third axis does not have the same direction then the resulting vector is flipped by 180° in order to have a direct orthogonal system.

The scale is simply handled by a bounding box which is the smallest box that can contain the model. The axes of the box are parallel to the principal axes of the tensor of inertia. Thus the scale is only defined by the size of the box. A rough description of the mass distribution inside the box is obtained by using the eigen values of the tensor of inertia (moment description).

As mentioned above, the shape is analyzed at three levels. The local level is defined by the normals. Assuming a triangular decomposition of the object and a normal for each triangle, the angles between the normals and the first two principal axes are computed using

$$\alpha_q = \cos^{-1}\left(\frac{\mathbf{n}\cdot\mathbf{a_q}}{\|\mathbf{n}\|\,\|\mathbf{a_q}\|}\right) \tag{4}$$

where

$$\mathbf{n} = \frac{[(\mathbf{r_2}-\mathbf{r_1})\times(\mathbf{r_3}-\mathbf{r_1})]}{\|(\mathbf{r_2}-\mathbf{r_1})\times(\mathbf{r_3}-\mathbf{r_1})\|} \tag{5}$$

The statistic of this description is then presented under the form of a histogram. We use three kinds of histograms called histogram of the first, second and third kind depending on the complexity of the description. The histogram of the first kind is defined as $h\ (\alpha_q)$ where q= 1 and 2. This histogram does not distinguish between the two angles and does not take into account the relation between them. Because of that, it has a low discrimination capability. The histogram of the second kind is made out of two histograms: one for each angle. Thus it distinguishes the angles but it does not establish any relation between them. The histogram of the third kind is a bidimensional histogram defined as $h\ (\alpha_1,\ \alpha_2)$. Not only does it distinguish between the angles but it also maps the relation between them as well.

Most of the time, normals are very sensitive to local variation of shape. In some cases, this may cause severe drawbacks. Let us consider an example: a pyramid and a step pyramid. In the case of the pyramid, the orientations of the normals are the same for all the triangles belonging to a given face while in the other case they have two orientations corresponding to those of the step. The histograms corresponding to these pyramids are very distinct although both models have a very similar global shape. In order to solve this problem, we introduce the concept of a cord measurement. A cord is defined as a vector that goes from the center of mass of the model to the center of mass of a given triangle. The cord is not a unit vector since it has a length. As opposed to a normal, a cord can be considered as a regional characteristic. If we take the pyramid and the step pyramid as an example we can see that the cord orientation changes slowly on a given region, while the normal orientation can have important variations. As for normals, the statistical distribution of the cord orientations can be represented by three histograms namely: histogram of the first, second and third kind. Since the cord has a length, it is also possible to describe the statistical distribution of the length of the cords by a histogram. This histogram is scale-dependent but it can be made scale-independent, by normalizing the scale: zero corresponding to the shortest cord and one to the longest.

Explicitly or implicitly, we are used to thinking about three-dimensional models as made out of surfaces. From a certain point of view this is right but at the same time we should not forget that a three-dimensional object is also a volume and consequently it might be interesting to analyze it as such. In a three-dimensional discrete representation, the building blocks are called voxels, a short hand for volume elements. Using such a representation, it is possible to binarize a three-dimensional model by loosing a small amount of information. The idea is simply to map the model's coordinates to the discrete voxel coordinates

$$\begin{bmatrix} x \\ y \\ z \end{bmatrix} \Rightarrow \begin{bmatrix} i\,\Delta_x \\ j\,\Delta_y \\ k\,\Delta_z \end{bmatrix} \tag{6}$$

where the Δ's are the dimensions of the voxel and i, j and k are the discrete coordinates. If the density of points in the original model is not high enough, it may be necessary to interpolate the original model in order to generate more points and to obtain a better description in the voxel space.

We have chosen to analyze the voxel representation with a wavelet transform [5-7]. There are many reasons for that: recent experiments tend to demonstrate that the human eye would perform a kind of wavelet transform [8]. This would also mean that the brain would perform a part of its analysis based on such a transform. Since we want our algorithm to mimic the human system, this is clearly a good direction to investigate. The wavelet transform performs a multiscale analysis. By multiscale we mean that the model is analyzed at different levels of details. There is a fast transform implementation of the wavelet transform which makes it possible to perform the calculation rapidly. The fast wavelet transform is an orthogonal transformation meaning that its base is orthogonal. The elements of the base are characterized by their scale and position. Each element of the base is bounded in space which means that it occupies a well defined region. This means that the analysis performed by the wavelet transform is local and that the size of the analyzed region depends on the scale of the wavelet. As an example, the one-dimensional wavelet is defined as

$$\sqrt{2^j}\,w\left(2^j q - n\right)_{n,\,j\in Z} \tag{7}$$

For our purpose we use DAU4 (Daubechies) wavelets which have two vanishing moments. The $N{\times}N$ (N being a multiple of two) matrix corresponding to the one-dimensional transform is

$$W=\begin{bmatrix} c_0 & c_1 & c_2 & c_3 & & & & & & \\ c_3 & -c_2 & c_1 & -c_0 & & & & & & \\ & & c_0 & c_1 & c_2 & c_3 & & & & \\ & & c_3 & -c_2 & c_1 & -c_0 & & & & \\ & & & & & \ddots & & & & \\ & & & & & & c_0 & c_1 & c_2 & c_3 \\ & & & & & & c_3 & -c_2 & c_1 & -c_0 \\ c_2 & c_3 & & & & & & & c_0 & c_1 \\ c_1 & -c_0 & & & & & & & c_3 & -c_2 \end{bmatrix} \tag{8}$$

where

$$c_0=\frac{(1+\sqrt{3})}{4\sqrt{2}} \quad c_1=\frac{(3+\sqrt{3})}{4\sqrt{2}}$$
$$c_2=\frac{(3-\sqrt{3})}{4\sqrt{2}} \quad c_4=\frac{(11\sqrt{3})}{4\sqrt{2}} \tag{9}$$

From equation (9) we can define

$$H\therefore\begin{bmatrix} c_0 & c_1 & c_2 & c_3 \end{bmatrix} \tag{10}$$

$$G\therefore\begin{bmatrix} c_3 & -c_2 & c_1 & -c_0 \end{bmatrix} \tag{11}$$

The doublet H and G is a quadrature mirror filter. H can be considered as a smoothing filter while G is a filter with two vanishing moments. The one dimensional wavelet transform is computed by applying the wavelet transform matrix hierarchically, first on the full vector of length N, then to the $N/2$ values smoothed by H, then the $N/4$ values smoothed again by H until two components remain. In order to compute the wavelet transform in three dimensions the array is transformed sequentially on the first dimension (for all values of its other dimensions) then on its second dimension and finally on its third dimension. The final result of the wavelet transform is an array of the same dimension as the initial voxel array.

The set of wavelet coefficients represents a tremendous amount of information. In order to reduce it, we compute the logarithm in base 2 of the coefficients in order to enhance the coefficients corresponding to small details. These usually have a very low value compared to those that have a large value and we integrate the signal for each scale. A histogram representing the distribution of the signal at different scales is then constructed: the vertical axis represents the total amount of signal at a given scale and the horizontal axis represents the "scale" or level of resolution. It is important to notice that each "scale" in the histogram represents in fact a triplet of scales corresponding to (s_x, s_y and s_z).

4. Analysis of the Color Distribution

In the following we assume that the model is either computer generated or that the colors have been corrected in order to take into account the illumination of the sensor. If we are not interested in the color location, the description of their distribution is very simple. We just need to compute the color histogram for the red, green and blue [9]. The vertical scale of these histograms corresponds to the occurrence of a given color while the horizontal scale represents the normalized intensity of the color.

If we are interested in the location of the color we can generalize the wavelet approach that we have presented in the previous section. The model has six dimensions: x, *y*, *z*, *R*, *G*, and *B*. The model is fist binarized

$$[x \quad y \quad z \quad R \quad V \quad B] \Rightarrow \lfloor i\Delta_x \quad j\Delta_y \quad k\Delta_z \quad l\Delta_R \quad m\Delta_V \quad n\Delta_B \rfloor \tag{12}$$

The six-dimensional wavelet transform is then computed. The array is transformed sequentially on the first dimension (for all values of its other dimensions) then on its second dimension up to the sixth dimension. The logarithm (base two) of the wavelet coefficient is computed and the total signal at each "scale" is integrated. The statistical distribution of the signal among the scales is presented under the form of a histogram. The horizontal axis of this histogram represents the scale and the vertical axis the corresponding signal. It is important to notice that each "scale" of the histogram is in fact an index that corresponds to a set of six scales in the wavelet space.

Three-dimensional color models can also be analyzed form another point of view [10]. It has been shown that the way geometry is analyzed by the human visual system is influenced by color and more precisely that separate colors are analyzed as distinct entities. Now a triplet red-green-blue represents each color. For each triplet we determine the dominant color and then we calculate the angle between the normal corresponding to that point and the first eigen vector. The statistical distribution of these angles is represented by a set of three histograms according to the dominant color. Thus three color-dependent distributions of the normals are obtained: one for the "red" normals, one for the "green" normals and one for the "blue".

One of the most widely used lighting model in computer graphic is the Phong model. It is generally defined as

$$I \approx I_s \left[r_d (\mathbf{s} \cdot \mathbf{n}) + r_s (\mathbf{r} \cdot \mathbf{v})^f \right] \tag{13}$$

where

$$\mathbf{r} = \frac{-\mathbf{s} + 2(\mathbf{s} \cdot \mathbf{n})\mathbf{n}}{\left\| -\mathbf{s} + 2(\mathbf{s} \cdot \mathbf{n})\mathbf{n} \right\|} \tag{14}$$

and where **n** is a unit normal at a given point, **s** is a unit vector from the same point to the light source, **v** is a vector in the direction of the observer, **r** is a unit vector that indicates the direction that would be taken by the reflected light if the material would have a perfect reflectivity, f is measure of the specularity of the surface, I_s is the source intensity, r_d is the diffuse reflectivity (or as it is often said the diffuse color) and r_s is the specular reflectivity. The diffuse reflectivity is in fact the Lambertian part of the reflected light while the specular reflectivity depends on the direction of observation. There are diffuse and specular reflectivity coefficients for the red, blue and green components of the light. This lighting model is more empirical than physical and consequently should be a crude approximation. Nevertheless, it has shown itself to be quite successful in realistically represent different kinds of materials. It is thus possible to identify a given material M by a set of seven coefficients

$$M \therefore \begin{bmatrix} R_d & G_d & B_d & R_s & G_s & B_s & f \end{bmatrix} \tag{15}$$

where R, G and B stand for red, green and blue. All these coefficients are defined on the interval [0,1] except for f, which is sometimes defined on the interval [0,128]. If the model is made out of one material, these coefficients constitute an acceptable definition of it. If the model is made of different kinds of materials, then it is possible to represent the statistical distribution of these materials by a set of seven histograms

$$M \therefore \left[h(R_d) \quad h(G_d) \quad h(B_d) \quad h(R_s) \quad h(G_s) \quad h(B_s) \quad h(f) \right] \tag{16}$$

The search engine uses these histograms in order to retrieve, in the database, similar models.

The texture can be seen in two different ways: as a pattern that is mapped on an object or as the complete color description of the object. Most of the time, the texture is an intensity, *RGB* or *RGBA* (alpha: transparency) image. We will not consider the transparency here. The texture is usually represented on a bidimensional orthogonal system. The coordinates, named S and T, are normalized. If the model also has a material description, it is possible to modulate the diffuse color of the model by the texture

$$\begin{bmatrix} R_{out} \\ G_{out} \\ B_{out} \end{bmatrix} = \begin{bmatrix} R_d R_{tex} \\ G_d G_{tex} \\ B_d B_{tex} \end{bmatrix} \tag{17}$$

In this case, the analysis of the texture does not need to be performed independently since we are interested in its resulting effect on the color distribution. If the texture is the sole color descriptor of the model then we can obtain a statistical description of the color distribution of the model by calculating the histograms of the red, green and blue distribution. If both material and texture are available, it would possible to analyze them independently although we do not recommend it because, generally speaking, the material and the texture are a twofold manifestation of a same entity: the color distribution.

5. Synergy

Up to now we have analyzed the scale, the shape and the color distribution as separate entities. Of course if somebody is performing a search in a database, he may be interested in combining them. For example the user can ask: "I would like an object of this size, looking like this one with a bit of red but the red is not that important". Our task is to translate such a query, into a simple "interface language". Basically the interface can have a checklist with scale, shape and color. The user just checks what they need. The last part of the sentence is about the relative weight of each parameter. According to this sentence, a low weight should be given to the color parameter. In general, the weights are determined according to type of search performed by the user.

Basically the weight can be used in two different ways: by error or by rank. In the first case, an error is computed for the scale, the shape and the color and a global error is defined. In the simplest case it can be

$$err_{global} = \frac{w_{scale}\, err_{scale} + w_{shape}\, err_{shape} + w_{color}\, err_{color}}{\left(w_{scale} + w_{shape} + w_{color}\right)} \tag{18}$$

The problem with that approach is that the sensitivity of the scale, shape and color algorithms may be quite different. Thus the parameter with the highest sensitivity may have the dominant effect. The other possibility is to replace the error by the rank. The classification is made separately for the scale, shape and color. Then the global rank is determined as follows:

$$rank_{global} = \frac{w_{scale}\, rank_{scale} + w_{shape}\, rank_{shape} + w_{color}\, rank_{color}}{\left(w_{scale} + w_{shape} + w_{color}\right)} \tag{19}$$

The errors are computed by comparing between them two sets of histograms. Since the number of channels is usually small (typically ten or forty) we compute the error as follow

$$err = \sqrt{\sum_{i=1}^{n}\left[h_{input}(i) - h_{ref}(i)\right]^2} \tag{20}$$

i.e. that we use the quadratic error.

6. Experimental Results

All of these algorithms are user transparent. In order to perform a search, a person only needs to choose the input model and decide if the search is based on scale, shape or color. The input model [11] can be taken directly from a database or from the outside world. In the last case the model can be edited in order to fit the user requirements. The material of the model can be modified by simply choosing a new one from a list (e.g. plastic, gold and so on). Editing can be performed directly on the diffuse and on the specular colors by picking them from a palette or tuning them to a suitable value by using a set of RGB-HSV sliders.

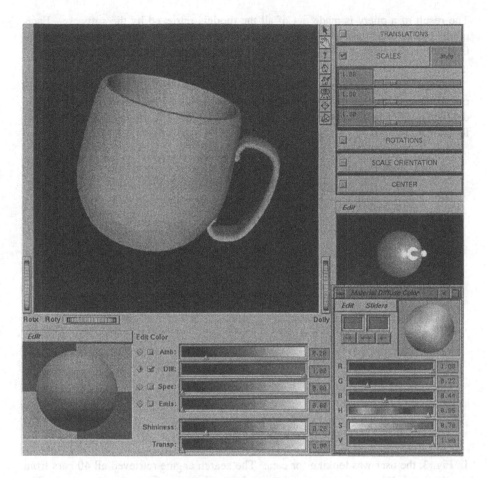

Fig. 1. User interface.

It is also possible to modify the shape of the input model in order to satisfy the user's requirements. For example, a coffee cup can be transformed easily into a pencil cap and a duck into a loony by clicking and dragging operations.

Fig. 2. Transformation of a cup into a pencil cap.

The result of a query is made out of all the models retrieved by the software. Each model can be examined with a viewer in three dimensions by using stereo glasses. A CGI program written in C++ handles each query on an applications server.

Results obtained with our VRML database are presented. This database is made out of more then 1000 models ; most of them are in color. Sample results from four distinct queries are shown below. The reference object is situated at the top-left location.

Fig. 3. Results for the cars.

In Fig. 3, the user was looking for cars. The search engine retrieved all 40 cars from the database; only the first six are shown. A search was performed with a plane. The search engine found all the planes belonging to the database; only three are shown.

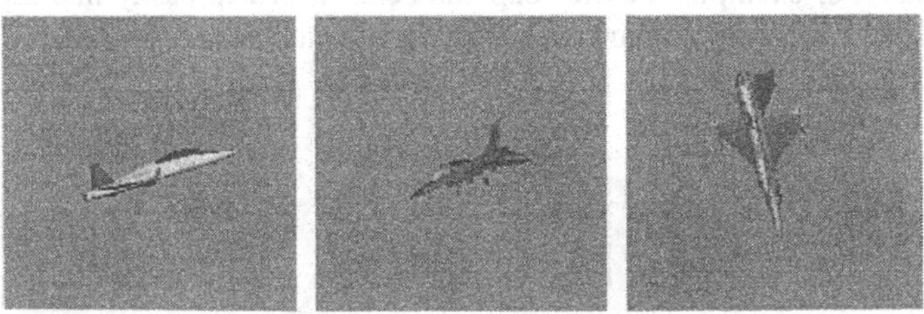

Fig. 4. Results for the planes.

A search was also performed with a street lamp. There were no other street lamps in the database but the search engine retrieved similar models.

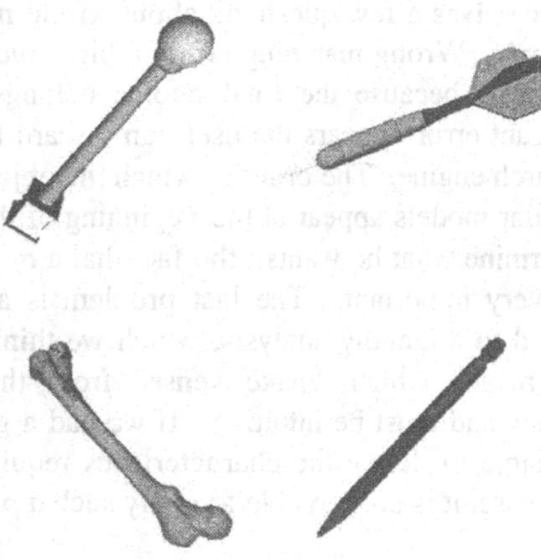

Fig. 5. Results for the street lamp.

In Fig. 6, the user was looking for brontosaurus. The search engine retrieved all of them even if they had different attitudes. This example indicates that our method can be applied to non-rigid object; at least to a certain extend.

Fig. 6. Results for the brontosaurus.

7. Conclusions

These examples clearly indicate that the proposed algorithms are valuable tools for performing queries by content in three-dimensional image and model databases. At this point we can ask ourselves a few questions about wrong matching, classification and definition of similarity. Wrong matching is not a big issue as long as it does not happen too often. This is because the final choice belongs to the user, not the algorithm. So if a flagrant error appears the user can discard the model as he would do with a text-based search engine. The order in which the objects appear is irrelevant as long as the most similar models appear at the beginning of the list because the user looks at this list to determine what he wants: the fact that a model appears in third of fourth position is not very important. The last problem is about the definition of similarity. This can lead to a lengthy analysis, which we think is unnecessary. The algorithm must yield results which "make sense" from the user and from the geometrical point of view and must be intuitive. If we had a global model of human vision it would be possible to define the characteristics required by the algorithms. But since this is not the case, it is not possible to apply such *a priori* approach.

The discrimination of the wavelet coefficients can be very high because they contain all the information about the original object. When a wavelet transformation is applied to an object no information is created nor lost. The data reduction scheme used provide us with an acceptable knowledge of the inter-scale signal distribution while the knowledge of the intra-scale distribution can be considered fragmentary. If two objects share the same inter-scale distribution they may be identified as similar even if their shape is not alike. If the system can process bigger descriptors, the problem can be solved by computing a set of moments for each scale in order to obtain a better description of their internal statistical distribution.

An important issue concerned the number of objects that the system can handle: the bigger the database, the higher the probability to find similar models and the higher the probability to have false detections. That comes from the fact that each descriptor contains only a small fraction of the total information about a given object. The classical solution to this problem is to increase the "resolution" of the descriptor by adding more channels to the histograms and by using more types of descriptors. Of course, there is a tradeoff between the total size of the descriptor and its discrimination capability, which is determined by the patience of the user waiting for the answer... To solve this problem we propose first to search the database with a compact version of the descriptor in order to obtain rapidly a short list of interesting models and then to clean up this short list using a more sophisticated version of the descriptor. One can also use simultaneously many descriptors and combine them with an equation similar to equation (19).

Before presenting our final conclusion, we would like to comment on the conditions of failure. In our case the most important condition of failure comes from the technique used to determine the reference frame. This technique is global because the reference frame is assimilated to the eigen vectors of the tensor of inertia. Let us suppose that we have two planes, the second being identical to the first except for a wing that is missing. If we compute their respective reference frames, we will end up with one reference frame centered on the fuselage and the other reference frame centered on the wing. Since the calculations of the descriptors are based on the reference frames, it is very unlikely that the descriptors be similar. For this reason, we are presently looking for a more robust way to define the reference frames.

We have presented a system and a set of algorithms for content-based three-dimensional models databases search. The software is intuitive and efficient and should be a valuable tool for three-dimensional model databases management. Future work will involve metadescriptions and neural network approaches.

References

[1] A. L. Ames et al., "The VRML Sourcebook", 1996, p. 1-7.

[2] I. Herman et al., "MADE: A Multimedia Application Development Environment", *Proc. of the IEEE International Conference on Multimedia Computing and Systems*, 1994, pp. 184-193.

[3] A. del Bimbo et al., "3-D Visual Query Language for Image Databases", *Journal of Visual Languages and* Computing, Vol. 3, 1992, pp. 257-271.

[4] C. Faloutsos et al., "Efficient and Effective Querying by Image Content", *Journal of Intelligent Information Systems: Integrating Artificial Intelligence and Database Technologies*, Vol. 3, 1994, pp. 231-262.

[5] S. Muraki, "Volume Data and Wavelet Transforms", *IEEE Computer Graphics and Application*, 1993, pp. 50-56.

[6] J. A. Schnabel and S. R. Arridge, "Hierarchical Shape Description of MR Brain Images Based on Active Contour Models and Multi-Scale Differential Invariants", *Proc. in Images Fusion and Shape Variability Techniques*, 1996, pp. 36-43.

[7] T. R. Downie et al., "A Wavelet Based Approach to Deformable Templates", *Proc. in Images Fusion and Shape Variability Techniques*, 1996, pp. 163-169 .

[8] D. J. Field, "Scale-Invariance and Self-Similar "Wavelet" Transforms: an Analysis of Natural Scenes and Mammalian Visual Systems", *Wavelet, Fractals, and Fourier Transforms*, Clarendon Press, Oxford, 1993, pp. 151-193.

[9] H. S. Sawhney and J. L. Hafner, "Efficient Color Histogram Indexing", *IEEE International Conference on Image Processing*, 1994, pp. 66-70.

[10] E. Paquet, "Recognition of Polychromatic Range Images based on Chromatic Sampling of Scalar Products and Logarithmic Histograms", *International Journal of Optoelectronics*, Vol. 10, 1995, pp. 251-255.

[11] K. Hirata and T. Kato, "Query by Visual Example: Content-Based Image Retrieval", *Proc. E. D. B. T. '92 Conf. on Advances in Database Technology*, Vol. 580, 1994, pp. 56-71.

Web-WISE: Compressed Image Retrieval over the Web

Gang Wei, Dongge Li & I. K. Sethi
Vision and Neural Network Lab.
Dept. of Computer Science
Wayne State University
Detroit, MI 48202
{gaw, dil, sethi} @cs.wayne.edu

Abstract

Web-WISE is a system designed to address the need for efficient content-based seeking and retrieval of images on the web. It supports searching by multi-features, including color and texture. Web-WISE contains three automatic components, which are 1): Internet Agent responsible for searching the Web to fetch images; 2): Analysis Agent, which extracts the color and texture features of color JPEG images directly from compressed domain and 3): Query Agent, which explores the image database using certain metrics and returns a set of images (thumbnails) visually similar to the query image. Due to its generic design, adding other features to the system will be very natural, including the ability to index and retrieve other types of visual information and retrieval based on edge and shape. This paper presents the theoretic and practical background of the system, the schemes of these three components and the performance of the system.

Keywords: WWW, content-based retrieval, DCT, feature extraction, image database

1. Introduction

The explosive growth of images and videos on the World Wide Web is making the Web a huge resource of visual information. Today efficient content-based retrieval of images over WWW is becoming increasingly desirable due to the needs motivated by various applications. A straightforward solution is to store the images and their content descriptions (image visual features) in a local database and use a query scheme to implement the content-based retrieval. However this is not always feasible because of the large size of image files and the huge amount of images on the Web. The practical way should fully take advantage of the Web as storage rather than merely a resource of images. After extracting the features of the original images, we can just keep a reduced version of the images (thumbnails) and their corresponding locations (URLs) on the Web. Then the original images are deleted to save disk space. Then URLs can be given with thumbnails as the query results, enabling the user to access the original images from the Web.

Web-WISE is such a system. The name "WISE" stands for "Wayne Image Seeker". It employs a search engine to download images from the Web. Then it extracts features from the images in compressed domain and stores feature vectors in the database. Through a graphic interface, it enables users to do content-based image retrievals by clicking thumbnails. The result thumbnails of a query are displayed in descending order of the similarity values and are presented to the user as well as URL's.

2. An Overview of Web-WISE

2.1 Structure

Web-WISE is in nature a content-based WWW image search engine. There have been a lot of research work and practical applications about search engines over the Web. Yahoo, AltaVista and Lycos are typical applications of this technique. Most of these kind of search engines are keyword-oriented, which enable users to retrieve information by typing in the subjects they are interested in. The documents are indexed by keywords extracted from them. The system developed by J.R. Smith and S.F.Chang supports both keywords and content-based retrieval [2]. It categorizes image by keywords in a hierarchical structure and extracts color regions of images and get their color histograms. Keyword-based systems work well for tasks like "Give me images of this type". But if we need a query like "Give me images that look like this", content-based retrieval has to be employed. Research carried out by B. Agnew et al. [3] and O. Munkelt et al. [4] index images by their visual contents however using different metrics. In Web-WISE, we emphasize on content-based retrieval and feature extraction directly from compressed domain.

Web-WISE consists of three modules: Internet Agent, Analysis Agent and Query Agent. Those modules work independently. The overall structure of the system is illustrated by Fig. 1.

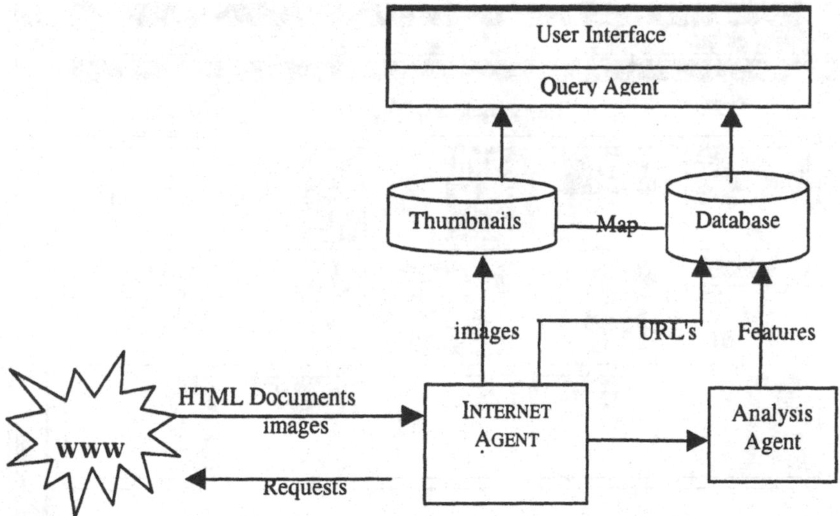

Fig. 1: The Overall structure of Web-WISE

- The Internet agent automatically traverses the Web by following hypertext links, fetching HTML documents and JPEG images to the local disk. The locations (URL's) of the images are also stored in a list in plain text format.

- Images are fed to the Analysis Agent which directly extracts color and texture features of the images in compressed domain. The feature values are imported to the database.
- Thumbnails are saved in a separate directory. There is a table which maps between the thumbnail filenames and the corresponding records in the database. URL list is also imported to the database.
- Query agent gets query image name from user interface, selects candidate images from the database, compares the query image with each candidate and assigns the candidate with a similarity value.
- The result thumbnails are displayed in descending order of their similarity values.

2.2 User Interface

Fig. 2 is the users interface of Web-WISE. To do an iteration of query, the user simply clicks the thumbnail. Randomly preview of the images in the database is allowed by clicking the "Random" button. As both global and local dominant histogramming retrievals are implemented, the user can try different schemes by choosing corresponding experiment numbers. Several parameters, which are to be discussed later, for similarity computations can also be set through the interface (however, each parameter has its default value).

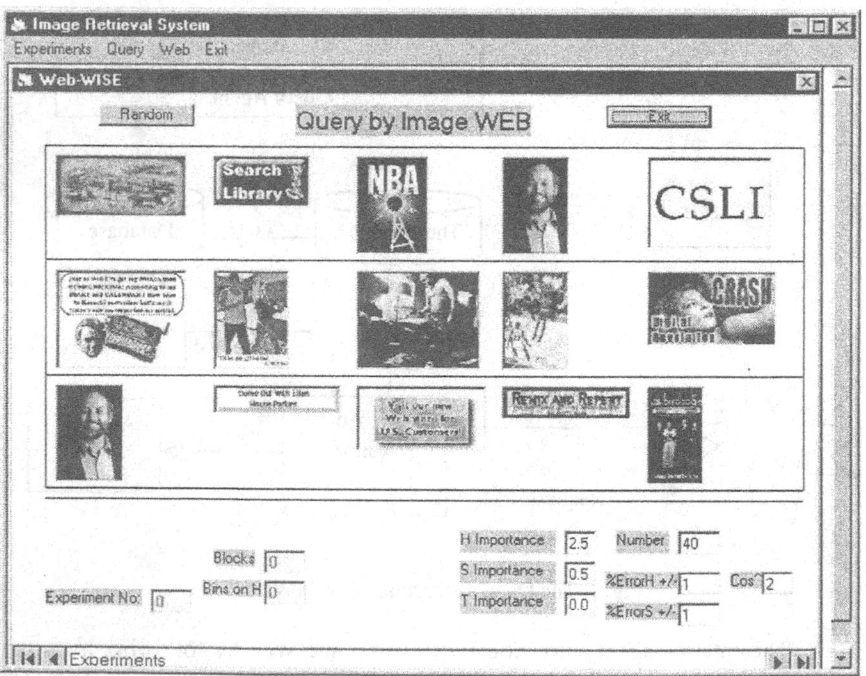

Fig. 2 Web-WISE interface

3. Searching Images on the Web

From a user's view, the Web is a large set of page documents, many of which contain reference to each other by hyperlinks. In this sense, the Web is a partially connected directed graph, with each page considered as a node and links as edges. Images on the Web are available through two forms: embedded and linked. To acquire images, a search engine should search the Web in some order from an arbitrary starting node and parse HTML documents to detect and download images.

The component in Web-WISE which explores the web for images is called Internet Agent. It is composed of three interactive sub-modules: GetLink, GetImage and Parsing Module.

The GetLink and GetImage modules are responsible for downloading HTML documents and images, respectively, from the URL's given by the Parsing Module. They hide the details of TCP/IP connections and HTTP requests from the Parsing Module. The Parsing Module analyzes the HTML documents fetched by GetLink module to extract URL's of JPEG images and links to more HTML documents, then send the URL's to the GetImage module and GetLink module.

The Internet agent traverses the Web in breadth-first search order. Starting from a designated URL, each time the GetLink Module fetches an HTML document and passes it to the Parsing Module. The Parsing Module is a push-down automaton model which analyzes HTML documents. When it detects the presence of a JPEG image, the URL is immediately sent to the GetImage Module to get the image. When a hyperlink to an HTML document is detected, however, it puts the URL to the rear of a queue. After Parsing Module finishes process a whole HTML, the GetLink Module takes the URL at the top of the queue to get another HTML document. The new document is sent to Parsing Module to start a new parsing. In implementation, several details deserve special attention.

- Certain algorithms need to be used to detect and avoid duplicate downloading.
- The depth of the search should be controlled (otherwise the searching will not terminate in some cases).
- The starting point of the agent should be delicately chosen to improve efficiency.

4. Feature Extraction and Representation

In content-based image retrieval, there should be some method to describe the content of the image. That is, some features should be extracted from the images so that the content similarity between two images could be evaluated. In Web-WISE, this is accomplished by the Analysis Agent. As have been mentioned before, the search engine fetches JPEG images, which are in compressed form. By extracting features directly from compressed data, we reduced the cost of computation significantly.

4.1 JPEG/DCT Model

Here we briefly review the JPEG/DCT compression algorithm [16] before describing how to extract feature information from compressed data. In baseline JPEG algorithm, the original image is transformed to frequency domain using forward discrete cosine transform (FDCT) on a block by block basis so that coefficients within the blocks are decorrelated and can thus be treated independently without loss of compression efficiency. The standard size of the block is 8*8. Then a weighted quantization is applied to each coefficient within the blocks to discard visually unimportant coefficients. This is a lossy process. Each quantized DCT block will become a sparse matrix. Then a lossless entropy coding is performed to obtain further compression. To decompress a JPEG image, entropy decoding is first applied to get quantized coefficients, then dequantization is performed, and the original image can be reconstructed by performing Inverse discrete consine transform (IDCT). FDCT and IDCT are the central parts of the JPEG compression/decompression schemes, which are defined as:

$$F_{uv} = \frac{C_u C_v}{4} \sum_{i=0}^{7} \sum_{j=0}^{7} \cos\frac{(2i+1)u\pi}{16} \cos\frac{(2j+1)v\pi}{16} f(i,j) \quad (1)$$

And

$$f_{ij} = \sum_{u=0}^{7} \sum_{v=0}^{7} \frac{C_u C_v}{4} \cos\frac{(2i+1)u\pi}{16} \cos\frac{(2j+1)v\pi}{16} F(u,v) \quad (2)$$

Where

$$c_u, c_v = \frac{1}{\sqrt{2}} \quad \mathit{for}\ u,v=0,\ \mathit{and}\ 1\ \mathit{othewise}$$

JPEG compression uses YCbCr color coordinate system, in which Y is the luminance component while Cb and Cr are chromatic components. Due to the aspects of human visual system, the sampling scale of Y is "finer" than that of Cb and Cr.

4.2 Properties of DCT coefficients

In JPEG decompression processes, IDCT is the most time-consuming part. Our approach extracts feature directly on the DCT coefficients without transforming them to spatial domain, thus considerable amount of time is saved.

By definition, a DCT coefficient F_{uv} is a linear combination of all pixel values in the block. Particularly, coefficient F_{00} (DC coefficient) is the average value of the original pixels in the block. If we extract the DC's of each block to construct a new image, it will be a downsampled version of the original one [9]. Fig. 3 compares the original image and the reconstructed image using only DC coefficients of component Y. Here we enlarged the reconstructed image. Although looking rather blocky, it bears a lot of information of the original one.

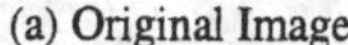

(a) Original Image (b) Reconstructed Image

Fig. 3 Original Image and its Reconstructed Image

Color histogram reflects the color distribution in an image. We observed that the histogram constructed on DC coefficients resembles closely that built on the original image. Fig. 4 indicates this fact. Thus we can use the reconstructed histogram as an estimation of the color distribution for the original one.

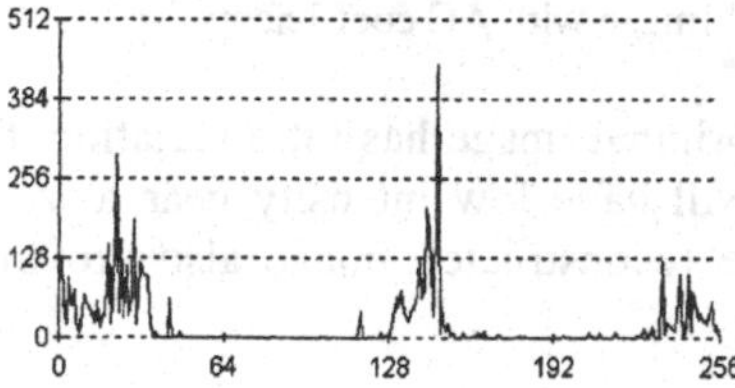

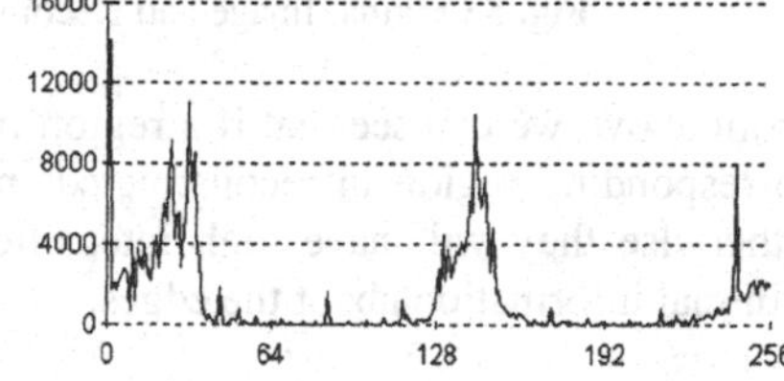

(a) Histogram of the Reconstructed Image (b) Histogram of the Original Image

Fig. 4 Histograms of the reconstructed and original images

The above two figures are the histograms of the hue component of the reconstructed image and original image, respectively. To get Fig. 4 (a), the DC coefficient in YCbCr should be transformed to HSI (hue, saturation and Intensity) color coordinate system first. The motivation of this transformation will be discussed in the next section.

The remaining coefficient (AC coefficients) reflects variations within a block [7]. For example, F_{10} essentially depends on the intensity difference in the vertical direction within a block. This relationship can be explained by a simple analysis of the definition of discrete cosine transform.

$$F_{10} = \frac{c_1 c_0}{4}\sum_{i=0}^{7}\sum_{j=0}^{7}\cos\frac{(2i+1)\pi}{16} f(i,j) = \frac{c_1 c_0}{4}\sum_{i=0}^{7}\cos\frac{(2i+1)\pi}{16}\sum_{i=0}^{7} f(i,j) \tag{3}$$

Considering cos $(\pi-\theta)=-\cos\theta$, the above equation can be expanded as

$$F_{10} = \frac{C_1 C_0}{4}\left[\cos\frac{\pi}{16}\left(\sum_{i=0}^{7} f(0,j) - \sum_{i=0}^{7} f(7,i)\right) + \cos\frac{3\pi}{16}\left(\sum_{i=0}^{7} f(1,j) - \sum_{i=0}^{7} f(6,j)\right)\right.$$
$$\left. + \cos\frac{5\pi}{16}\left(\sum_{i=0}^{7} f(2,j) - \sum_{i=0}^{7} f(5,j)\right) + \cos\frac{7\pi}{16}\left(\sum_{i=0}^{7} f(3,j) - \sum_{i=0}^{7} f(4,j)\right)\right] \quad \textbf{(4)}$$

Similarly, F_{01} and F_{11} depend on the variation in the horizontal and diagonal direction, respectively. Fig. 5 shows the image constructed with the maximum value of F_{01}, F_{10} and F_{11} within blocks of Y component.

(a) Original Image

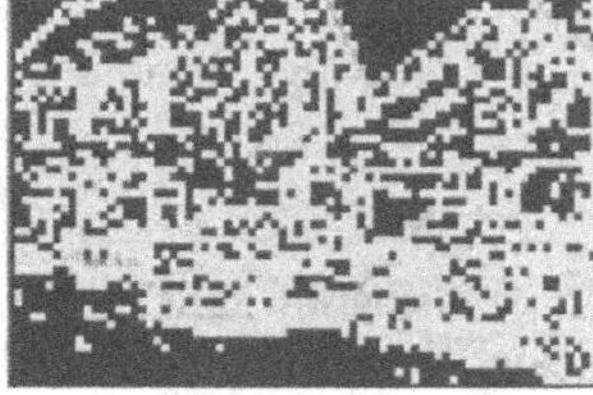

(b) Reconstructed Image

Fig. 5 Original Image and Reconstructed Image with AC coefficients

From above we can see that if a region in the original image has little variation, the corresponding region in reconstructed image will have low intensity near to zero. Otherwise they will have high intensities. The reconstructed image also provides minimal information about the edges.

4.3 Feature Extraction and Representation

All content-based image retrieval systems employ a set of features to describe images in the database. Color is maybe the most widely used feature. Reflecting the distribution of color within images, color histograms prove to be an efficient feature in content-based retrieval. I.K. Sethi et al. employed localized dominant hue and saturation values [1], which can achieve better performance than the use of global or local histogramming because it contains the information of the spatial distribution of color. J.R. Smith and J.K.Chang isolated color regions using back-projection and extracted properties for each region.

In Web-WISE, histograms are built on images reconstructed with only DC coefficient. Those images have already been reduced in size by 8*8 (Y component) or 16*16 (Cb and Cr components). If we further divide them using fixed image partition or color region extraction scheme, the number of pixels in each region may be too small to give a description of the feature with desired precision. Therefore, global histogram is the default criteria in determining the similarity although local histogramming retrieval also supported in our system.

Color features are represented in HSI (Hue, saturation and intensity) color coordinate system, which correlates human visual system better than YCbCr space. Hue and saturation are color components in which hue refers to the average spectral wavelength and saturation reflects the purity of the color. Intensity stands for brightness. After transforming the YCbCr components of the JPEG images to HSI color space, the Analysis agent calculates the global histograms of H and S channels.

Since each hue or saturation value is represented with eight bits, there are 256 bins in each histogram, which makes the number of dimensions of the feature vector very large for each image. To get compact representation of image feature for efficient storage and retrieval, quantization of the histograms is required. Noting that human visual system is more sensitive to of hue information, the quantized hue histogram has 16 bins while saturation histogram has 8 bins. Non-uniform quantization is applied to get better performance [17]. The distribution of hue and saturation values of all the images in database are calculated to decide the range of each bin so that all bins contain the same number of pixels as far as the all images in the database are concerned. This ensures that for any given pixel, its probability of falling into any bins is equal.

The texture feature is used as an auxiliary measure in Web-WISE. The maximum values of F_{01}, F_{10} and F_{11} of each DCT block are used to reconstruct an image as Fig.5 (b). Relying on a fixed image-partitioning scheme, we divide the reconstructed image into 4*4 blocks and compute the average intensity of each block. Each value reflecting the extent of variation in that block, the spatial distribution of texture can be represented by the 16-dimension vector.

Thus we represent an image with a 40-dimensional vector, in which 16 dimensions and for hue, 8 for saturation and 16 for texture.

5. Similarity Metrics and Image Query

The results of the retrieval are determined by the similarity between the query image and target images in the database. This section describes how we compute the similarity between two images are how the result of a query is generated.

5.1 Similarity Computation

As each image is represented by a 40-dimensional vector, the whole image database could be seen as a 40-D space where all images are distributed. In a query like "find k images that is most similar to this image", actually we are trying to decide the k nearest neighbor of the query image in this 40-D space, using some distance metrics. The smaller the distance is, the closer two objects are and thus the larger the similarity value.

The Query Agent gets the name of the query image from the user interface, then gets the feature vectors for the query image and the candidate images from the database. Distance and similarity values between query image and each candidate image are calculated. The interface displays the thumbnails of result images in descending order of the similarity values. Since the sensitivity of human visual system may vary a lot to different features, 3 coefficients for the three sections (hue, saturation and texture) of the feature vector can be specified through the user interface to indicate the significance of each feature in determining the similarity between images. The similarity between two images Q and T is defined by the following formula:

$$s(Q,T) = \frac{1}{(1+(aD_h(h_Q,h_T)+bD_s(s_Q,s_T)+cD_t(t_Q,t_T)))} \quad (5)$$

Here D_h, D_s and D_t are functions to calculate the vector section distances between Q ant T in hue, saturation and texture, respectively. Their parameters are feature vectors defined in section 4.3. The coefficient a, b and c are importance weights for the vector section distances, which are to be specified by the user. By default, a > b (we assume more importance to hue than to saturation), c is set to zero (texture is used as an auxiliary feature). Let M be the number of dimensions of the vectors, D_h, D_s and D_t are defined as:

$$D(u,v) = 1 - \frac{\sum_{i=0}^{M-1} \min(u(i),v(i))}{\min(\sum_{j=0}^{M-1} u(j), \sum_{k=0}^{M-1} v(k))} \quad (6)$$

5.2 Image Filtering: Reduce the number of candidates

The similarity metrics described have some drawbacks. First, the computation complexity is high. If the similarity computation is applied to all the images in the database, there would be no surprise that the system is slow. Secondly, all bins within one section have the same impact on the distance measure. However, human visual system is usually more sensitive to the dominant color of images. In another word, to get better performance, the bins with larger values should assume greater significance than those with smaller values in determining the distance of vectors.

In query agent, a simple but effective method is used to solve the above two problems. For each image record in the database, an extra field, max_H_bin, is added, which represents the number of the bin with the maximum value (this is the dominant color bin) in the hue feature vector. In the query process, this field can be used as a filter to kick out images that do not seem promising so that the number of candidate images to be applied the similarity computation can be greatly reduced. For each image in the database, if the difference between its max_H_bin and that of the query image is greater than a threshold, then their dominant color is very likely to be so different that it is not worth further analysis as a potential result. In this case, this image can be discarded and the similarity value is simply set to 0. Otherwise it is chosen as a candidate to which the similarity calculation described above will be applied. For further filtering, one more field, max_S_bin could be used with similar method.

After the filtering process, we have a set of candidate. Apparently, applying similarity computation only on these images can achieve great speedup. Each candidate gets a value through the similarity computation. Images with the largest similarity values are returned as the query results.

6. Performance Evaluation

Presently there are about 1000 images in the database of Web-WISE. As two retrieval schemes (global and local dominant histogramming) are supported, there are two tables in the database. An image is represented with one record in each table. For a default query, the global histogramming is used. The record contains a 40-D feature vector, the URL of the image, and two additional fields representing the numbers of bins with maximum values for hue and saturation, for the purpose described in Section 5.2. Fig. 6 is an example of the result of a default query. In this query only global hue and saturation are compared, with the importance coefficient of each being 2.5 and 0.5, respectively. The upper-left image is the query image and the numbers above each image are the similarity values. Note that the value for query image itself is 1.

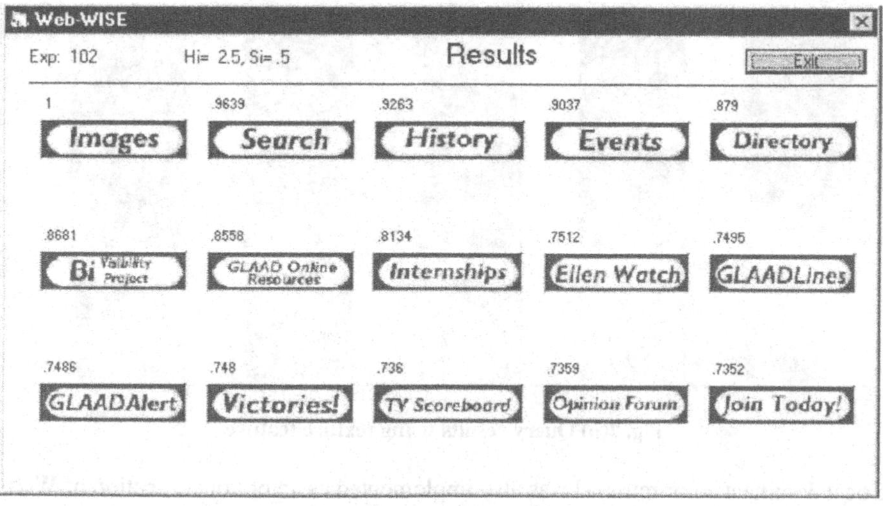

Fig. 6 The set of result images

Texture is an optional feature used in Web-WISE. Its effect on similarity value can be enabled or disabled by setting its weight to 1 or 0 from the user interface. Fig. 7 compares the performances of queries with and without using texture for the same query image.

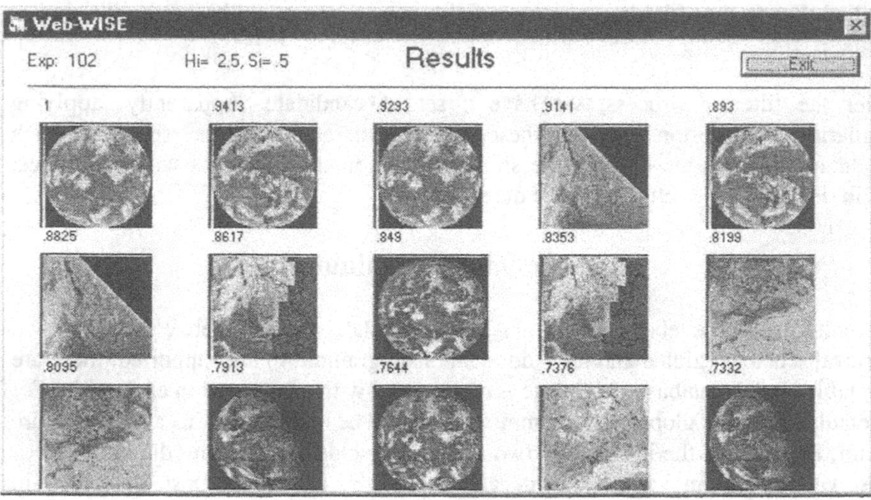

Fig. 7 (a) Query results without using texture feature

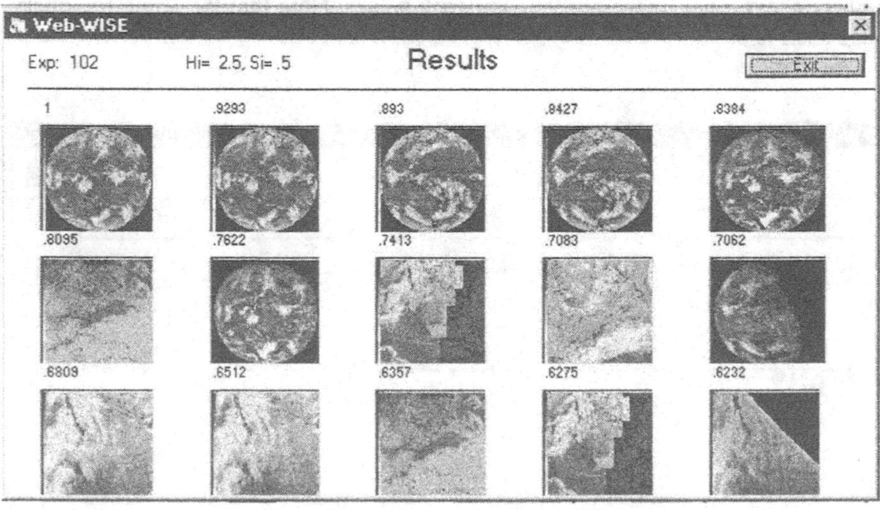

Fig. 7(b) Query results using texture feature

Local dominant color retrieval was also implemented as an optional function of Web-WISE. In this scheme, each image is partitioned into 4*4 blocks. For each block we extract the dominant hue and saturation value (peak area value, as described in [1]). By doing this we represent an image with a 32-D vector, 2 dimensions for a block. Fig. 8 compared this scheme with the global histogramming retrieval. Due to the reason explained in Section 4.3, this scheme outperforms global histogramming scheme only occasionally. Moreover, for image less than 32 pixels wide or high, the local scheme will fail in the feature extraction stage (an image constructed with DC coefficients is 8*8 downscaled, thus an image smaller than 32*32 pixels can not be partitioned 4*4 properly). These images have to be discarded.

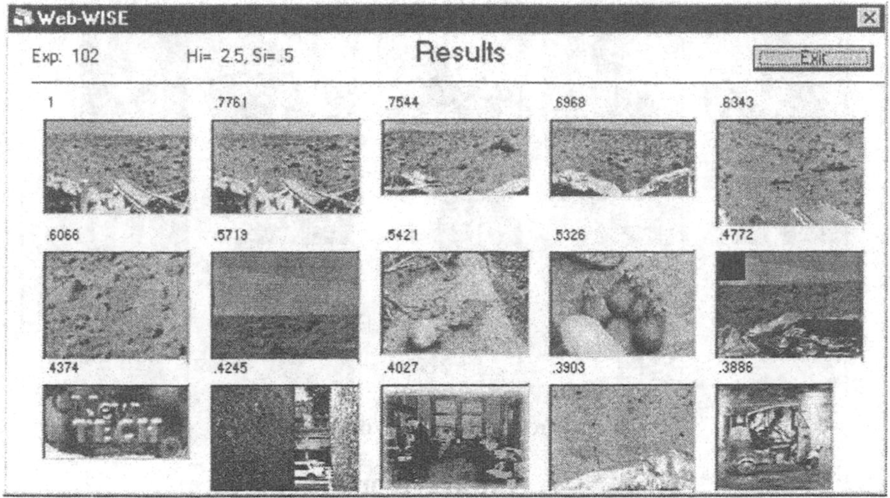

Fig. 8 (a) Query result using global histogram

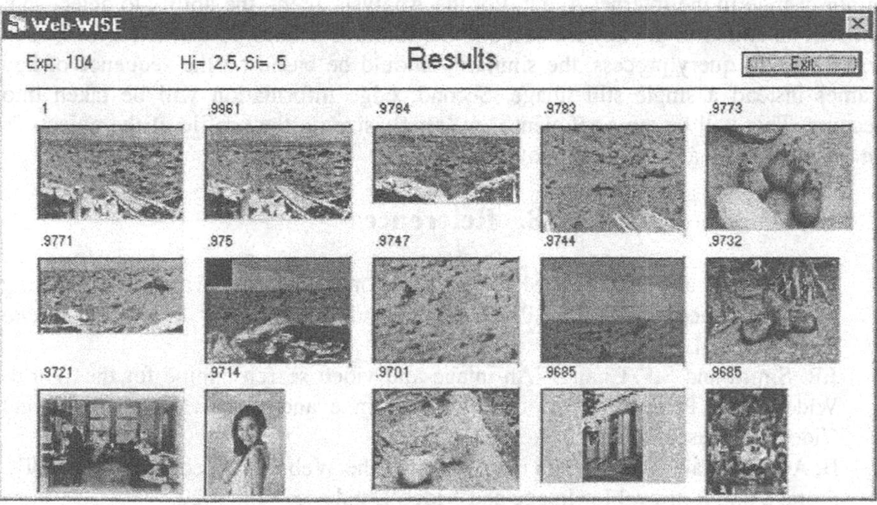

Fig. 8 (b) Query result using local dominant features

However, we observed that local scheme is efficient in retrieving images with salient color regions, provided the images are sufficiently large. This fact is illustrated in Fig. 9.

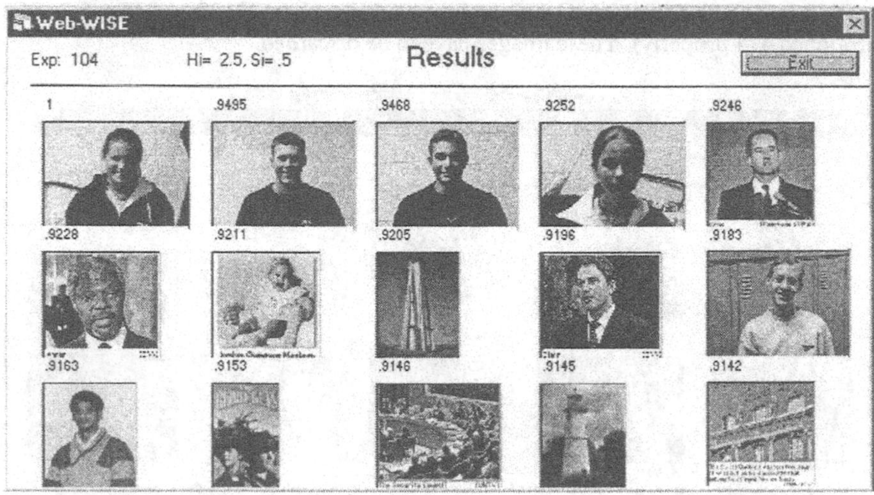

Fig. 9 Query results using local dominant color

7. Future Work

Web-WISE is a highly adaptive system. Two extensions will be explored in our future research. First, we will enable it to retrieve videos on the Web. This requires a minor change in the Internet Agent. For the Analysis agent, the ability to detect shot cut will be added to get key frames, and the features should be extracted from each key frame. In query process, the similarity should be based on the sequence of key frames instead a single still image. Second, edge information will be taken into account. This will be more efficient if we emphasize on the profile of the objects in an image rather than the distribution of color.

8. References

1. I.K. Sethi et al., "Color-WISE: A system for image similarity retrieval using color," Proceedings of the SPIE: Storage and Retrieval for Image and Video Database, Vol. 3312.
2. J.R. Smith and S.F. Chang, "An image and video search engine for the World-Wide-Web," Proceedings of the SPIE: Storage and Retrieval for Image and Video Database, Vol. 3022.
3. B. Agnew et al., "Multimedia indexing over the Web." Proceedings of the SPIE: Storage and Retrieval for Image and Video Database, Vol. 3022.
4. O. Munkelt et al., "Content-based image retrieval in the World Wide Web: a web agent for fetching portraits," Proceedings of the SPIE: Storage and Retrieval for Image and Video Database, Vol. 3022
5. M. Waston, "Programming Intelligent Agent for the Internet", Computing McGraw-Hill
6. B.L. Yeo, "Efficient Processing of Compressed images and video," Ph.D. dissertation, Dept. of Electrical Engineering, Princeton University

7. B. Shen, I.K. Sethi, "Direct feature extraction from compressed images," Proceedings of the SPIE: Storage and Retrieval for Image and Video Database, Vol. 2670
8. N.V.Patel, I.K. Sethi, "Compressed video processing for cut detection ", IEE Proceedings, Vis. Image Signal Process., Vol. 143
9. X. Wan, C.J.Kuo, "Image Retrieval based on JPEG Compress Data," SPIE Proceedings, Multimedia Storage and Archiving Systems, Vol. 2196
10. J.R.Smith, S.F.Chang, "Tools and Techniques for color image retrieval," Proceedings of the SPIE: Storage and Retrieval for Image and Video Database, Vol. 2670
11. A. Berman, L. Shapiro, "Efficient image retrieval with multiple distance measures, " Proceedings of the SPIE: Storage and Retrieval for Image and Video Database, Vol. 3022
12. V.N. Gudivada, V.V. Raghvan, "Content-based Image Retrieval System," IEEE Computer, Vol. 28, No. 9.
13. H. Lu, B. Ooi and K. Tan, "Efficient Image Retrieval by Color Contents," Proceedings International Conference on Applications of Databases, 1994
14. S. Santini, R Jain, "Similarity Queries in image databases," IEEE International Conference on Computer Vision and Pattern Recognition, pages 646-651, San Francisco, CA, USA, June 1996.
15. M.Stricker and M. Orengo, "Similarity of color images, " Proceedings of the SPIE: Storage and Retrieval for Image and Video Database, Vol. 2420
16. G.K.Wallace, "The JPEG Still Picture Compression Standard", Communications of the ACM, vol. 34, no. 4, Apr. 1991
17. R.C.Gonzalez, R.E.Woods, "Digital Image Processing", Addison-Wesley Publishing Company

Embodying Semiotic Cues in Video Retrieval

J. Assfalg[1] C. Colombo[2] A. Del Bimbo[1] P. Pala[1]

[1] DSI — Università di Firenze, Firenze, ITALY I-50139
[2] DEA — Università di Brescia, Brescia, ITALY I-25123

Abstract. Video information processing and retrieval is a key aspect of future multimedia technologies and applications. Commercial videos encode several planes of expression through a rich and dense use of colors, editing effects, viewpoints and rhythms, which are exploited together to attract potential purchasers. Databases of commercials can be accessed in order to analyze how a commercial has been developed, retrieve commercials similar to an example, catalogue commercials according to the kind of message conveyed to the user. In this paper, we present a system allowing the retrieval of commercial streams based on their salient semantics. Semantics is regarded from the semiotics perspective: collections of signs and semantic features like colors, editing effects, motion etc. are used as basic blocks with which the meaning of a commercial is constructed. In our system, it is possible to retrieve commercials according to both the meaning they convey and to their similarity to examples.

1 Introduction

Modern multimedia archiving systems must be capable of handling huge collections of distributed and heterogeneous digital data, encompassing audio, video, text, graphics, etc. Both the size and the eclectic nature of the information involved call for the definition of novel strategies to process, store and retrieve data in a multimedia database. This raises in particular the problem of how to represent the data content in an automatic and unsupervised way.

That of content representation is one of the major issues in multimedia engineering. Building up a representation is tantamount to defining a model of the world through a *description language*, whose semantics captures only a few significant aspects of the information content. For instance, in an image database system where images are internally represented only by color, images of objects with the same color but different shapes would share the same internal description.

Diverse world representations are induced by different languages and semantics. In the case of text, for example, the meaning of single words is specific yet limited, and an aggregate of several words—i.e. a phrase—is needed to produce a higher degree of significance and expressivity. Hence, the rules for the syntactic composition of signs in a given language not only generate a language more complex syntactically, but also a new world representation, provided with a richer semantics in the hierarchy of signification.

Since each language sign can induce diverse semantics in different representations, a message equivocation may arise if the sender and the receiver do not share the same representation. In information retrieval, the user and the system exchange messages — in the form of queries and responses, respectively — in a common interface language. To avoid equivocation, a retrieval system should embed a semantic level reflecting as much as possible the one humans refer to during the interrogation — of course, it would be pointless for a user to search for dogs in a database which classifies all animals in a same category. In practice though, it is the user to fill the *semantic gap* by adapting his semantic level to the one embedded in the retrieval system. As an example, consider again color-based image retrieval, and suppose the user's task is to obtain images of country landscapes. Such a (high level) task will need to be translated by the user into the (lower level) system query language by providing the system with a set of color examples of typical country landscapes, obtaining in response an unsatisfactory set of retrieved images, refining the query set by a selection of the retrieved images, and so on — an iterative procedure which is usually both effort intensive and time consuming.

Examples such as the one above stress the difficulty of characterizing high level semantics of pictorial information (either in still images or video databases) through plain image perceptual features such as color, shape, spatial relationships, editing effects, motion, and so on. Yet, virtually all the systems based on automatic storage and retrieval of visual information proposed so far use such *perceptual representations* of the pictorial world which, when considered as language signs, have a quite limited semantics.

Much more than text, video conveys information through a multiplicity of planes of communication which encompass what is represented in the images, how the images are linked together, how the subject is imaged and so on.

The extraction and manipulation of the information embedded in video data is a key challenge for future multimedia technologies and emerging applications such as digital libraries and interactive video analysis. However, the advances of computer systems towards becoming "true" multimedia applications largely depend on the availability of tools for manipulating video semantic information.

Several recent papers have addressed aspects and problems related to the access and retrieval by content of video streams. Research on automatic segmentation has been presented by several authors [1], [2], [3], [4], [5]. All of them analyze inter-frame differences to automatically detect sharp and gradual shot transitions (cuts, fades and dissolves) and other special transition effects (such as mattes and wipes), and to estimate the motion of camera and objects in the scene. Automatic segmentation into higher level aggregates has also been addressed by some authors. In [6], a set of rules to identify macro segments of a video was proposed. Algorithms to extract story units from video are described in [7]. Also in [8], [9] the specific characteristics of a video type are exploited to build higher level aggregates of shots.

While significant research efforts have been devoted to the analysis of movies, news reports, and sport videos, the analysis of *commercial videos* has been virtually neglected by the research community. Only quite recently a few works explicitly addressing commercials investigate the possibility of detecting and extracting advertising content from a stream of video data [10]. The automatic characterization of a commercial video is complicated by the intrinsic and peculiar features of its time-varying content. First, the effectiveness of commercials is mainly related to its perceptual impact than to its mere content or explicit message: the way colors are chosen and modified throughout a spot, characters are coupled and shooting techniques are selected create a large part of the message in a commercial, while the extraction of canonical contents (e.g. imaged objects) has less conceptual relevance than in other contexts. Second, strict time requirements compel the director to make a condensed use of color, rhythm, camerawork, sound etc. Finally in commercials, traditional editing effects are augmented with novel and specific artifacts (e.g. computer graphics, cartoons, etc.), so as to draw at best the audience's attention and emotions to the product being promoted.

Until recently, commercials design and production has been a discipline based on a set of usually effective yet empirical rules. Formalized by professionals in the marketing field, such rules associate each single induced impression and emotion with given combinations of editing effects [11]. Quite recently, marketing companies have introduced semiotic methodologies in the process of spot making, so as to ground the principles of advertising into a solid scientific context, and to better combine together the artistic quality of a commercial with its communication effectiveness. According to semioticians, the analysis (and production) of commercials must be focused on the same narrative mechanisms and structures fairy-tales are based on: this leads to characterize a commercial according to four broad semiotic classes, or categories [12].

In this paper, we address the problem of retrieval by semantic content of commercials based on its vehiculated message as related to the semiotic level of the hierarchy of signification. In the presented system, a four-category semiotic characterization of commercials is produced by extracting a number of *perceptual level features of visual content* — colors, motion, cuts, dissolves etc. — and then combining them together according to an established set of rules. The semiotic categories are then used to access a database of commercial videos.

2 Semiotics and Commercials

Semiotics is the science of signs as carriers of *sense*. Semiotics suggests that signs are related to their meaning by social conventions (or, in the semiotics jargon, *codes*), i.e. by a specific cultural context.

Semiotic principles represent a reference and useful framework for the analysis of videos. Codes in a video include genre, editing effects (cuts, fades, dissolves, cutting rate and rhythm), camerawork (shot size, focus, camera movement, angle, slope of framing), manipulation of time (compression, flashbacks, flashforward, slow motion), and well defined choices of lighting, color, sound, graphics (text or cartoons) and narrative style. Through a semiotic analysis, the nature and use of each code can be highlighted, thus making it explicit how signs are properly organized for the construction of sense. According to semiotics studies, the narrative structure of commercials conforms to a four-element morphology closely related to the one introduced by Propp for fairy-tales [12]. Therefore, commercials can be classified into four different categories, related to the narrative element which is relevant w.r.t. the others. *Practical* commercials emphasize the qualities of a product according to a common set of values. These commercials represent everyday life scenes, commonplaces that are recognized by the audience. The product is described in a familiar environment so that the audience naturally perceives it as useful in everyday life. *Critical* commercials introduce a hierarchy of reference values . In this kind of commercials, the product is the subject of the story. It is a real story, allowing to focus on the qualities of the product through an apparently objective description of its features. *Utopic* commercials provide the definite evidence that the product is able to succeed in critical tests. In this kind of commercials, the story doesn't follow a realistic plot: rather, situations are presented as in a dream. Wide scenarios are used to present the product, which is shown to succeed in critical conditions often in a fantastic and unrealistic way. *Playful* commercials emphasize the match between user's needs and product's qualities. These commercials represent a manifest parody of the other typologies of commercials: it is clearly stated to the audience that they are watching advertising material. Situations and places are visibly different from everyday life, and deformed in such a caricatural and grotesque fashion that the agreement between product qualities and purchaser's needs is often remarked in an ironical way (e.g. an old woman driving a Ferrari at 30 Km/hour).

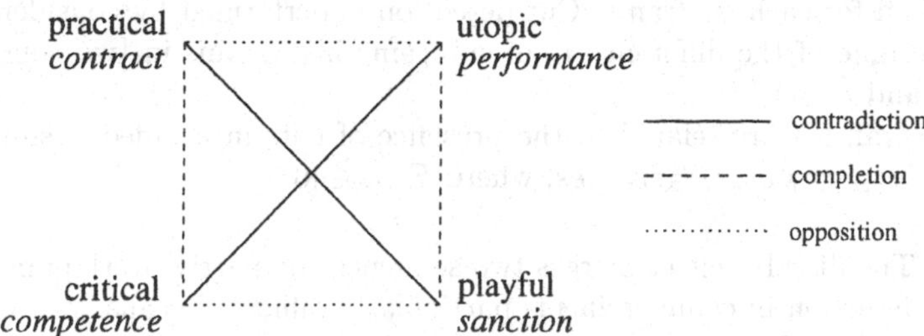

Fig. 1. The semiotic square for commercials.

A common representation of commercials categories uses the *semiotic square.* This square combines in pair each out of four semiotic objects with a same semantic level according to three basic relationships: opposition, completion, contradiction (see Fig. 1). Of these, only the last one has a quantitative characterization: this implies that the objects placed at opposite sides of the square diagonals are strongly related each other, being complementary.

In Sect. 4 it will be shown how all these somewhat abstract categories and statements can come into play in the creative process of spot making — the spot language.

3 Feature Extraction

In this section we introduce the perceptual-level features which are significant for a semiotic-level characterization, and summarize the algorithms used to extract such features automatically in a video.

3.1 Video Segmentation

The primary task of video analysis is its segmentation, i.e. the identification of the start and end points of each shot that has been edited, in order to characterize the entire shot through its most representative *keyframes*. The automatic recognition of the beginning and end of each shot implies solving two problems: *i)* avoiding incorrect identification of shot changes due to rapid motion or sudden lighting change in the scene; *ii)* identification of sharp transitions (*cuts*) as well as gradual transitions (*dissolves*).

Cuts Rapid motion in the scene and sudden change in lighting yield low correlation between contiguous frames especially in the case in which a high temporal subsampling rate is adopted. To avoid false cut detection, a metric has been studied which proves highly insensitive to such variations, while maintaining reliable in detecting "true" cuts. Each frame has been partitioned into nine subframes. The histogram $\mathcal{H}(H,S) : \Re^2 \rightarrow \Re^+$ of hue H and saturation S properties is considered for each subframe. Cut detection is performed by considering the average volume of the difference v_i of subframe histograms in two consecutive frames (i and $i+1$).

The overall feature related to the presence of cuts in a video is simply obtained as $F_{cuts} = \#cuts/\#frames$, where $F_{cuts} \in [0,1]$.

Dissolves The dissolve effect merges two sequences by partly overlapping them. Dissolves detection in commercials is particularly difficult because of their very limited duration. Due to this peculiarity, existing approaches to dissolve detection (developed for movie segmentation purposes) have shown a very poor performance. We use instead *corner statistics* as a means to detect dissolves. Indeed, during the dissolve, the first sequence gradually fades out (i.e. is darkened) while the second sequence fades in. Therefore, during the editing effect, corners associated to the first sequence gradually disappear and those associated with the second sequence gradually appear. This yields a local minimum in the number of corners detected during the dissolve.

Corner detection is based on the algorithm presented in [13]. An image location (x,y) is defined as a corner if the intensity gradient ∇I in a patch $(x+u, y+v)$ around it is not isotropic, i.e. it is distributed along two preferred directions.

Operationally speaking, a corner is characterized by large and distinct values of $\lambda_1(x, y)$ and $\lambda_2(x, y)$, the eigenvalues of the gradient auto-correlation matrix.

The feature $F_{dissolves} \in [0, 1]$ related to the presence of dissolves is evaluated as $F_{dissolves} = \#dissolves/\#frames$.

Rhythm Another relevant video editing characteristic is the *rhythm* of a sequence of shots, as related to shot duration and to the use of cuts and dissolves to join shots. For instance, a sequence of short shots can be used to keep continuously alive the audience's attention and emphasize dynamism, modernity, etc., while a sequence of gradually shorter shots can induce an increase of tension. The rhythm $r(i_1, i_2) \in [0, 1]$ of a video sequence over a frame interval $[i_1, i_2]$ is defined as

$$r(i_1, i_2) = \frac{\#cuts + \#dissolves}{i_2 - i_1 + 1} \quad , \tag{1}$$

where $\#cuts$ and $\#dissolves$ are measured in the same interval. A simple feature measuring the internal rhythm of an entire sequence is the average rhythm, as related to the overall number of breaks: $F_{breaks} = r(1, \#frames)$. This is a normalized quantity, and it holds $F_{breaks} = F_{cuts} + F_{dissolves}$. The absence of breaks is obviously described by the dual feature $F_{continuous} = 1 - F_{breaks}$.

3.2 Shot Content

Once a video has been fragmented into shots and video editing features have been extracted, the content of each shot needs to be internally described. To this end, features are extracted from each shot keyframe describing characteristics such as the presence and distribution of relevant colors in the scene, and the distribution and orientation of lines highlighting specific camera takes.

Colors A description of the shot chromatic content is obtained by performing keyframe color segmentation, thus highlighting its main color regions. In our system, image segmentation is carried out by color cluster analysis: segmentation is then achieved by back-projecting cluster centroids (feature space) onto the image. The use of the HSI color space allows that small feature distances correspond to similar colors in the perceptual domain. Clustering in the 3-dimensional feature space is obtained using an improved version of the standard K-means algorithm, which avoids convergence to non-optimal solutions. Competitive learning is adopted as the basic technique for grouping points in the color space as in. The chromatic content of a video sequence is expressed using a set of eight numbers $c_i \in [0, 1]$ with i denoting one color out of the set {*red, orange, yellow, green, blue, purple, white, black*}, each number quantifying the presence in the keyframe of a region exhibiting the i-th color. Color related features used in the system are $F_{recurrent} \in [0, 1]$ (expressing the ratio between colors which recur in a high percentage of keyframes and the overall number of significant colors in the video sequence) and $F_{saturated}$ (expressing the relative presence of saturated colors in the scene). Their dual are respectively $F_{sporadic} = 1 - F_{recurrent}$ and $F_{unsaturated} = 1 - F_{saturated}$.

Lines The detection of significant line slopes in the keyframe is accomplished by using the Hough transform [14] to generate a line slope histogram. The feature $F_{slanted} \in [0, 1]$ gives the ratio of slanted (i.e. with a slope neither horizontal nor vertical) lines with respect to the overall number of lines in the sequence. Its dual is $F_{hor/vert} = 1 - F_{slanted}$.

4 Mapping Features onto Semiotic Categories

We summarize below the mapping between the set of perceptual features (lower semantic level) and each of the four semiotic categories (higher semantic level) of commercials. Such a mapping allows to organize a database of commercials on the basis of their semiotic content and provide content-based access facilities. The idea is to express the degree of similarity $\mathcal{S}$ of each video with a given query by a simple weighted average of the partial scores expressing the match with each individual semiotic category:

$$\mathcal{S} = w_1\, S_{practical} + w_2\, S_{critical} + w_3\, S_{utopic} + w_4\, S_{playful} \quad , \tag{2}$$

the query being expressed in terms of the set of weights $\{w_i\}$. The way partial scores are obtained highlights the language link for each semiotic category. **Practical** commercials have a linear narrative style: everything in the video must appear real and close to everyday experience. Camera takes are usually frontal, and care is taken that all transitions take place in a smooth and natural way. This implies choosing long dissolves for merging shots (short dissolves are deliberately interpreted by the system as cuts), and the prevalence of horizontal and vertical lines — giving respectively the impression of relax and solidity — over slanted lines:

$$S_{practical} = F_{dissolves} + F_{hor/vert} + F_{unsaturated} \quad . \tag{3}$$

In **playful** commercials, the presence of the camera is always emphasized, and all possible effects are used to stimulate the active participation of the audience in the creation of sense. Everything looks strange and "false" (colors are unnatural, camera takes are usually unprobable, etc.). Hence

$$S_{playful} = F_{cuts} + F_{slanted} + F_{saturated} \quad . \tag{4}$$

The main characteristic of **utopic** commercials is to present the product as part of a world which looks real but doesn't resemble everyday life (i.e. it is a realistically rendered ideal world). For this reason, care is taken to produce a movie-like atmosphere, with a set of dominant colors defining a closed chromatic world and with all of the traditional editing effects (cuts, dissolves) possibly taking place:

$$S_{utopic} = F_{cuts} + F_{dissolves} + F_{recurrent} \quad . \tag{5}$$

Critical commercials spend most of the spot time displaying the product (typically in central and frontal views), while the audio comment continues listing its qualities: the scene has to appear "more realistic than reality itself." For this reason, the number of breaks is kept low, while the ever changing colors in the background due to smooth camera motions contribute to draw the attention to the (constant) color of the product:

$$S_{critical} = F_{continuous} + F_{sporadic} + F_{hor/vert} \quad . \tag{6}$$

The discussion above validates the usual semiotic square representation of commercials categories (see again Fig. 1). Specifically, the pairs practical-playful and utopic-critical appear to be strongly complementary: notice for instance that the features characterizing the "playful" category (eq. (4)) are virtually dual to those of the "practical" (eq. (3)).

5 Experimental Results

The system includes over 150 commercial videos digitized from several Italian TV channels. Each video was processed to extract the perceptual level features as described in Sect. 3. Features extracted from all the database videos were stored in an index signature file associating each set of features with the video they refer to. A graphic interface (see Fig. 2) was designed to support video browsing (upper left part of the interface) and two retrieval modalities:

- the user can select the degree to which the video should conform to the four basic semiotic categories (upper central part of the interface);
- the user can select one of the videos from the database and query for similar videos (upper right part of the interface).

In the first case, a set of four weights is extracted according to the values selected by the user for the degree of conformity to each category. Categories are arranged according to the semiotic square of Fig. 1: the relevance of each category, ranging from 0% to 100%, can be selected through a scroll-bar. The matching score S is computed for each video in the database according to eq. (2), and videos are presented in decreasing order of match in the lower left part of the interface. In the case of search by global similarity, matching scores are computed by considering the correlation between the characteristic features of the video used as reference and those of the rest of the database. Videos are then presented to the user in decreasing order of similarity. Also, by selecting a video from the output list, the video can be either viewed at full or in its most salient keyframes through a movie player application.

Fig. 2 shows the output of the retrieval system in response to a query of purely practical commercials. At the right of the list of retrieved items, a bar-graph displays the degree by which each shot of the best ranked video — cuts and dissolves are represented by white thin vertical lines — belongs to the practical category. The best ranked videos obtained in response to the query all advertise typical family products (the first three retrieved commercials advertise a soup, a soap and a kind of rice, respectively): this again highlights the fact that the kind of promoted product often drives the choice of the semiotic category to use. Figs. 3–5 show some relevant frames for the two best ranked commercials in Fig. 2. From a rapid inspection to these frame sequences, it is evident that practical spots have a quite linear narrative structure, making it is easy to reconstruct the story told in the spot and fill in the "semantic gaps" from frame to frame. Figs. 4–6 show the feature distributions for the two best ranked commercials in Fig. 2. Notice how in practical commercials long dissolves prevail over cuts (see again Fig. 3), horizontal/vertical lines are dominant, and colors are non-saturated.

We conducted a test to measure the extent to which system results conformed to human expectations. The tests were conceived to estimate the agreement between system results and human expectations in the case of reference queries. We considered a set of 5 experts in the semiotic and marketing field. A database of 20 commercial videos was created and each expert was asked to rank videos according to their conformity with the four reference categories. According to this, each expert provided four ordered lists $\{l_k^{play}\}_{k=1}^{20}$, $\{l_k^{uto}\}_{k=1}^{20}$, $\{l_k^{prac}\}_{k=1}^{20}$ and $\{l_k^{crit}\}_{k=1}^{20}$. Each entry in a list can be one out of three values: $\{no, med, high\}$. Hence, the entry l_k^c represents the extent to which the k-th video conforms to the category c – *no* if it doesn't conform at all, *high* if it matches the category and *med* if it matches only to a limited extent.

Fig. 7 evidences how database commercials were classified by the semiotic experts. In this representation considers the semiotic square has been divided into 25 regular regions (Fig. 7(a)). Semiotic categories that are in relation of contradiction are located on opposite sides of the square (practical–playful and critical–utopic).

For a generic commercial, the two most representative categories are considered. Hence, the representation of a commercial is expressed through two labels (l_1, l_2) each label belonging to the set $\{no, med, high\}$. In the square, each region is characterized by a unique combination of labels: for instance, region 20 is characterized by $(Playful = high, Critical = med)$. This allows us to represent the distribution of database commercials with a set of bars. The height of each bar is proportional to the percentage of database commercials exhibiting the combination of labels associated with the region. Fig. 7(b) shows the distribution of database commercials as it is devised according to experts classification. It can be noticed how many database commercials conform to the playful category, thus evidencing a general trend of nowadays commercials to prefer an unconventional way of presenting the product. Fig. 7(c) shows the distribution of database commercials as it is devised according to the system classification. It can be noticed the match between the 'view' of database commercials that is presented by the system and by the experts.

Acknowledgments

The authors warmly thank Bruno Bertelli, Laura Lombardi, Mauro Caliani and Jacopo M. Corridoni for their help in the development of this research.

References

1. S. Smoliar, H. Zhang. Content-based video indexing and retrieval. *IEEE Multimedia*, 2(1):63–75, Summer 1994.
2. Y. Tonomura, A. Akutsu, Y. Taniguchi, and G. Suzuki. Structured video computing. *IEEE Multimedia*, (3):35–43, Fall 1994.
3. A. Hampapur, R. Jain, and T. Weymouth. Digital video segmentation. In *2nd Annual ACM Multimedia Conference and Exposition*, San Francisco, CA, Oct. 1994.
4. J.M. Corridoni and A. Del Bimbo. Film editing reconstruction and semantic analysis. In *Proc. CAIP'95*, Prague, Czech Republic, Sept. 1995.
5. F. Arman, A. Hsu, and M. Chi. Feature management for large video databases. In W. Niblack, ed., *Conf. on Storage and Retrieval for Image and Video Databases*, pages 2–12, San Jose, CA, May 1993.
6. P. Aigrain, P. Joly, P. Lepain, and V. Longueville. Medium Knowledge-based macro segmentation of video sequences. In M. Maybury, ed., *Intelligent Multimedia Information Retrieval*, 1996.
7. M. Yeung, B.L. Yeo, and B. Liu, Extracting story units from long programs for video browsing and navigation. In *Proc. IEEE Int'l Conf. on Multimedia Computing and Systems*, pages 296–305, Hiroshima, Japan, June 1996.
8. D. Swanberg, C.F. Shu, and R. Jain. Knowledge guided parsing in video databases. In W. Niblack, ed., *Conf. on Storage and Retrieval for Image and Video Databases*, pages 13–24, San Jose, CA, May 1993.
9. J.M. Corridoni and A. Del Bimbo. Structured digital video indexing. In *Proc. 13th Int'l Conf. on Pattern Recognition ICPR'96*, pages (III):125–129, Wien, Austria. August 1996.
10. R. Lienhart, C. Kuhmünch, and W. Effelsberg. On the detection and recognition of television commercials. In *Proc. Int'l Conf. on Multimedia Computing and Systems*, pages 509–516, Ottawa, Canada, June 1997.
11. C.R. Haas. *Pratique de la publicité.* Bordas, Paris, France,1988.
12. J.-M. Floch. *Sémiotique, marketing et communication. Sous les signes, les stratégies.* Presses Universitaires de France, Paris, France, 1990.
13. J.M. Pike and C.G. Harris. A combined corner and edge detector. In *Proc. Fourth Alvey Vision Conference*, pages 147–151, 1988.
14. D.H. Ballard and C.M. Brown. *Computer Vision.* Prentice-Hall, Engelwood Cliffs, NJ, 1982.

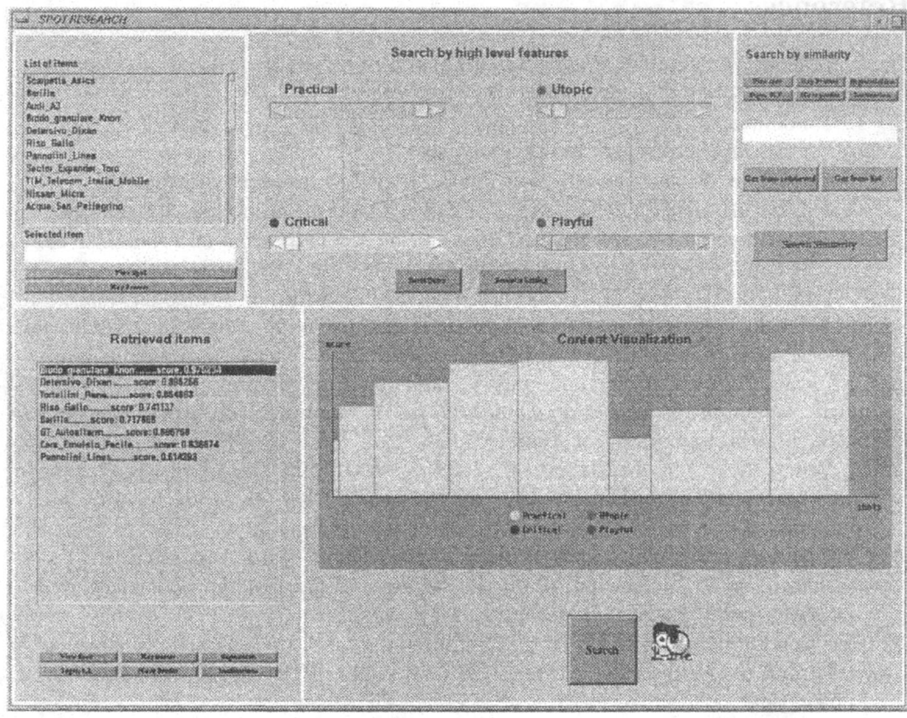

Fig. 2. Retrieval of practical commercials.

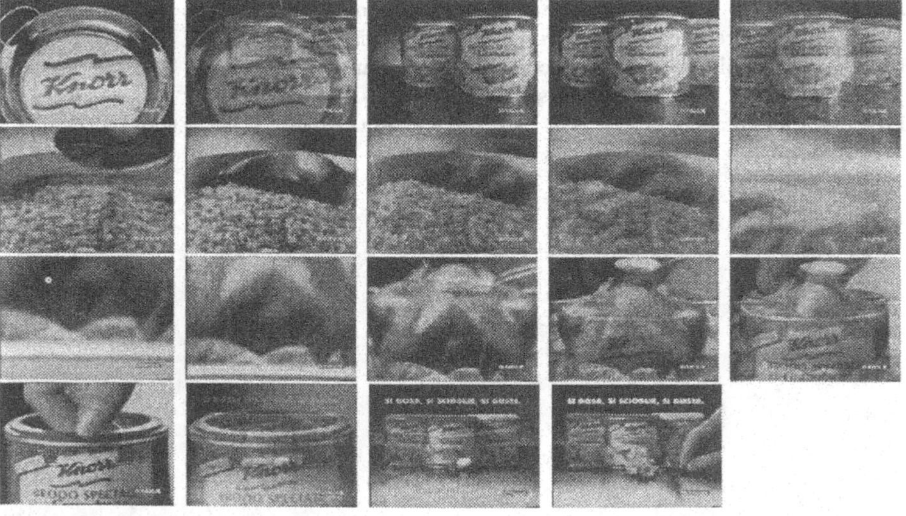

Fig. 3. Some relevant frames for the first ranked spot in Fig. 2 ("Knorr").

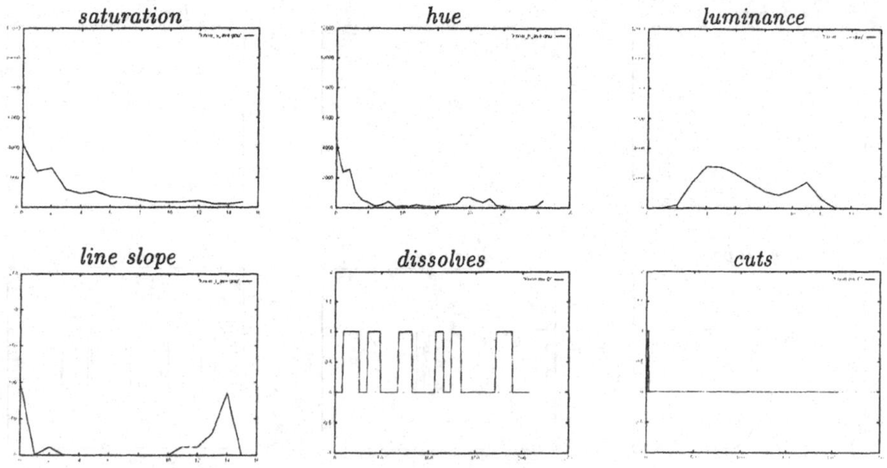

Fig. 4. Feature values for the "Knorr" commercial.

Fig. 5. Some relevant frames for the second ranked spot in Fig. 2 ("Dixan").

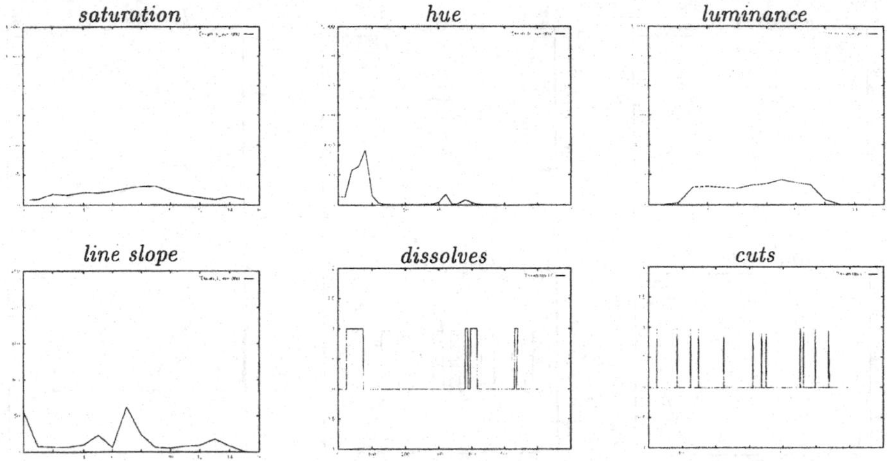

Fig. 6. Feature values for the "Dixan" commercial.

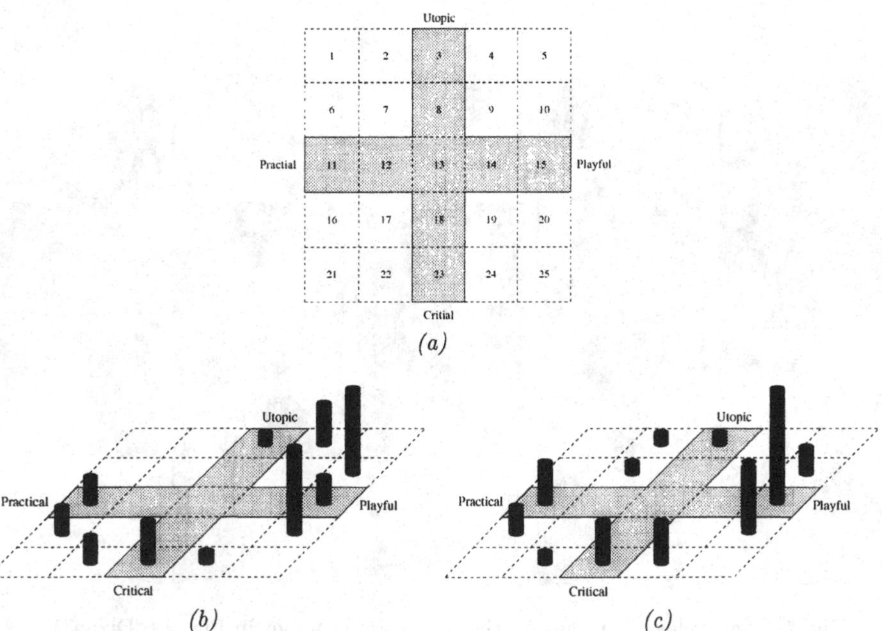

Fig. 7. Representation of database commercials semiotic features. (a) Partionment of the semiotic square, (b) expert's and (c) system's classification of commercials.

Supporting Video Applications through 4DIS Temporal Framework*

Rynson W.H. Lau[1] Hong Va Leong[2] Qing Li[2] Antonio Si[2]

[1] Department of Computer Science, City University of Hong Kong, Hong Kong
[2] Department of Computing, Hong Kong Polytechnic University, Hong Kong

Abstract. Video has become an essential component of multimedia applications nowadays. To support multimedia and video applications efficiently, video objects must be modeled and structured in secondary storage effectively, allowing flexible, easy, and efficient retrieval. Existing video database systems are quite ad hoc, each being designed for a specific kind of application. In this paper, we illustrate that a temporal database system is crucial in providing powerful capabilities for supporting both storage and manipulation requirements of a generic video manipulation framework. We present the design of our experimental prototype which is developed upon the 4DIS temporal database system. Utilizing the query capability of the 4DIS system, abstraction primitives are developed to provide a core set of operations for various kinds of application semantics. In particular, the temporal database system provides persistent storage for video objects and query capabilities for retrieving video objects. Functional capabilities of various video applications could then be easily developed upon this core set of operations.

1 Introduction

Video has become an essential component of today's multimedia applications, including VCR, video-on-demand, virtual walkthrough, etc. [21]. To support video applications efficiently, video objects must be modeled and structured in secondary storage effectively, allowing flexible specification and efficient retrieval [6, 8, 15, 30, 32, 36]. Existing video database systems are quite ad hoc, each of which being designed for a specific kind of application. What is of great necessity is a generic framework that supports the specifications and operations on video objects so that the functional capabilities of various applications can be defined and supported via the correct specification of system behaviors.

In [27], we have shown that a video object could be modeled as a collection of frame objects, which can be related in a temporal sequence. Such temporal structures are demonstrated using the temporal modeling constructs and temporal constraints of the *Four Dimensional Information System* (*4DIS*), an object-oriented temporal database system [28]. An advantage of modeling video

* This research is supported in part by the Hong Kong Polytechnic University central research grant numbers 353/066 and 351/217 and RGC grant number CityU 77/95E.

objects in a temporal database system is that video applications could be easily implemented using the query capabilities of the temporal database system. For instance, in [27], we demonstrate that VCR operations, including Play, Pause, Resume, Fast Forward, Fast Backward, Slow Motion, Search, and Stop, are supported by implementing each function as a temporal query on 4DIS.

In this paper, we extend and generalize our video database framework to support functional capabilities of various video applications, rather than specifically for VCR operations as in [27]. The framework provides abstraction primitives upon which different video applications could be implemented.

The architecture of our framework is depicted in Fig. 1. We define a layered architecture utilizing the 4DIS temporal database system as persistent storage. The storage defines the *kernel layer* of our framework and constitutes the lowest layer. It provides the core manipulating capabilities to video objects such as modeling constructs and temporal constraints for describing video objects and query languages for retrieval of video objects.

Since not all functional capabilities of a temporal database system are needed for video applications, the *driver layer* provides a set of driver routines, to be utilized by individual video applications, as a layer of abstraction on the kernel layer, hiding low level implementation details and unused features from video applications. Each driver is implemented as a temporal query or a simple set of queries, operating on the kernel temporal database layer. They act as the set of core functions upon which operational semantics of various applications could be established.

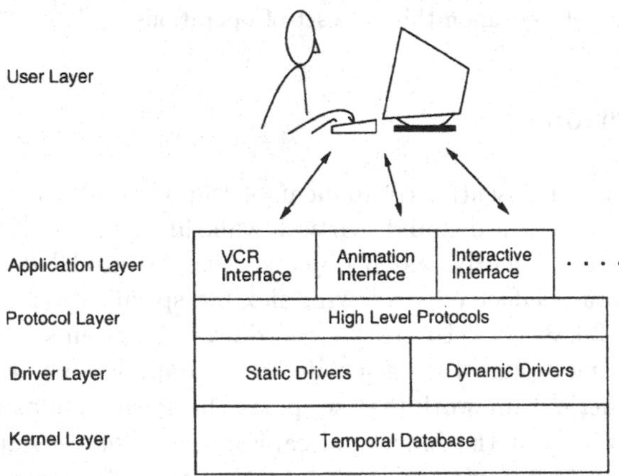

Fig. 1. Architecture of a video database framework.

The *protocol layer* provides a level of communication transparency between video applications and the database engine. Applications might be executed locally, within the same machine as the database engine, or they might be executed remotely, resulting in a client/server architecture. The protocol layer receives instructions from a video application, either locally or remotely, and invokes the driver routines with the corresponding arguments. This allows a video application to be developed in a platform independent manner.

The *application layer* includes different video applications developed for specific purposes. A good example will be the VCR player, providing the presentation functions for video objects. Another example might be a video editing application, allowing new video objects to be dynamically constructed from existing ones. The application layer is similar in spirit to the external layer or view of a conventional database system [29]. Each video application is implemented based on the driver routines provided.

This paper is organized as follows. In Section 2, we give a survey of relevant research on video database objects and their manipulation, as well as basic research in the context of temporal database models, which will form the basis of the work described in this paper. Section 3 summarizes the key features of the 4DIS temporal database system. The core driver routines and their implementations in supporting video applications are presented in Section 4. The warping of drivers within a communication independent framework is supported by the protocol layer, as delineated in Section 5. The utilization of driver routines in implementing sample video applications is illustrated in Section 6. Finally, we summarize with a brief discussion on the results and offer future research remarks in Section 7.

2 Related Work

In this section, we briefly review some related work, which is generally divided into two areas: temporal data modeling, and management of video data.

2.1 Temporal Data Modeling

Adding time support into databases has been an active area of research for the last fifteen years or so. Broadly speaking, the following time dimensions have been introduced and/or exploited in temporal databases: (i) *Valid time* concerns the time a fact was true in reality. (ii) *Transaction time (an interval)* concerns the time the fact was stored in the database. (iii) *User-defined time* refers to the fact that the semantics of the data values are known only to the user and are not interpreted by the DBMS. However, most of the existing work has been focused on the first two dimensions, as they are orthogonal to each other, and are of more general interests to databases.

Valid time has been added to many data models. In particular, the relational model mostly supports only valid time, with a few only supporting transaction time or supporting both time dimensions. Valid times can be represented with single chronon identifiers (i.e., event timestamps), with intervals (i.e., as interval timestamps), or as valid time elements which are finite sets of intervals. Valid time can be associated with entire tuples or with individual attribute values. As surveyed in [22], there are six basic alternatives to representing attribute values:

(1) Atomic valued – values do not have any internal structure; (2) Set valued – values are sets of atomic values; (3) Functionally atomic valued – values are functions from the valid-time domain to the attribute domain; (4) Ordered pairs – values are an ordered pair of a value and a timestamp; (5) Triplet valued – values are a triple of attribute value, valid-from time, and valid-until time; (6) Set-triple value – values are a set of triplets. In the context of object-oriented (OO) data models, there have been three general approaches to adding valid-time support to an OO data model and query languages [31], namely, using the object model directly (as exemplified by OODAPLEX [10]), using other extensions to support time (such as Sciore's *annotations* [26]), and direct incorporation of time into the data model (as exemplified by the TEER model [12] and the OSAM*/T [34]).

Transaction time support is necessary when one would like to restore the state of the database to a previous point in time, or to support versions of data objects. Hence, transaction time is often branching, as opposed to the linear time model underlying valid time. An important distinction to be made in support for transaction time is the granularity of versioning. In particular, *extension versioning* refers to the fact that object instances or their attribute values are versioned, whereas *schema versioning* refers to the definitions of objects are versioned. Prominent examples supporting extension versioning include OODAPLEX [10], Postgres [33], Matisse OODBMS [1], and the Chou-Kim versioning model [7]. Only a few systems and/or models support schema versioning, notably Matisse OODBMS [1] adopting the *eager* approach, and Orion [3] adopting the *lazy* approach (which is mainly based on Kim-Chou versioning model [17]). To some extent, Postgres also supports schema versioning by specifying system catalog to consist of transaction-time relations. The transaction time, however, is only linear [33].

2.2 Video Data Management

In parallel with the research work on temporal databases in recent years, there have been significant interests and considerable amount of research in developing management systems for video objects. Here, we review some existing work on video data management, with an attempt to compare and contrast different approaches of modeling and/or managing video data, emphasizing the necessity of incorporating temporal data modeling into video data management. While there have been several research projects on video databases initiated, the following are what we regard as representative ones which have explicit temporal semantics support in their models and systems.

Bimbo *et al.* [4] used Spatio-Temporal Logic to support the retrieval by content of video sequences through visual interaction. Temporal Logic is a language for the qualitative representation of ordering properties in the execution sequences of temporal systems. In their database, video sequences are stored along with a description of their contents in Spatio-Temporal Logic. Retrieval is supported through a 3D iconic interface.

Day *et al.* [9] introduced a graph based model called Video Semantic Directed Graph (VSDG) to conceptually model video data. They suggest that their VSDG representation, when superimposed on an object-oriented hierarchy, leads to an architecture for handling video. The main deficiencies of this approach are that it does not allow for searching and direct access to video segments containing specified items of interest, nor does it provide adequate support for declarative and associative search.

Duda and Weiss [11] proposed an algebraic video data model, which allows users to (1) model nested video structures such as shot, scene and sequence, (2) express temporal compositions of video segments, (3) define output characteristics of video segments, (4) associate content information with logical video segments, (5) provide multiple coexisting views and annotations of the same data, (6) provide associative access based on the content, structure, and temporal information, and (7) specify multi-stream viewing. However, their model does not support the exploration of interactive movies and home video editing.

Gibbs *et al.* [14] presented a model for the interpretation, derivation, and temporal composition of media objects. A media object is defined as a timed audio/video/audio-video stream or a sequence of temporally related timed audio/ video/audio-video streams. While this model provides adequate supports for the editing and presentation of audio-video content, it lacks in an easy retrieval of an audio-video stream based on inherent items of interest or meta-data.

Jain and Hampapur [16] proposed a video model (ViMod) based on studies of the applications of video and the nature of video retrieval requests. The features of their model include content dependence, temporal extent, and labeling. A feature is said to be content independent if the feature is not directly available from the video data. Certain aspects of a video can be specified based on viewing a single frame in temporal intervals, whereas other features like motion can be specified only based on a time interval. The changes that occur in a video can be tracked over the extent of a time interval.

On the *video structuring* aspect, related research has been conducted at multiple levels. At the first level, some prototype systems allow users to access desired images from image databases by directly making use of visual cues. For example, in [23], Niblack *et al.* reported that the QBIC system at IBM allows images to be retrieved by a variety of image content descriptors including color, texture, and shape. However, even though they address the issues regarding the exploitation of spatial structure in images for effective indexing for retrieval, they do not deal with video sequences where the *temporal* information also has to be considered. At the next level, video data structuring is viewed as segmenting the continuous frame stream into physically discontinued units, usually called *shots*. In general, these physical units need to be grouped/clustered [19] to form semantically meaningful units, such as scenes. The shots and/or scenes are described by one or several representative frames, known as *key frames* [18, 35]. Still, very few facilities of supporting temporal semantics have been explicitly provided in so-called *story-based* video structuring work [19, 35]; rather, the main focus at this level has been primarily on the support of video information browsing. As a consequence, declarative queries and associative search related to the temporal ordering of scenes can hardly be accommodated.

3 The 4DIS Temporal Database System

In this section, we briefly describe the 4DIS temporal data model and its querying capabilities [28]. We will only describe those features of 4DIS that are relevant for video applications here. We next, illustrate how the temporal modeling constructs of 4DIS could be employed for modeling and querying video objects.

3.1 The 4DIS Data Model

The 4DIS model provides uniformity in modeling data and meta-data in an extensible framework so that they can be manipulated by a single set of basic operators. For the purpose of this paper, we will only focus on the capability of 4DIS in modeling data. 4DIS supports two basic modeling constructs: *objects* and *mappings*. Every fact in the real-world corresponds to an object in a 4DIS database. The 4DIS model supports *atomic*, *class*, and *composite objects*. Atomic objects represent the symbolic constants in a database. Class objects specify the descriptive and classification information. Composite objects describe entities of application environments. Mappings belong to a special kind of composite objects. They are mainly used for modeling relationships between two objects.

Each relationship among objects is modeled as a four-valued tuple: ⟨`domain-object`, `mapping-object`, `range-object`, `chronon-object`⟩. Let D denote the set of all domain objects, M the set of all mapping objects, R the set of all range objects, and T the set of all chronon objects. Each relationship, $r_{4dis} = \langle\, d \in D,\ m \in M,\ r \in R,\ t \in T\,\rangle$, in a 4DIS database is a member of $D \times M \times R \times T$, representing a relationship, m, between objects d and r at chronon t. The chronon object could be a transaction time chronon or it could be a valid time chronon. In the context of this work, we only focus on transaction time.

A 4DIS database, $\mathcal{DS}$, contains a collection of inter-related objects at various time. It is a collection of ⟨`domain-object`, `mapping-object`, `range-object`, `chronon-object`⟩ four-valued tuples:

Definition 1 (Database) *:*
$\mathcal{DS} = \{r_{4dis} = \langle d, m, r, t\rangle | d \in D \wedge m \in M \wedge r \in R \wedge t \in T\}$. □

Definition 2 (Group-precede) *:*
A group of time chronons $T_1 = \{t_{11}, t_{12}, ..., t_{1m}\}$ *precedes another group of time chronons* $T_2 = \{t_{21}, t_{22}, ..., t_{2n}\}$*, denoted* $T_1 \preceq T_2$*, if and only if* $(\forall i, j : 1 \leq i \leq m \wedge 1 \leq j \leq n : t_{1i} \leq t_{2j})$. □

Two different mapping objects, member and subclass, are predefined to support three different abstraction primitives, which include classification/instantiation, aggregation/decomposition, and generalization/specialization. The member mapping is used to define both classification and aggregation abstraction primitives while the subclass mapping is used to define the generalization abstraction primitive.

Several class objects are predefined: meta-class, String, Integer, and Real. Meta-class is a system predefined class object which contains all class objects.

String, Integer, and Real are system predefined class containing atomic string, integer, and real respectively. An object instantiated to be a member of meta-class will be of type class object; it could contain other objects as its members. String, Integer, and Real class objects are thus, members of Meta-class.

Properties of a class object is modeled as a relationship between two class objects in 4DIS, defined via a user-defined mapping-object. We require that a class object could only be related to another class object, as specified in Constraint 1.

Constraint 1 (Class Relation) :
$\langle C_1, m, C_2, t\rangle \in \mathcal{DS} \wedge \langle C_1, \mathsf{member}, \mathsf{meta\text{-}class}, t_1\rangle \in \mathcal{DS} \wedge t_1 \leq t \Rightarrow$
$(\exists t_2 : t_2 \leq t : \langle C_2, \mathsf{member}, \mathsf{meta\text{-}class}, t_2\rangle \in \mathcal{DS})$. □

To illustrate, at chronon t, the definition of a class object Person might be modeled by the following set of tuples:

$\langle$ Person, member, meta-class, t $\rangle$
$\langle$ Person, name, String, t $\rangle$
$\langle$ Person, age, Integer, t $\rangle$
$\langle$ Person, address, String, t $\rangle$.

Here, Person is defined as a class object by defining Person as a member of meta-class. Name, age, and address are user-defined mapping objects which relate the class object Person to the predefined class objects of String, Integer, and String respectively. Similarly, class object Employee could be defined as a subclass of class object Person using the predefined subclass mapping object as follows:

$\langle$ Employee, subclass, Person, t $\rangle$
$\langle$ Employee, work-for, Company, t $\rangle$.

Here, Employee is defined as a class object by defining it as a subclass of the class object Person. Company is another class object defined by another similar set of tuples and is related to class object Employee via the mapping object work-for. Relationships defined in a class will be inherited into its subclasses recursively as defined in Constraint 2.

Constraint 2 (Inheritance) :
$\langle C_1, m, C_2, t_1\rangle \in \mathcal{DS} \wedge \langle C_3, \mathsf{subclass}, C_1, t\rangle \in \mathcal{DS} \Rightarrow \langle C_3, m, C_2, t\rangle \in \mathcal{DS}$. □

For instance, since Employee is defined as a subclass of Person at chronon t, relationships defined for Person, such as name are inherited into Employee at chronon t as well.

An instance object is instantiated when the instance object is related to a class object via the member mapping object. By instantiating an instance object, o, as a member of a class object C, the properties of C are inherited into o. In other words, relationships could be defined on o using the mapping objects defined on class object C. Furthermore, only the mapping objects defined on C could be defined on o. In addition, the object related to o through a mapping object, m, must be a member of the class object related to C via m.

Constraint 3 (Instance Relation) :
$\langle C_1, \mathsf{member}, \mathsf{meta\text{-}class}, t_1\rangle \in \mathcal{DS} \wedge \langle C_2, \mathsf{member}, \mathsf{meta\text{-}class}, t_2\rangle \in \mathcal{DS} \wedge$

$\langle o_1, \text{member}, C_1, t_4\rangle \in \mathcal{DS} \wedge \langle o_1, m, o_2, t\rangle \in \mathcal{DS} \wedge t_1 \leq t_4 \leq t \wedge t_2 \leq t$
$\Rightarrow (\exists t_3, t_5 : \{t_1, t_2\} \preceq \{t_3\} \preceq \{t\} \wedge t_2 \leq t_5 \leq t : \langle C_1, m, C_2, t_3\rangle \in \mathcal{DS} \wedge$
$\langle o_2, \text{member}, C_2, t_5\rangle \in \mathcal{DS})$. □

For instance, the definition of an object John at t_1 might be modeled by:

⟨ John, member, Person, t_1 ⟩
⟨ John, name, "John", t_1 ⟩
⟨ John, age, 3, t_1 ⟩
⟨ John, address, "Los Angeles", t_1 ⟩.

A very useful integrity constraint is that each tuple inserted into the system would remain valid until its value is modified: the *until-changed* semantics [24], used in many temporal database systems. It is formally defined in Constraint 4.

Constraint 4 (Until-changed) :
$\langle d, m, r, t_1\rangle \in \mathcal{DS} \wedge \neg(\exists t_2 : t_1 < t_2 \leq t : \langle d, m, r', t_2\rangle \in \mathcal{DS}) \Rightarrow \langle d, m, r, t\rangle \in \mathcal{DS}$.
□

Based on this until-changed semantics, we define the duration within which a tuple is valid, the *validity duration*.

We denote a snapshot of the database, $\mathcal{DS}$, at time t by $\mathcal{DS}^t$ and is defined as the set of four-valued tuples valid on or before t, as defined by the until-changed semantics. This is formalized in Definition 3.

Definition 3 (Database Snapshot) :
$\mathcal{DS}^t = \{\langle d, m, r, t'\rangle | \langle d, m, r, t'\rangle \in \mathcal{DS} \wedge d \in D \wedge m \in M \wedge r \in R \wedge t' \in T \wedge$
$\neg(\exists t'' : t' < t'' \leq t : \langle d, m, r, t''\rangle \in \mathcal{DS})\}$. □

Similarly, we denote a snapshot of an object, d, at time t by d^t and is defined as the set of valid four-valued tuples defined on or before t, with object d acting as the domain of the tuples, as specified in Definition 4.

Definition 4 (Object Snapshot) :
$d^t = \{\langle d, m, r, t'\rangle | \langle d, m, r, t'\rangle \in \mathcal{DS} \wedge m \in M \wedge r \in R \wedge t' \in T \wedge$
$\neg(\exists t'' : t' < t'' \leq t : \langle d, m, r, t''\rangle \in \mathcal{DS})\}$. □

We could define an extension of a class object, C, at time t based on the definition of snapshot. Denoting the extension of C at time t as $\mathcal{E}(C^t)$, it is defined as the set of snapshots of its member instance objects at time t, formally defined as Definition 5.

Definition 5 (Class Extension) :
$\mathcal{E}(C^t) = \{o^t | \langle o, \text{member}, C, t'\rangle \wedge t' \in T \wedge$
$\neg(\exists t'' : t' < t'' \leq t : \langle o, \text{member}, C, t''\rangle \in \mathcal{DS})\}$. □

A 4DIS database could be viewed graphically by a 4 dimensional (4D) geometric representation. The four axes in the 4D space represent the domain, mapping, range, and chronon objects. This is illustrated in Fig. 2. Since a 4D space is intuitively difficult to present, we depict a 4D space by a sequence of database snapshots along the temporal dimension. The mapping axis holds only mapping objects. The temporal dimension holds only the chronon objects. All objects other than chronon objects appear on both the domain and range axes. Each relationship is represented by a point in the 4D space.

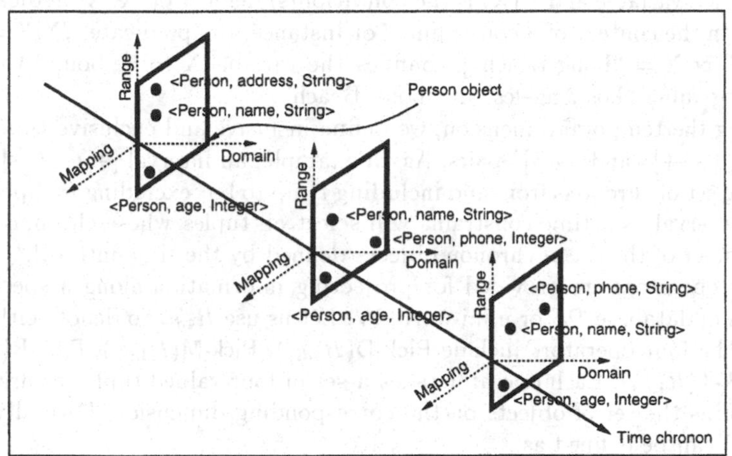

Fig. 2. Four dimensional representation of 4DIS database.

3.2 Query Specification in 4DIS

The specification of queries is based on a simple query-by-example (QBE) like syntax [29]. Each 4DIS query is specified as a list of sub-queries: $\{\langle D$ constraint, M constraint, R constraint, T constraint$\rangle, \cdots\}$. Each sub-query is a four-valued constraint: $\langle D$ constraint, M constraint, R constraint, T constraint$\rangle$, indicating the constraint for each of the four different dimensions. Each constraint can be a predicate, or the wildcard matching pattern of "?" symbol.

Each sub-query specification is based on three geometric components: *line*, *plane*, and *space* components. A line component is specified with a "?" symbol in place of one of the constraints. Similarly, a plane and a space component is specified with two and three "?" symbols in place of two and three of the constraints respectively. In addition, a "-" symbol in place of any constraint denotes no information on that particular dimension. Therefore, $\langle$-, -, -, -$\rangle$ denotes no information. Lastly, a "." in place of the T constraint denotes the most current chronon. Therefore, $\langle$-, -, -, .$\rangle$ denotes no information at now.

Each sub-query, $\langle D$ constraint, M constraint, R constraint, T constraint$\rangle$, applies on the database, $\mathcal{DS}$, and returns a set of four-valued tuples which satisfy the constraints, representing a *mini-world*, $\mathcal{W}$, of the database.

Definition 6 (Mini-world) *:*
$\mathcal{W} = \{w_{4dis} = \langle d, m, r, t\rangle | d \in D \wedge m \in M \wedge r \in R \wedge t \in T \wedge w_{4dis} \in \mathcal{DS}\}$. □

The mini-worlds of different sub-queries will be unioned together to form the results of the query. One could bind a variable with a constraint which will be used for unification in a way similar to QBE for relating attributes.

A constraint could be specified in a form of predicate. Each predicate is a well-formed formula, involving some matching variables, of the form: ?X $[P(\mathrm{X})]$. Here, X is a variable and $P(\mathrm{X})$ is the condition(s) defined on X. A predicate can be used in the context of a constraint. For instance, the predicate, ?X[X = "Los Angeles" or X = "Long Beach"], confines the variable X to be bound to either the string value "Los Angeles" or "Long Beach".

Along the temporal dimension, we define inclusive and exclusive time intervals by "[+, +]" and "[-, -]" pairs. As an example, an interval [+ t_1, t_2 -] represents the set of chronons from and including t_1 up to but excluding t_2. Specifying a time interval as a time constraint will select all tuples whose chronon object is a member of the set of chronon objects defined by the time interval.

Four operators are provided for projecting information along a specific dimension of database $\mathcal{DS}$ or mini-world $\mathcal{W}$. Let us use R_{4dis} to denote either $\mathcal{DS}$ or $\mathcal{W}$. The four operators include Pick-D(R_{4dis}), Pick-M(R_{4dis}), Pick-R(R_{4dis}), and Pick-T(R_{4dis}). Each operator takes a set of four-valued tuples as argument and returns the set of objects on the corresponding dimension. Formally, Pick-D(R_{4dis}) can be defined as:

$$\text{Pick-D}(R_{4dis}) = \{d \mid \exists \langle d, m, r, t\rangle \in R_{4dis}\}.$$

Pick-M(R_{4dis}), Pick-R(R_{4dis}), and Pick-T(R_{4dis}) can be defined similarly. The syntax of the query language can be expressed in the Backus Naur Form. Details of the syntax can be found in [28].

3.3 Temporal Operators

Several temporal operators are defined on chronon objects, with respect to the database $\mathcal{DS}$ or perhaps the mini-world $\mathcal{W}$:

- The FIRST operator takes in a set of four-valued tuples as argument and returns the oldest chronon object of the set.
  FIRST(R_{4dis}) $= t$ which is the lone element in the set
  $\{t|\langle d,m,r,t\rangle \in R_{4dis} \wedge \neg(\exists t' : t' < t : \langle d,m,r,t'\rangle \in R_{4dis})\}$.
- The LAST operator takes in a set of four-valued tuples as argument and returns the last (most current) chronon object of the set.
  LAST(R_{4dis}) $= t$ which is the lone element in the set
  $\{t|\langle d,m,r,t\rangle \in R_{4dis} \wedge \neg(\exists t' : t' > t : \langle d,m,r,t'\rangle \in R_{4dis})\}$.
- The BEFORE operator takes in a four-valued tuple as argument and returns the previous snapshot of the relationship (tuple). There is an optional argument R_{4dis}, which is the mini-world or database to be used for the operation, and it is $\mathcal{DS}$ when the argument is left out.
  BEFORE(r_{4dis}, R_{4dis}) $= t$ which is the lone element in the set
  $\{t|t = \langle d,m,r',t'\rangle \wedge r_{4dis} = \langle d,m,r,t\rangle \wedge$
  $(\exists t' : t' < t : \langle d,m,r',t'\rangle \in R_{4dis} \wedge$
  $\neg(\exists t'' : t' < t'' < t : \langle d,m,r'',t''\rangle \in R_{4dis}))\}$.

- The NEXT operator takes in a four-valued tuple as argument and returns the next snapshot of the relationship (tuple). Again, an optional argument R_{4dis} could be specified.
  $\mathsf{NEXT}(r_{4dis}, R_{4dis}) = t$ which is the lone element in the set
  $\{t|t = \langle d, m, r', t'\rangle \wedge r_{4dis} = \langle d, m, r, t\rangle \wedge$
  $(\exists t' : t' > t : \langle d, m, r', t'\rangle \in R_{4dis} \wedge$
  $\neg(\exists t'' : t < t'' < t' : \langle d, m, r'', t''\rangle \in R_{4dis}))\}$.
- The PRECEDE operator takes in a four-valued tuple, r_{4dis}, as argument, with an optional R_{4dis} argument. It returns the set of chronon objects that precedes the chronon object of r_{4dis} within the space of R_{4dis}. If the R_{4dis} argument is missed out, the operator applies on the database, $\mathcal{DS}$.
  $\mathsf{PRECEDE}(r_{4dis}, R_{4dis})$ is the set
  $\{t'|t' < t \wedge t \in \mathsf{Pick\text{-}T}(r_{4dis}) \wedge \langle d, m, r, t'\rangle \in R_{4dis}\}$.
- The SUCCEED operator takes in a four-valued tuple, r_{4dis}, as argument, with an optional R_{4dis} argument. It returns the set of chronon objects after the chronon object of r_{4dis} within the space of R_{4dis}.
  $\mathsf{SUCCEED}(r_{4dis}, R_{4dis})$ is the set
  $\{t'|t' > t \wedge t \in \mathsf{Pick\text{-}T}(r_{4dis}) \wedge \langle d, m, r, t'\rangle \in R_{4dis}\}$.

As an example, one may retrieve the employment history of John in chronological order, using the temporal operators defined above as follows:
{⟨John, work-for, ?,
[+FIRST(⟨ John, work-for, ?, ?⟩), LAST(⟨ John, work-for, ?, ?⟩)+]⟩}.

3.4 Representing Video Objects

One way to represent a video object is to model individual frames of a video object as a relationship to the video object at various chronons. Each video object belongs to the class Video and each frame object belongs to the class Frame. The frames can be created, instantiated, and related to appropriate video object via the current-frame relationship as follows:

⟨Video, member, meta-class, t_1⟩
⟨Frame, member, meta-class, t_1⟩
⟨Video, current-frame, Frame, t_1⟩
⟨Frame1, member, Frame, t_1⟩
⟨Frame2, member, Frame, t_1⟩
⟨Frame3, member, Frame, t_1⟩
...
⟨Video1, member, Video, t_1⟩
⟨Video1, current-frame, Frame1, t_1⟩
⟨Video1, current-frame, Frame2, t_2⟩
⟨Video1, current-frame, Frame3, t_3⟩
...

Here the objects Frame1, Frame2, and Video1 are merely object identifiers that occupy very little storage. Only the relationship themselves are modeled. This also leads to a saving of storage, if the frames are later being copied from one place to another by higher level primitives and operations. The chronon objects at which subsequent frames are related to a video object via the **current-frame** mapping object could be defined according to a temporal constraint. For instance, given a constraint that 30 frames have to be clustered within one second in order to provide continuous motion, the validity duration of each frame will be $\frac{1}{30}$, i.e., $t_{k+1} = t_k + \frac{1}{30}$.

4 The Driver Layer

The 4DIS temporal database system provides some driver routines as programming interfaces for developing video applications, apart from those for other user applications. Each driver routine is implemented as a 4DIS query. The set of core driver routines includes the following:

- COPY(Video, t_1, t_2): send the segment of the input video between times t_1 and t_2 to the output channel. Here, the output channel may be a file or a hardware video output interface. Alternatively, the output can be stored in another video for future use. This is illustrated in Fig. 3*a*. This routine is implemented as a 4DIS query:
  $\{\langle \text{Video, current-frame, ?, } [+t_1, t_2+] \rangle\}$.

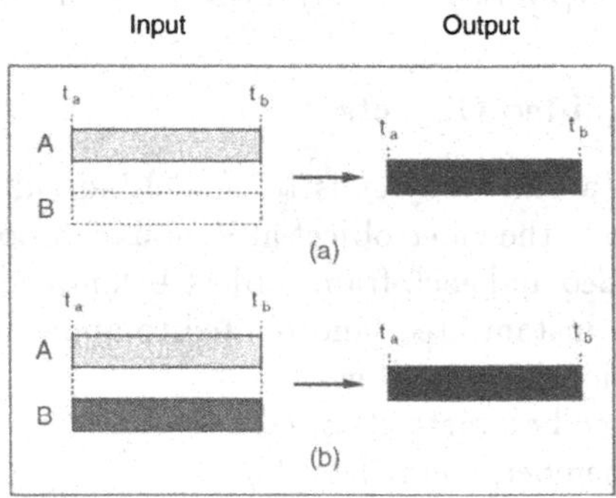

Fig. 3. Routines COPY and COMPOSE.

  Notice that when $t_2 < t_1$, the video stream will be played in reserve since the frames are retrieved in a reverse order.
- COMPOSE(Video$_1$, t_{1a}, t_{1b}, **operator**, Video$_2$, t_{2a}, t_{2b}): compose two input video segments into a single output video segment as its output. The segment of Video$_1$ is specified by two chronons t_{1a} and t_{1b} ($t_{1a} \leq t_{1b}$), and those of Video$_2$ is specified by two chronons t_{2a} and t_{2b} ($t_{2a} \leq t_{2b}$). Each frame within segment Video$_1$ is composed with a frame along the same relative position within segment Video$_2$. This routine assumes that the length of the two intervals are identical, i.e., $t_{1b} - t_{1a} = t_{2b} - t_{2a}$. If $t_{1a} = t_{1b}$ and $t_{2a} = t_{2b}$, it

will degenerate into one that operates on a particular frame. It is illustrated in Fig. 3*b*.

The parameter **operator** defines the operation to be performed to compose two frames in producing the output frame. The set of possible operators includes set, and, or, etc. [13, 25]. Thus, one may provide a variety of cross fading on the two input video segments or the "dissolving" effect. Other functions may include superimposition of the videos. It is implemented thus:

```
t = 0
while t ≤ t_1b − t_1a do
    f_1 = Pick-R(⟨Video_1, current-frame, ?, t_1a + t⟩)
    f_2 = Pick-R(⟨Video_2, current-frame, ?, t_2a + t⟩)
    f_3 = f_1 operator f_2
    ⟨Video_3, current-frame, f_3, .+t⟩
    t = NEXT(⟨Video_1, current-frame, ?, t_1a + t⟩, DS) − t_1a
end.
```

- BEGIN(Video): return the chronon at which the first frame of a video object is associated. It is implemented using the FIRST temporal operator:
  FIRST(⟨Video, current-frame, ?, ?⟩).
- END(Video): return the chronon at which the last frame of a video object is associated. It is implemented using the LAST temporal operator:
  LAST(⟨Video, current-frame, ?, ?⟩).
- GET-CHRONON(Video, frame): return the chronon object at which the frame is related to Video via the current-frame relationship. It is implemented as:
  Pick-T(⟨Video, current-frame, frame, ?⟩).
- SET-FRAME(Video, cframe, d): perform a positioning function. It returns the frame related to Video which is d chronon from that relating cframe to Video. If it is beyond the last frame of the video, the last frame is returned. Similarly, if it is beyond the first frame of the Video, the first frame is returned. It is implemented as follows:

```
t = Pick-T(⟨Video, current-frame, cframe, ?⟩) + d
if (t > END(Video)) then t = END(Video)
if (t < BEGIN(Video)) then t = BEGIN(Video)
Pick-R(⟨Video, current-frame, ?, t⟩).
```

- SKIP(Video, n, m): return another video object which is composed of a subset of frames of the original Video object, by skipping m frames after retrieving and outputting every n frames in the chronon order. This routine is implemented as 4DIS queries:

```
W = {⟨Video, current-frame, ?, ?⟩}
W_1 = ϕ
r = ⟨Video, current-frame, ?, FIRST(W)⟩
while r ≠ NULL do
    for i = 1 to m do r = NEXT(r, W)
    for i = 1 to n do
        r = NEXT(r, W)
        W_1 = W_1 ∪ {r}
```

end
end
return $\mathcal{W}_1$.

5 The Protocol Layer

In our video database application, we are employing a client/server design, since it can be foreseen that there will be various types of applications running on various types of machines. For instance, the specific graphics processor and the video database server will be running on some dedicated machines supporting a high I/O throughput by employing such as parallel I/O technology.

In order to provide a transparent interface for applications to utilize the relatively low level drivers locating and executing at the video database server, the protocol layer is introduced to hide the communication details, in much the same way that remote procedure call (RPC) [5] and transport/network layer are hiding low level details of communication in a client/server application.

In the protocol layer, we provide function calls with similar forms as the driver routines, with the same set of parameters. The outputs from the driver routines will be passed back to the RPC caller. Since there can be interactions between the client and server, we allow the server to invoke callbacks to the client for some dedicated functions defined only by the client application. The callback interaction is more like a reverse RPC from the server back to the client. The standard functionalities of RPC such as argument marshaling, object identifier translation, and timeout/error recovery will be supported in the protocol layer, where there is a set of utility routines for these functionalities. The client stub procedures sending RPC for the remote driver functions will make calls to these utility routines as well. There is also a similar set of utility routines located at the server to be used by server stubs, which call the driver routines.

We will briefly illustrate the events that will happen after an application invokes a driver, say, SLOW MOTION. A stub procedure named SLOW MOTION will be invoked at the client machine, which looks up a table to locate the IP address of the database server and the port number of the RPC server for the functionality. Dynamic binding is also possible here. The server will create a separate connection with the client, which is normally a TCP connection. The object identifier of the video object will be passed to the server along the TCP connection. The server will invoke functions to retrieve the video object identifier and perform the 4DIS operations. Furthermore, the output of the driver routine will be redirected from the default output, i.e., **outputchannel**, to the server stub procedure, which formats the video data stream like marshaling for transmission over the TCP connection. Specific information, such as a callback from server to client, will be served by means of out-of-band data along the TCP connection, besides the regular video data from the server to the client. When the driver completes its execution, the server closes the connection with the client.

Additional functions dedicated to the video database application may also be supported. A good example is the handling of the high transmission requirement for video objects, since they are of large size. It might be worthwhile to compress the information before transmission and restoring the information after transmission, if the compression can be executed almost real-time and does not involve disk overhead. These functions are also located at the protocol layer.

6 The Applications Layer

The driver routines provide a means for building video applications, hiding the details of the temporal database system. Due to space limitations, we depict only the implementation of two kinds of video applications: VCR and Video Editing.

6.1 VCR Application

A VCR application contains a set of presentation operators for displaying a video object at various constraints. We define eight presentation operators for presenting a video object, including PLAY, SLOW MOTION, FAST FORWARD, FAST BACKWARD, PAUSE, RESUME, SEARCH, and STOP.

1. PLAY(Video):
   PLAY simply plays back the video object of interest. It is implemented using the static driver COPY, by providing it with the validity duration of the video object:
   COPY(Video, BEGIN(Video), END(Video)) >> `outputchannel`.
2. SLOW MOTION(Video):
   SLOW MOTION is implemented identically to PLAY. The only difference between the two operators is when each frame is rendered, SLOW MOTION will display it for a much longer duration than PLAY will. This requirement is enforced at the video application, rather than at the temporal database system, since the data requirements of both operators are the same.
3. FAST FORWARD(Video):
   The operational behavior of FAST FORWARD is governed by the speed of viewing. It could be implemented using the driver SET-FRAME by continuously setting the current frame to the one after a certain duration. For example, a three-time fast forward can be implemented as:
   while (true) do
   f = displayed frame obtained from the video application
   SET-FRAME(Video, f, $\frac{3}{30}$)
   end.
   An alternative is to implement FAST FORWARD using the SKIP routine:
   PLAY(SKIP(Video, 1, 2)).
4. FAST BACKWARD(Video):
   Again, the operational behavior of FAST BACKWARD is governed by the speed of viewing. Therefore, FAST BACKWARD is also implemented using the static driver SET-FRAME. However, FAST BACKWARD needs to present the video object in a reverse direction:

while (true) do
f = displayed frame obtained from the video application
SET-FRAME(Video, f, $-\frac{3}{30}$)
end.

Again, an alternative implementation is to implement FAST BACKWARD using the SKIP routine:

v = SKIP(Video, 1, 2)
COPY(v, END(v), BEGIN(v)) >> **outputchannel**.

5. PAUSE(Video):
The PAUSE operator could be implemented by repeatedly copying the frame presented at the current moment to the output channel:
f = displayed frame obtained from the video application
t = GET-CHRONON(Video, f)
while true do
COPY(Video, t, t) >> **outputchannel**.
6. RESUME(Video):
The RESUME operator could be implemented by playing the video from the moment that the play back of the video is paused to the last frame of the video object:
f = displayed frame obtained from the video application
t = GET-CHRONON(Video, f)
COPY(Video, t, END(Video)) >> **outputchannel**.
7. SEARCH(Video, *time*):
The SEARCH operator locates the frame of Video at a particular time *time* and begins the play back action starting from that frame. It is easily implemented using COPY:
COPY(Video, *time*, END(Video)) >> **outputchannel**.
8. STOP(Video):
The STOP operator terminates the copying of the video object to the output channel. It could be implemented by copying the frame at the null chronon:
COPY(Video, -, -) >> **outputchannel**.

6.2 Video Editing

Video editing applications construct a new video object by defining a temporal relationship between two existing video objects. Seven temporal relationships are supported. They include before, meets, dissolve, over, insert, starts, finishes, and equals [20]. These kinds of temporal construction could be readily implemented using our COPY and COMPOSE driver routines. Due to space limitation, we will only illustrate the implementation of three sample temporal constructors: dissolve, over, and insert. The semantics of these three temporal relationships are depicted in Fig. 4*a*, 4*b*, and 4*c* respectively.

The insert constructor inserts a video segment within the middle of another video segment. It is one of the most common operator in video editing. The over constructor allows the superimposition of a video segment on top of the middle of another video segment, as is useful in bringing together two scenes. This is

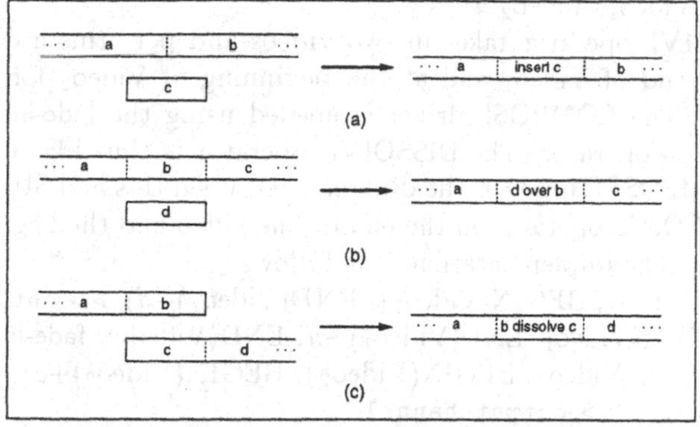

Fig. 4. Temporal relations in video editing.

generalized into SUPERIMPOSE operator using **over** as the executed operator on the video objects. The **dissolve** constructor illustrates the common cinematic effect of "dissolving". The end of one video segment will be dissolved into the beginning of another. The implementation of these operators are as follows:

1. INSERT(Video_1, Video_2, t):
   The INSERT operator takes two video objects as argument and insert Video_1 into Video_2 at the point of time t. It can be implemented as:
   COPY(Video_2, BEGIN(Video_2), t) >> `outputchannel`
   COPY(Video_1, BEGIN(Video_1), END(Video_1)) >> `outputchannel`
   COPY(Video_2, t, END(Video_2)) >> `outputchannel`.
2. SUPERIMPOSE(Video_1, Video_2, t, `s-op`):
   The SUPERIMPOSE operator takes in two video objects and performs a superimposition of Video_1 onto the middle part of Video_2 starting at time t, based on the superimposition operator. Video_1 is assumed to be of shorter duration than Video_2, i.e., (END(Video_2) − BEGIN(Video_2)) > ((END(Video_1) − BEGIN(Video_1)). The behavior of superimposition is defined by the superimposition operator, `s-op`. We support six superimpose operators, including **and**, **or**, **xor**, **set**, **over**, and **fade-in-fade-out** [13, 25]. One of the most useful operators in movie editing is the use of the **over** operator as depicted in Fig. 4*b*, which is to place some transparent image onto another image, for example, to create the effect of someone flying in the sky. The general SUPERIMPOSE operator can be implemented using the COMPOSE driver using the `s-op` composition operator, as follows:
   COPY(Video_2, BEGIN(Video_2), t) >> `outputchannel`
   COMPOSE(Video_2, t, t+END(Video_1)−BEGIN(Video_1), `s-op`, Video_1,
   BEGIN(Video_1), END(Video_1)) >> `outputchannel`
   COPY(Video_2, t+END(Video_1)−BEGIN(Video_1), END(Video_2))
   >> `outputchannel`.

   Notice that the superimpose operators such as **and** are supported in hardware and are readily available to be utilized by the 4DIS temporal database system and in turn, by the driver routines.

3. DISSOLVE(Video_1, Video_2, l):
The DISSOLVE operator takes in two videos and performs a dissolving effect at the end of Video_1 and at the beginning of Video_2 for a period of duration d. The COMPOSE driver is needed using the **fade-in-fade-out** superimposition operator. The DISSOLVE operator is thus like a special case of SUPERIMPOSE. However, the difference between this and SUPERIMPOSE is that DISSOLVE operates on the end of one video and the beginning of the other video. The implementation is as follows:

```
COPY(Video1, BEGIN(Video1), END(Video1)-d) >> outputchannel
COMPOSE(Video1, END(Video1)-d, END(Video1), fade-in-fade-out,
        Video2, BEGIN(Video2), BEGIN(Video2)+d)
        >> outputchannel
COPY(Video2, BEGIN(Video2)+d, END(Video2)) >> outputchannel.
```

7 Discussion and Future Research Directions

In this paper, we have presented a video database architecture developed upon the 4DIS temporal database system. The temporal system provides persistent and querying supports for video objects, allowing application abstractions to be defined according to the their semantics. We have demonstrated the feasibility of our framework via the implementation of several kinds of video applications.

Several issues constitute our current research directions. First, we would like to support quality of service (QoS) for video objects in 4DIS. In general, QoS can be measured by two factors: the number of frames presented per second and the degree of resolution of each frame. We have illustrated in Section 3 that the number of frames presented per second could be modeled by the validity duration of the current-frame relationship between a frame object and a video object. However, to successfully provide this service, the temporal database system should properly enforce this constraint during run-time. To this end, we are currently incorporating the concept of real-time transactions into 4DIS. However, the environment may not be capable of supporting the required number of frames per second, such as in a high traffic network. One alternative is to reduce the resolution of each frame. This could reduce the transmission and presentation time for rendering each frame. A video frame object must be properly structured to incorporate information for presentation at various levels of resolution. This is not supported in the current 4DIS implementation, where each frame is modeled as an atomic object.

Another issue that we are investigating is the modeling and support of audio objects. Audio objects exhibit a large degree of resemblance with video objects. Both media are composed of a collection of components, related in a temporal sequence. Presentation operators on both media require timing relationships among their components. Due to the striking similarities between these two media, we believe that our approach to manage video objects could be fruitfully applied on audio objects as well.

We are also modeling the synchronization process between video and audio objects at various levels of granularity. Synchronization between video and audio objects can be at a coarse level such as aligning the start and end of an audio annotation with the first and last frame of a video frame sequence, both of which last for a duration of several seconds. In contrast, synchronization could be achieved at a fine level such as alignment of a lip movement which usually requires synchronization between video frame sequence and audio annotation every $\frac{1}{30}$ seconds [20]. We are currently modeling synchronization with temporal constraints. Finally, we are implementing a prototype based on the Ode object-oriented database system [2].

References

1. ADB. MATISSE Versioning. Technical Report, ADB/Intellitic., 1993.
2. AT&T Bell Laboratories. *Ode 4.1 User Manual.*
3. J. Banerjee, W. Kim, H.K. Kim, and H.F. Korth. Semantics and Implementation of Schema Evolution in Objec-Oriented Databases. In *Proc. of the ACM Int'l Conf. on Management of Data*, 1987.
4. A. Bimbo, E. Vicario, and D. Zingoni. Sequence Retrieval by Contents Through Spatio Temporal Indexing. In *Proc. of IEEE Symposium on Visual Languages (Cat. No.93TH0562-9)*, pages 88–92, 1993.
5. A. D. Birrell and B. J. Nelson. Implementing remote procedure calls. *ACM Transactions on Computer Systems*, 2(1):39–59, February 1984.
6. C. W. Chang, K. F. Lin, and S. Y. Lee. The Characteristics of Digital Video and Considerations of Designing Video Databases. In *Proc. of ACM International Conference on Information and Knowledge Management*, pages 370–377, 1995.
7. H. Chou and W. Kim. A Unifying Framework for Version Control in a CAD Environment. In *Proc. of the Int'l Conf. on Very Large Data Bases*, 1986.
8. T. S. Chua and L. Q. Ruan. A Video Retrieval and Sequencing System. *ACM Transactions on Information Systems*, 13(4):373–407, 1995.
9. Y.F. Day, S. Dagtas, M. Lino, A. Khokhar, and A. Ghafoor. Object-oriented conceptual modeling of video data. In *Proceedings of the 11th Int'l Conference on Data Engineering*, pages 401–408, Taipei, Taiwan, 1995.
10. U. Dayal. Queries and Views in an Object-Oriented Data Model. In *Proc. of the 2nd Workshop on Database Programming Languages.* ACM Press, 1989.
11. A. Duda and R. Weiss. Structured Video: A Data Type with Content-Based Access. Technical Report, MIT/LCS/TR-580, 1993.
12. R. Elmasri and G. Wuu. A Temporal Data Model and Query Language for ER Databases. In *Proc. of the 6th Int'l Conf. on Data Engineering.* IEEE Computer Society, 1990.
13. R. Fielding. *The Technique of Special Effects Cinematography.* Focal Press, fourth edition, 1985.
14. S. Gibbs, C. Breiteneder, and D. Tsichritzis. Data modelling of time based media. In *Proc. of the ACM Int'l Conf. on Management of Data*, pages 91–102, 1994.
15. R. Hjelsvold and R. Midstraum. Modelling and Querying Video Data. In *Proc. of the Int'l Conf. on Very Large Data Bases*, pages 686–694, 1994.
16. R. Jain and A. Hampapur. Metadata in Video Databases. *ACM SIGMOD Record*, 23(4):27–33, 1994.

17. W. Kim and H. Chou. Versions of Schema in OODB. In *Proc. of the Int'l Conf. on Very Large Data Bases*, 1988.
18. J.C.M. Lee, Q. Li, and W. Xiong. Automatic and Dynamic Video Manipulation. In B. Furht, editor, *Handbook of Multimedia Computing*, 1998 (to appear).
19. Q. Li and J.C.M. Lee. Dynamic Object Clustering for Video Database Manipulations. In *Proc. IFIP 2.6 Working Conference on Visual Database Systems*, pages 125–137, Lausanne, Switzerland, 1995.
20. T. D. C. Little and A. Ghafoor. Interval-Based Conceptual Models for Time-Dependent Multimedia Data. *IEEE Transactions on Knowledge and Data Engineering*, 5(4):551–563, August 1993.
21. B. Mannoni. A Virtual Museum. *Communications of the ACM*, 40(9):61–62, 1997.
22. E. McKenzie and R. Snodgrass. An Evaluation of Relational Algebra Incorporating the Time Dimension in Databases. *ACM Computing Surveys*, 23(4):501–543, 1991.
23. W. Niblack, R. Barber, W. Equitz, M. Flickner, E. Glasman, D. Petkovic, P. Yanker, C. Faloutsos, and G. Taubin. The QBIC Project: Query Images by Content Using Color, Texture and Shape. In *SPIE Proc. Storage and Retrieval for Image and Video Databases, vol. 1908*, pages 173–186, 1993.
24. G. Özsoyoğlu and R. Snodgrass. Temporal and Real-Time Databases: A Survey. *IEEE Transactions on Knowledge and Data Engineering*, 7(4):513–532, 1995.
25. T. Porter and T. Duff. Compositing Digital Images. *ACM Computer Graphics (SIGGRAPH'84)*, 18(3):243–259, July 1984.
26. E. Sciore. Using Annotations to Support Multiple Kinds of Versioning in an Object-Oriented Database System. *ACM TODS*, 16(3):417–438, 1991.
27. A. Si, R.W.H. Lau, Q. Li, and H. V. Leong. Modeling Video Objects in 4DIS Temporal Database System. In *Proceedings of ACM Symposium on Applied Computing, Multimedia Systems Track*, pages 525–531, February 1998.
28. A. Si, H. V. Leong, and P. Wu. 4DIS: A Temporal Framework for Unifying Meta-Data and Data Evolution. In *Proceedings of ACM Symposium on Applied Computing, Database Technology Track*, pages 203–210, February 1998.
29. A. Silberschatz, H. F. Korth, and S. Sudarshan. *Database System Concepts*. McGraw-Hill, 1996.
30. S. Smoliar and H. Zhang. Content-Based Video Indexing and Retrieval. *IEEE Multimedia*, 1(2):62–72, 1994.
31. R. Snodgrass. Temporal Object-Oriented Databases: A Critical Comparison. In W. Kim, editor, *Modern Database Systems*, pages 386–408. Addison-Wesley, 1995.
32. R. Srihari. Automatic Indexing and Content-Based Retrieval of Captioned Images. *IEEE Computer*, 28(9):49–56, September 1995.
33. M. Stonebraker, L. Rowe, and M. Hirohama. The Implementation of POSTGRES. *IEEE TKDE*, 2(1):125–142, 1990.
34. S.Y.W. Su and H.M. Chen. A Temporal Knowledge Representation Model OSAM*T and Its Query Language OQL/T. In *Proceedings of the Int'l Conf. on Very Large Data Bases*, 1991.
35. Y. Tonomura. Video Handling Based One Structured Information for Hypermedia Systems. In *Proceedings of the ACM Int'l Conf. on Multimedia Information Systems*, pages 333–344, New York, USA, 1991.
36. Y. Tonomura, A. Akutsu, Y. Taniguchi, and G. Suzuki. Structured Video Computing. *IEEE Multimedia*, 1(3):34–43, 1994.

Video Sequence Similarity Matching

D. A. Adjeroh, I. King, and M.C. Lee

e-mail: [donald, king, mclee]@cse.cuhk.edu.hk
Department of Computer Science and Engineering,
The Chinese University of Hong Kong,
HONG KONG.

Abstract. Content-based retrieval of multimedia data with temporal constraints, such as video and audio sequences, requires a consideration of the temporal ordering inherent in such sequences. Video sequence-to-sequence matching is therefore an important step in realizing content-based video retrieval. This paper provides an overview of the general issues involved in video sequence matching, points out the immediate problems that must be addressed before video sequence matching becomes practical, and proposes some general methods to address the problems. Implications of the video sequence matching problem on other areas of multimedia information retrieval are also highlighted.

1 Introduction

Content-based access to image data collections has received a relatively considerable attention. In comparison, not much have been done on content-based access to multimedia data collections with temporal constraints, such as video or audio sequences. At best, the considerations so far have mainly been limited to the initial step of sequence segmentation - video partitioning [11] and audio segmentation [27]. Content-based access to multimedia information with temporal constraints however requires access beyond the initial segmentation. This requires a paradigm-shift in the representation and manipulation of these data types, calling for a special consideration of the spatio-temporal dynamics of the information contents within and between the individual segments, and of the inherent temporal ordering of the information carried by such data types.

Generally speaking, content-based video retrieval is concerned with the indexing and retrieval of video information based on the actual contents of the video sequence. The first stage of this process is the initial partitioning of the video data into its constituent scenes, and the identification of the various editing effects in the video [11]. The next stage involves the analysis of the individual scenes to provide further information for fine-grained access to the digital video. One such analysis results in the generation or selection of representations for each video scene, such as the video key frame or other higher-level compact representations [13, 20]. Further stages of the retrieval process could involve the classification or clustering of the partitioned video based on certain characteristics of the scenes, such as motion complexity or the activity in the scene [2, 7].

From these, it could then become possible to address the more difficult problem of semantic access to the video contents [7].

Though most of the representations are typically generated based on some motion and temporal information in the video, the representations themselves generally do not have these information, and hence the temporal information are generally no longer available at the time of comparison. Thus, while such methods could be adequate for exact matching of the video sequences, they are generally inappropriate when similarity is required. Further, it is well known that the ordinary matching of text sequences is a computationally intensive problem. With the various features with which the visual information in the video can be described, video sequence matching is bound to stretch the available computational resources to their limit. In spite of these problems, a few methods have been proposed for video sequence matching [4, 7, 12, 29]. In this paper, we present an overview of the general issues involved in video sequence matching, and describe some methods that can be used to address some of the problems.

2 Video Sequence Searching

The *video sequence searching problem* can be stated as follows: given a query video sequence, and a database of video sequences, find one or all the occurrences of the query video sequence in the database. The problem then is to search the entire database of video sequences for the requested query video sequence, producing a list of the positions in the database sequence where a match starts (or ends). Solution to this problem depends on a variant of the problem - the *video sequence matching problem*: given two video sequences, determine whether they are matches or not. Since the problem here is exact matching of the sequences, it is not difficult to compute some simple statistics with which the matching can be performed. It is known however that exact matching is not very appropriate in multimedia information systems, especially those that involve visual data. A more useful variant of the problem is that of video sequence *similarity searching* which will in turn depend on video sequence *similarity matching.*

2.1 Video Sequence Similarity Matching

Given two video sequences, the *video sequence similarity matching problem* is to determine whether the two sequences are similar or not, based on some stated acceptable levels of similarity. Unlike in the case of exact matching, since we are interested only in similarity between the sequences, the V_1 and V_2 need not be of the same length, and the size of the frames in each sequence need not be the same. Conceptually, V_1 can represent the entire database, while V_2 is just a short query sequence. Here, we would be interested in knowing if there exists any subsequence of V_1 that is similar to V_2. The notion of similarity also implies that, in choosing the portion of the database to match with the query sequence, the similarity threshold may need to be used to determine possible limits on the size of the portion to be considered at each iteration step. The problem of exact matching is thus easily handled by just an appropriate choice of the threshold.

In video sequence matching, three basic types of matching can be identified:

1. *scene-to-scene*: check if two scenes are similar;
2. *scene-to-sequence*: check if a scene similar to the query scene occurs in the database sequence;
3. *sequence-to-sequence*: check if a sequence similar to a query sequence occurs in the database sequence. The query can contain more than one scene.

The case of sequence-to-scene matching is simply handled by taking the database sequence as the longer of the two sequences. We observe that (2) is a generalization of (1), and will make use of methods for (1). Similarly, (3) is a generalization of (2) and its solution will depend on solutions to (2). The basic problem then is finding solutions to (1): the scene-to-scene matching problem.

2.2 General Approaches and Key Problems

The technique used to address the video sequence similarity matching problem generally depends on the methods used to represent the sequences. For instance, if the video information in a given sequence is treated as a time series signal, then ideas from time series analysis can be used to compare two such signals. Three key issues can be identified as needing immediate attention before video sequence matching can become feasible and reliable. These are: (i) providing suitable representations for describing video sequences along with their temporal constraints; (ii) devising appropriate measures or metrics with which the similarity between the sequences can be assessed; and (iii) finding ways of speeding up the different aspects of the matching process. These issues are treated in more detail in the subsequent sections.

3 Representations for Video Sequence Matching

The particular technique used to represent the video sequence determines the methods for the matching, and possibly, the techniques with which the matching can be speeded up. The reliability of the match results also depends heavily on the representation formalism used. A video sequence is primarily a time ordered sequence of images (called frames). At the first instance, a video sequence V can the be represented as an ordered set of frames: $V = F_1, F_2, ..., F_n$, where F_i represents the i-th frame in the sequences. The problem of video sequence representation for similarity matching then deals with how to extract some information from V, with which matching can be performed. Three basic aspects of the problem can be identified: (i) representing the original video sequence, (ii) selecting appropriate features and their extraction from the adopted representation, and (iii) transcribing the features into forms suitable for sequence matching. In all cases, consideration will have to be made of the special characteristic of video sequences that make them significantly different from other types of sequence data - repetitions, periodicity, and redundancies; multidimensional features; video scene breaks; incorporation of motion and spatial information; time codes and temporal information, etc.

3.1 Spatio-Temporal Representation

By spatio-temporal representation, we refer to how the original video sequence can be represented prior to extracting the required features for matching.

Frame-by-Frame Representations Here, all the frames in the sequence are used, and features are extracted from each frame. The features can then be viewed as a sequence of features values, with the sequence determined from the temporal ordering of the frames in the video. The advantage here is accuracy, since no information is lost before the features are extracted. The major problem is the time that may be required to analyze each frame to extract the required features, bearing in mind the usually large number of frames, and the fact that multiple features may be needed. Frame skipping is one basic method used to handle the problem, though this may lead to some loss in accuracy, and the choice of an appropriate skip factor still remains an issue.

Compact Representations Usually, rather than using all the frames in a sequence, some surrogate for the sequence is used, from which the index features are extracted. Such surrogates are typically more compact than the original sequence, and are assumed to provide a typical representation (i.e. summary) of the contents of the scene. With appropriate canonical representations, video sequence similarity matching becomes more feasible, although at a possible loss in accuracy. A simple example here is the video key frame. A simple representation similar to the key frame can be generated by computing an average frame using some of (or all) the frames in a scene.

Some other forms of compact representation are generated by considering the motion in the video, and by applying some transformations on the original sequence [13, 20]. These representations are generally larger than the original size of the frames, and the size could vary depending on the contents of the video. The video mosaics, layered representations, or the super resolution frames are typical of this type of representation. Though the compact representations are usually generated based on the motion in the video, the motion information itself is no longer available in these representations. Thus, when the objective is video sequence matching, motion-based features may be difficult to be derived from such representations. Another problem is that two different scenes could produce very similar compact representations (example the video mosaics) - depending on the transformations used.

Multi-resolution/pyramidal representations Another form of representation that can be used for the video is the pyramidal representation [3]. Depending on the type of scene, the frames can be represented at different resolutions. Conceptually, the reduced frames can be used just like the full frames in the frame-by-frame representation. Similarly, key frame selection, generation of compact representations, and extraction of the required features can all be carried out at this reduced scale. An extension of the idea is the generation of the multi-scale representation in three dimensions, with the temporal axis forming the third dimension. Thus, like in the compact representation, an entire scene or a set of frames can be reduced to their 3D multi-resolution form, from which some features can be computed for the purpose of sequence matching. This may also leed to loss of information about the motion in the video.

3.2 Features to be Used[1]

After choosing an appropriate spatio-temporal representation for the sequence, we still have to decide on some specific features that can be used to capture the information in the sequence. Clearly, the particular features chosen will depend highly on the representation adopted. Typical features could be the color, texture, frame differences, motion information, etc. For color and texture, some statistics can be calculated for each frame (or representation). A sequence of such statistics then forms a sequence of color or texture feature values, which can then be used for matching. More accurate descriptors may be obtained by subdividing the frames into subframes, and then performing the matching based on sequences of feature values computed for each subframes. Here, each subframe will yield a sequence of feature values, and thus for each feature, we may have multiple sequences.

For motion features, the motion vectors from adjacent frames can be used as features. The motion vectors indicate both directional information and the amount of motion between frames. To avoid the problem of object segmentation, block motion vectors, rather than object motion vectors could be used. The vectors can be clustered based on some neighborhood criteria to obtain a reduced set of vectors, which can then be used to form a feature sequence. In [12, 7], motion-based features were used as the basic cue in video scene retrieval.

3.3 Representing the Features - Feature Transcription

In its simplest form, the features (such as statistics from color or texture, motion vectors, etc.) will be real-valued. Depending on the similarity measure to be used for comparing the sequences, (for instance, simple correlation function), the index features can be stored as just their original real-valued data sequences. Generally, the feature values may need to be further transcribed into forms more suitable for sequence matching. The first problem will be normalizing the feature values to within some given range, for instance [0 1], for easy and uniform

[1] For simplicity, further discussions assume the frame-by-frame representation, though they are also applicable to the other representations.

assessment of the differences or similarity between two feature values. Such normalization can be performed at the time of feature extraction. Below, we consider three methods with which the original real-valued feature values can be represented for the purpose of sequence matching.

Bounding-Box Representation The bounding-box representation for video sequences was proposed in [4], based on the concept of minimum bounding hyper-rectangles used by Faloutsos et al [8] for the representation and matching of time series data. The basic idea is to transform the original real-valued time series data sequence into a set of rectangles in feature space. A sliding window is used on the original sequence to extract a sequence of features - called trails - from the original sequence. Each trail is further subdivided into smaller regions (sub trails), and each is subsequently represented by its minimum bounding rectangle (MBR). Each MBR thus corresponds to a sub-trail, with possible overlaps between the rectangles. The advantage of the MBR representation is that, rather than storing all the original data points, only few points - the coordinates of the rectangles, and some identifier are stored - resulting to improvements in the storage efficiency and the amount of time needed to match two sequences. With multiple features, the idea is easily extended into the concept of minimum bounding hyper-rectangles.

The major differences between the MBR used in [8] and the bounding-box used in [4] are that the bounding boxes allowed no overlap, and are based on prominent index points, rather than sliding windows. The prominent index points are determined based on certain constrains (thresholds) on the feature values, such as the temporal difference between two prominent points, the difference between successive feature values, etc. One problem with the bounding box method is the difficulty in choosing appropriate prominent points. This is due to the inherent problem in defining effective rules or constraints based on the feature values, since these could vary significantly between video scenes. A slight change in the actual value of a given feature at a given temporal position would lead to different dimensions for the bounding box or rectangle, and thus to a different description of the same sequence. A more difficult problem is how to device methods for reliable similarity matching using the box representation.

From Video Sequences to Video Strings Another form with which the information in a video sequence can be captured is the string representation. Here the sequence of feature values is transformed into a sequence of symbols, with the symbols coming from a defined alphabet. For a given feature, this will involve the initial transformation of the real-valued feature values into some discrete classes. Each symbol in the string represents a class, and the set of all symbols form the alphabet. Thus, the total number of symbols will depend on the number of classes used. We call such a string representation of the video sequence a *video string*, or *vstring* for short. For multiple features, we may have different classifications, leading to multiple alphabets, with possibly different cardinality. In such a case, we will have multidimensional video strings, with the strings from each feature forming a dimension. In [29], the string approach was used to match image sequences, with the longest common subsequence as the similarity measure.

Usually, symbols in traditional text strings are taken as presence/absence symbols - that is, the symbols either appear in the string or they do not appear, and two different symbols are assumed not to have much in common. For instance, the symbols "*a*", and "*b*" in the English alphabet are assumed not to have much in common. For the video string, when the symbols are taken from an alphabet obtained from the classification of real valued features, we assume that nearby classes are related. That is, a feature value that belongs to class I is nearer to another feature value that belongs to class II, than to one that belongs to say class III. This implies that the symbols in the video string will be multi-valued. This modification is needed to improve the accuracy of the similarity measurement using video strings, and will have some important implications in defining distances between video strings. For some other types of classification (e.g. semantic classification of the sequences), the symbols can be treated as the traditional presence/absence symbols. The example below gives a string representation of some of the basic video frame transitions: fast forward, slow motion, reverse, partial reverse.

Example: Video scene transitions represented by video strings. The symbols a, b, c, etc. represent the different classes to which the feature values are grouped. A video frame is represented by a symbol. $ stands for video scene break.

Original Sequence: *aaabbbcccdddeee$vvvxxxyyy*;

Fast forward:

skip=1: speed=×2 : *aabccdee$vxxy*

skip=2: speed=×3 : *abcde$vxy*

skip=3: speed =×4 : *abce$vy*

Slow motion:

speed=×0.5 : *aaaaaabbbbbbccccccddddddeeeeee$vvvvvvxxxxxxyyyyyy*

speed=×0.25 : *aaaaaaaaaaaabbbbbbbbbbbb* ... *eeeeeeeeeeee$vvvvvvvvvvvv* ...

Reverse: *yyyxxxvvv$eeedddcccbbbaaa*

Partial reverse (due to video editing):

Case 1: *bbbccc* and *dddeee* transposed: *aaadddeeebbbccc$vvvxxxyyy*

Case 2: *bbb* and *vvv* transposed: *aaavvvcccdddeeebbbxxxyyy*

Note: For partial reverse, the first, second and fourth scene breaks in Case 2 are required since the edit operation involved two different scenes. In Case 1, the editing involved only frames in the same scene. Fast reverse is very similar to fast forward.

The problem with the video string representation is the possible loss of accuracy, since the quantization needed for the classification will necessarily introduce some error. The problem can be reduced by use of more classes (larger alphabet size), though at the expense of more classification time. The advantage of the string representation is that it is fairly general - we can represent different types of video scene classifications using strings sequences. For instance, we can easily classify using quantitative features from the video - such as motion vectors, angles, color, etc. The semantic classes used for video scene classification can also form the alphabets for the video string, e.g. special effects: {fade-in, fade-out, dissolve, pan, zoom, normal}; scene contents: {news, sports, documentary, adverts}; spatial motion direction: {north, south, east, west, north-west, north-east, south-west, south-east}; etc. More importantly, the video string representation provides an intuitive method to model various characteristics of the video sequence - such as repetitions, reverse, fast forward, video scene breaks, etc.

Tree (Hierarchical) Representations Generally, the use of trees in image representation is based on some structural (for instance, containment) relationships between image and objects or sub-regions in the image. In representing video sequences with trees, we can consider the case when the trees are used in their traditional form - based on structural information, including objects in each frame. Here each object is just a node in the tree, and the root node corresponds to the entire sequence or a given scene. An object could contain some other object, which will be represented as a child of the previous object.

A second approach could be to classify the feature values and then each class forms a node in the tree representation. Conceptually, the members of a given class could further be classified into different subclasses. The tree is then extended by using each new subclass as a new child node, and the parent class as a parent node. The process can continue ad infinitum, with the height of the tree only limited by the application requirements or available resources. Such a hierarchical representation provides a more general form of the video string representation, and a more accurate representation of the video sequence. Also, both the semantic classification and the classification based on feature values can easily be integrated in the tree representation.

While the tree representation is conceptually appealing, its realization could be considerably more difficult than the video string. Determining the number of levels to use could be difficult, and will typically be application dependent. Matching with trees is also considerably more difficult than string matching [30].

4 Similarity Measures for Video Sequences

After transcribing the sequence data into an appropriate representation, the actual matching using the selected representation will still have to be performed. Since we are often more interested in similarity (rather than exact) matching, there is need for some measure to indicate the degree of similarity between two sequences. Formally, given two video sequences A and B, represented by their p-feature values at each temporal indices t and r: $A = [a_1(t), a_2(t), ..., a_p(t)]^T$; $B = [b_1(r), b_2(r), ..., b_p(r)]^T$ where, $r = 0, 1, 2, ..., n; t = 0, 1, 2, ..., m$, x^T stands for transpose of x. The problem is to find a mapping: $f : A \times B \rightarrow \Re$, such that f is independent of n and m, the sequence lengths. Even with some possible variations in A and/or B, the function is still expected to provide a reliable measure of the similarity between them. That is, $f : A \times B \times \theta \rightarrow \Re$, where θ

represents some variations on the p features from A or B. The above formulation assumes that each of the p features are robust (or should have been normalized) against certain variations in the frames, such as those due to scale changes, illumination, rotation, etc. Similarly, we expect that the resulting values from f can be normalized to some given ranges, say [0 1], such that the degree of similarity increases uniformly from perfect mismatch (0), to perfect match (1).

Different measures (functions) could be used to achieve the required mapping. The measure could be a distance function or a similarity function, and need not necessarily be a metric. Whether the two sequences are similar (matches) or not will then depend on some pre-defined threshold of similarity. The threshold can be chosen based on the application, or could be stipulated by the user. Below, we consider some measures that can be used to compute the similarity between two video sequences - based on the specific representation used.

4.1 Cross Correlation

When the sequence data is viewed in its original real-valued form, without transcription into some other representation, the simple cross correlation function can be used as the similarity matching function. Let a_t and b_t be single-features (one dimensional signals): $a_t = a_0, a_1, \ldots, a_n; b_t = b_0, b_1, ..., b_m$; To obtain the correlation between the two signals, we simply glide one signal (typically the query) over the other (the database), and integrate the product from $-\infty$ to $+\infty$, for each value of the time shift, say s. The similarity mapping function is then given by the cross correlation function: $f(a_t, b_t, s) = \sum_{t=0}^{n} a_t b_{t+s}; a_t = 0, t > n; b_t = 0, t > m;$

For each time shift s, we obtain a different value for the similarity between the signals. The closest match is then given by the b_{t+s} for which the correlation $f(.)$ is maximum. Thus, if we align a_t with this b_{t+s}, we obtain the best match. The major problem here is the enormous computational cost. Cheever et al [6] used the cross correlation method to compare DNA sequences. To avoid the huge computational effort required, they proposed computing the correlation in the frequency domain via the Fast Fourier Transform (FFT). Still, it is known that even the use of FFT algorithms may not always afford much help, depending on the size of the query sequence relative to the database sequence [10]. Thus, speeding up calculations involving the correlation function remains a major problem.

4.2 Bounding Box Approach

With the bounding box method, we need special measures to compare two bounding boxes (or sequence of bounding boxes) for possible similarity. The simplest and obvious method could be to calculate some statistics such as the area (or volume), aspect ratio, etc. using the dimensions of the bounding box. Faloutsos et al [8] proposed the use of the overlaps between bounding rectangles. In this case, the similarity function can be defined using the amount of intersection between the MBR from the two sequences. Chang and Lee [4] used the difference between the "box density" of two given boxes as a similarity measure.

For a given feature, the box density was defined as the average accumulated difference between two consecutive feature values within the same box. They also used some other features such as the aspect ratio, dimensions of the box, etc., and the final similarity between bounding boxes is determined by the combination of these various measures using a cost function.

4.3 Statistics-Based Measures

These are similarity measures that depend on statistical information from the sequences, without actually performing point-by-point matching. Simple examples here include the ratio of the sequence lengths, the frequency of occurrence of the symbols, number of breaks, etc. For instance, $|A| + |B| - 2.common(A, B)$ is a simple measure of similarity; where $common(x, y)$ is a function that returns the number of symbols common to strings x and y, and $|x|$ is the length of x. Sankoff [19] defined a number of statistics on genetic sequences, and combined these with the traditional edit distance measures to present some non-local measures of similarity between sequences.

More complex statistics-based measures have also been proposed. Milosavljevic [15] proposed the use of *algorithmic mutual information* between sequences as a measure of their similarity. The algorithmic mutual information between two sequences A and B was defined as the difference between the encoding lengths of A, when A is encoded independently, and when A is coded relative to B, by making use of their similarity. The mutual information was further used to compare DNA sequences. Ohya [17] also used concepts related to entropy to define a measure of similarity between genetic sequences. In [28], various definitions for local complexity for biological sequences were provided, and information entropy was proposed as a possible measure of similarity.

One major drawback with the statistics-based measures is that they ignore the order constraint inherent in sequences. It is clear that since these methods do not use the temporal ordering in the sequence - an important information in video sequences, they will not produce very accurate measures for their similarity. They however can provide methods for obtaining a fast overview of the degree of similarity. Thus they can be combined with some other measures for possibly focussing the match process on those areas of the database sequence with a better promise of having a match.

4.4 Video String Edit Distance

One advantage of the string representation is that, we can borrow ideas from previous results in related areas (such as in theoretical computer science - string pattern matching, or in molecular biology -DNA and protein sequence alignment) for their efficient storage and matching. However, we will have to make appropriate considerations with respect to the special characteristics of video sequences, and the different types of transitions that can occur in such sequences.

Edit Distances The distance between two strings is traditionally calculated using the string edit distance. Given two strings $A : a_1a_2 \ldots a_n$, and $B : b_1b_2 \ldots b_m$, over an alphabet Σ, and a set of allowed edit operations, the edit distance indicates the minimum number of edit operations that will be required to transform one string into the other. Three basic types of edit operations are used - *insertion* of a symbol, $(\varepsilon \rightarrow a)$; *deletion* of a symbol, $(a \rightarrow \varepsilon)$; and *substitution* of one symbol with another $(a \rightarrow b)$. ε represents the zero-length empty symbol, and $x \rightarrow y$ indicates that x is transformed into y). The cost of each edit operation is determined by use of a weighting function. The edit distance between A and B is usually determined using some recurrence equations [21]:

initializations:

$d_{0,0} = 0; d_{i,0} = d_{i-1,0} + \alpha_{del}(a_i); d_{0,j} = d_{0,j-1} + \alpha_{ins}(b_j); 1 \le i \le n; 1 \le j \le m$

main recurrence:

$d_{i,j} = min\{d_{i-1,j}\alpha_{del}(a_i), d_{i-1,j-1} + \alpha_{sub}(a_i, b_j), d_{i,j-1} + \alpha_{ins}(b_j)\}$

where $\alpha_{del}, \alpha_{ins}, \alpha_{sub}$ are the respective costs of deletion, insertion, and substitution edit operations. Approximate pattern matching (i.e. the *k-difference problem*) is then performed by checking if there exists a substring A_s of A, such that the edit distance between A_s and B is less than k. Another form of approximate string matching - the *k-mismatch problem* checks for a substring of A having only a maximum of k mismatches with B. Different algorithms have been proposed for the string edit distance problem, primarily based on dynamic programming [21, 24], and various extensions and improvements have been made [23, 9, 14]. The edit distance forms the basis for various algorithms used in text pattern matching and in DNA sequence alignments [25].

New Video String Edit Operations We can use a similar method to define the similarity mapping function between video strings. However, due to the special types of transitions that can occur in a video sequence, the three basic edit operations will be inadequate in handling the unique characteristic of video sequences. For the case of video strings, we extend the traditional edit distance by defining some new edit operations, namely: transposition, break, fusion/fission.

Transposition: interchange two symbols in one of the strings: $abdc \rightarrow adbc$; (transpose b and d), $abdc \rightarrow cbda$ (transpose a and c). For video, apart from the temporal positioning, the actual contents of the frames in the sequence are also important is assessing the similarity between sequences. Thus the transposition operation will be useful in handling the special transitions and possible partial occlusions that may occur in a video sequence, or the cases of partial reverse. Though the temporal ordering may not be the same in these cases, the contents of the scene may still be viewed as similar by the human observer, since the frames basically contain the same information.

Fusion/fission: fusion: merge a consecutive stream of the same symbol into a single symbol: $aaa \rightarrow a$; fission: convert a single symbol into a stream of symbols all of the same type: $a \rightarrow aaa$. This operation is needed to deal with the repetitive nature of symbols in a video string. It also provides a natural way of handling special video frame transitions, such as fast forward, and slow motion, since these are usually achieved by dropping or repeating frames. The cost of such an operation could be different from that of the repetitive use of the insert or delete operations.

Break: insert or delete a scene break between two adjacent symbols: insert break: $ab \to a\$b$; delete break: $a\$b \to ab$; \$ stands for a scene break. This is a special edit operation that allows the insertion or deletion of video scene breaks at any point in the sequence. Because of the significance of scene breaks in video sequences, they may not be very accurately modeled as an ordinary concatenation, or by mere insertion and deletion operations.

Video String Edit Distance Using the new operations, we can derive a general similarity measure (actually distance measure in this case) for video strings as follows: Let O_p be the set of edit operations: $O_p = \{ins, del, sub, bre, tra, fus\}$, representing insertion, deletion, substitution, break, transposition, and fusion, respectively, and α_p be another set containing the respective cost of each edit operation: $\alpha_p = \{\alpha_{ins}, \alpha_{del}, \alpha_{sub}, \alpha_{bre}, \alpha_{tra}, \alpha_{fus}\}$. Also, let S_E denote a sequence of edit operations that transforms A into B: $S_E = \{{}^{a_1}_{b_1}O_1, {}^{a_2}_{b_2}O_2, \ldots, {}^{a_l}_{b_l}O_l\}$ where ${}^{a_i}_{b_i}O_i$ indicates that $a_i \to b_i$ by edit operation O, $(O \in O_p)$ at edit step i; a_i and b_i are the two symbols[2] involved in i-th edit operation. We can the determine the distance between A and B as the cost of using this particular edit sequence:

$$c(S_E) = \sum_{O_i=ins} \alpha_{ins}\delta({}^{a_i}_{b_i}O_i) + \sum_{O_i=del} \alpha_{del}\delta({}^{a_i}_{b_i}O_i) + \sum_{O_i=sub} \alpha_{sub}\delta({}^{a_i}_{b_i}O_i)$$

$$+ \sum_{O_i=tra} \alpha_{tra}\delta({}^{a_i}_{b_i}O_i) + \sum_{O_i=fus} \alpha_{fus}\delta({}^{a_i}_{b_i}O_i) + \sum_{O_i=bre} \alpha_{bre}\delta({}^{a_i}_{b_i}O_i)$$

$\delta({}^{a}_{b}O)$ is a distance function, whose result depends on a, b, and the edit operation O. Since we can have various sequences of edit operations that can transform A into B, we will actually have a set of edit sequences: $S_{A\to B} = \{S_E^1, S_E^2, \ldots, S_E^q\}$ where S_E^i is the i-th edit sequence that transforms A into B. The edit distance is then given by the minimum cost edit sequence:

$$D(A, B) = min\{c(S_E) | S_E \in S_{A\to B}\}$$

The distance measure above is quite general, incorporating facilities to handle cases with different cost functions for the edit operations; and more importantly for multi-valued symbols. More details on the video string edit distance can be found in [1].

4.5 Tree Matching

When the data is represented as trees, we can use traditional methods for tree matching to compare the sequences. Expectedly, since the tree is a generalization of the string, the tree matching problem is very similar to the problem of string

[2] They could also be strings for certain edit operations, such as the fusion operation.

matching, but decidedly more difficult [22, 30]. The idea is still based on the general concept of edit distances. Given two labeled trees A and B, the tree edit distance is the minimum number of edit operations that will transform one tree into the other. The basic edit operations are *substitution*: change the label of a node; *deletion*: remove a node, and make its children become the children of its parent, and *insertion* which is the complement of deletion.

As with string edit distances, different costs can be allocated to each edit operation. The overall distance is then determined by the cost function and the number of times each edit operation is applied in transforming one tree into the other. When the video string is represented as a tree, there may be need to define some new tree edit operations to account for the special transitions that may occur in a video sequence. Tree matching is treated in more detail in [22, 30].

4.6 Combining Different Similarity Metrics

Because of the diversity of the information content in a typical video sequence, it may be difficult to determine one single metric that provides the best similarity measure in all situations. Better results may be obtained if different metrics are involved in assessing the similarity between sequences. If we can afford the additional computation and space requirements, one may wish to use more than one representation for the sequences, and hence different similarity measures, each for each representation. Conversely, different similarity measures can be defined based on the same representation (for instance the statistical measures can be combined with other measures), and the problem is how to combine the results returned from each measure.

Various methods can be used to combine the multiple metrics. For instance, each metric can be treated independently, with the results of one stage forming the input to the next stage. Alternatively, each measure can be used independently, and the final results are combined by some weighting function. Another method could be to compare different parts of the sequence using adaptively selected metrics. This will require some form of pre-analysis of (parts of) the sequences to determine which metrics will be most appropriate for its comparison. This further implies that some method of pre-classification may be needed.

5 Efficiency Issues

With the huge volume of data involved in a typical video database, efficiency becomes a major consideration in matching the sequences. In fact, the efficiency problem pervades all aspects of the matching process - from the initial spatio-temporal representation for the video sequence to feature extraction to the final matching using the transcribed features.

5.1 Efficiency Problem for the Representation Phase

Efficiency issues here are concerned with methods for speeding up the representation phase. It may be recalled that the problem of efficiency primarily motivated the search for different representation schemes (the compact representations) to capture the video content. Feature extraction based on these representations also need to be speeded up, since this could affect the response time to a query (we still have to extract the equivalent features from the query sequence at the time of query request). Beyond this level, there may also be need to find faster methods for transcribing the features into the required representations for sequence matching. Clearly, at each stage of the representation phase, the methods to be used for improving the efficiency will surely depend on the particular representation under consideration. For now, we ignore the other methods of representation, and concentrate on efficient sequence matching based on the video string representation.

5.2 Speeding Up Video String Matching

With the incorporation of new edit operations for the video string, the challenge is to minimize the additional computation that may be required on account of such new operations. The efficiency problem can be considered from two angles: the edit distance computation, and matching using the edit distance. The edit distance can be computed using a matrix (table) whose entries $D(i, j)$ represent the minimum number of edit operations required to transform the length j prefix of B into a substring of A ending in the i-th letter. A number of methods have been proposed to speed up the computation of the traditional edit distance [23, 9, 5]. For the overall matching process, two basic methods can be used to speed up the matching: *pre-indexing* and *pre-selection*. Pre-indexing (or pre-processing) usually involves the description of the database strings using a pre-defined index. The indices are usually generated by use of some hashing function or a scoring scheme [16]. Pre-selection methods generally divide the matching problem into two stages [18]. In the first stage, an initial filtering is performed to select candidate regions of the database sequence that are likely to be matches to the query sequence. In the second stage, a detailed analysis is made on only the selected regions to determine if they are actually matches. The introduction of new edit operations however, imply that the definition of the string edit distance will have to be modified (see section 4.4), and thus different edit tables and new considerations will have to be made [14].

5.3 Compressed Domain Methods

Compressed domain operations have been proved to provide huge speed ups for traditional frame based video indexing. It may be a useful endeavor to investigate if corresponding improvements can be obtained when the video sequence matching problem is performed in the transform domain. Once again, depending on the representation used, it may be possible to extend the similarity measures to video represented in the compressed form. For instance, compressed video is typically coded as a sequence of symbols using run length encoding. Thus, the video string concept can be easily adapted to handle such string sequences.

6 Discussion and Conclusion

The foregoing sections have presented a number of issues relating to sequence-to-sequence matching of video data. It is clear that before the dream of content-based access to video information can be realized, the video data will have to be considered in terms of a sequence, with its inherent temporal ordering of the information contents. The major drawbacks in video sequence-to-sequence matching remains the difficulty in finding appropriate representations for the video sequence when the objective is similarity matching (the representational problem), and the huge computational effort required for the different stages of the sequence matching process (the computational problem).

A majority of the indicated problems will have to be addressed before sequence matching of video data becomes practical. Some other aspects of the video sequence matching problem are also worthy of some attention, since they could be used to improve the overall matching process - in terms of efficiency and/or reliability of results. With the video string representation, such issues could include a probabilistic analysis of video string characteristics; study of the effect of alphabet size; and analysis of video strings for possible semantic information. Another issue has to do with determining appropriate methods with which the results from the similarity measures can be normalized, and how they can be related to human perception of similarity. This also brings up the question of evaluation methods for video sequence matching algorithms.

The question of video sequence matching is of both practical and theoretical significance. The problem is primarily motivated by the practical need for content-based video retrieval. Results from such an endeavor will equally be of interest in other applications areas where data come with some temporal constraints, such as in music and audio retrieval [27], medical imaging and biomedical monitoring [26], crime investigation (copyright infringement), TV broadcasting, etc. Initial attempts at the video sequence matching problem may have to rely on methods developed in other related areas such as, time series analysis, biological sequence analysis, or traditional text string matching. It is however expected that algorithms developed for video sequence matching (e.g. those that address the still open problem of efficiency) can also be useful in such areas.

References

1. Adjeroh D.A., Lee M.C. and I. King, "A distance measure for video sequence similarity matching", Proc., IW-MMDBMS'98, Ohio, August 1998, to appear.
2. Adjeroh, D. A. and Lee, M.C., "Adaptive Transform Domain Video Scene Analysis", Proc., IEEE Multimedia97, Ottawa Canada, June 3-6, 1997.
3. Burt P. and Adelson E., The Laplacian pyramid as a compact code", IEEE Trans. on Communications, 31, 523-540, 1983.
4. Chang C-W, and Lee S-Y, "Video content representation, indexing, and matching in video information systems", J. Visual Comm. & Image Repr., 8, 2, 107-120, 1997.
5. Chang W.I. and Lawler E.L., "Sublinear approximate string matching and biological applications", Algorithmica, 12:327-344, 1994.

6. Cheever E.A., et al, "Using signal processing techniques for DNA sequence comparison", Proc., 5th Annual Northeast Bioengineering Conf., pp. 173-174, MA, 1989.
7. Dimitrova N. and Golshani F., "Motion recovery for video content classification", ACM Trans. on Information Systems, 13, 4, 408-439, 1995.
8. Faloutsos C, Ranganathan M, and Manolopoulos Y, "Fast subsequence matching in time-series databases", Proc., ACM SIGMOD, pp. 419-429, Minneapolis MN, 1994.
9. Galil Z and Park K., "An improved algorithm for approximate string matching", SIAM J. of Computing, 19, 6, 989-999, 1990.
10. Gonzalez R.C. and Woods R.E., Digital Image Processing. Addison-Wesley Publishing Co., Readings Massachusetts, 1992.
11. Idris F. and Panchanathan S., "Review of image and video indexing", J. Visual Comm. & and Image Repr., 8,2,146-166, 1997.
12. Iorka M. and Masato Kurokawa, "Estimation of motion vectors and their aoolication to scene retrieval", Machine Vision and Applications, 7:199-208. 1994.
13. Irani M. and Anandan P., "Video indexing based on mosaic representations", Tech. Report, Dept. of Applied Maths. & Comp. Sc., The Weizmann Institute, Israel, 1997.
14. Lee J-S, et al, "Efficient algorithms for approximate string matching with swaps", LNCS 1264, Combinatorial Pattern Matching, pp. 28- 39, 1997.
15. Milosavljevic A., "Discovering dependencies via algorithmic mutual information: a case study in DNA sequence comparison", Machine Learning, 21, 35- 50, 1995.
16. Myers E.W., "A sublinear algorithm for approximate keyword searching", Algorithmica, 12: 345-374, 1994
17. Ohya M, "Information treatment of genes", Trans., Japanese IEICE, E72. 5, 556-560, 1989.
18. Pevzner P.A. and M.S. Waterman, "A fast filteration algorithm for the substring matching problem", LNCS 684, Combinatorial Pattern Matching, pp. 197- 214, 1993.
19. Sankoff D., "Edit distance for genome comparison based on non-local operations", LNCS 644, Combinatorial Pattern Matching, pp. 121-135, 1992.
20. Sawhney H.S., and Ayer S., "Compact representation of videos through dominant and multiple motion estimation", IEEE Trans. PAMI, 18, 8, 814-830, 1996.
21. Sellers P.H, "The theroy of computation of evolutionary distances: pattern ecognition", J. of Algorithms, 1, 359-373, 1980.
22. Tai K.C., "The tree-to-tree correction problem", J. of ACM, 26: 422-433, 1979.
23. Ukkonen E., "Finding approximate patterns in strings", J. of Algorithms, 6, 132-137, 1985.
24. Wagner A. and Fischer M.J, "The string-to-string correction problem", J. of ACM, 21: 168-173, 1974.
25. Waterman, M.S. (ed.), Mathematical Methods for DNA Sequences, CRC Press, Boca Raton, Florida, 1989.
26. Wendling F.,et al, "Extraction of spatio temporal signatures from depth EEG seizure signals based on objective matching in warped vectorial observations", IEEE Trans. Biomedical Engineering, 43, 10, 990-1000, 1996.
27. Wold W., et al, "Content-based classification, search and retrieval of audio", IEEE Multimedia, 3, 3, 27-36, 1996.
28. Wooton J.C. and Federhen S., "Statistics of local complexity in amino acid sequences and sequence databases", Computers and Chemistry, 17, 2, 149-163, 1993.
29. Yazdani N and Ozsoyoglu Z.M., "Sequence matching of image", Proc., 8th Int'l Conf. on Scientific and Statistical Database Management, pp. 53-62, 1996.
30. Zhang K, and Shasha D., "Simple fast algorithms for the editing distance between trees and related problems", SIAM J. of Computing, 18, 6, 1245-1252, 1989.

Image Retrieval by Multi-scale Illumination Invariant Indexing

Theo Gevers and Arnold W.M. Smeulders

Intelligent Sensory Information Systems, University of Amsterdam
Kruislaan 403, 1098 SJ Amsterdam The Netherlands
{gevers, smeulders}@wins.uva.nl

Abstract. The purpose is to arrive at image retrieval invariant to a substantial change in illumination.
We will extend the theory that we have recently proposed on illumination invariant color models [6]. Then, a multi-scale image representation is produced by applying Gaussian derivatives at different scale levels on the illumination invariant color models. In this way, a multi-dimensional multi-scale image index is obtained which is illumination-independent and invariant under the group of rotations in the image domain. The multi-scale image representation is taken as input for image retrieval by query by example (i.e. an example image is given by the user) and image retrieval by arranging the image database as a binary tree (i.e. no example image is given is available).
Experiments have been conducted on a database consisting of 500 images taken from multicolored man-made objects in real world scenes. From the experimental results it can be observed that image retrieval by multi-scale invariant indexing provides high retrieval accuracy even under spatially and spectrally varying illumination.
Key-words: image retrieval, dichromatic reflection, reflectance properties, photometric color invariants, Gaussian derivatives, scale-space, multi-scale invariant indexing, query by example, decision trees

1 Introduction

Color provides powerful information for image retrieval. A simple and effective retrieval scheme is to represent and match images on the basis of color histograms as proposed by Swain and Ballard [16]. The work makes a significant contribution in introducing color for image indexing. However, it has the drawback that when the illumination circumstances are not equal, the retrieval accuracy degrades significantly. This method is extended by Funt and Finlayson [5], based on the retinex theory of Land [10], to make the method illumination independent by indexing on illumination-invariant surface descriptors (color ratios) computed from neighboring points. However, it is assumed that neighboring points have the same surface normal. Therefore, the derived illumination-invariant surface descriptors are negatively affected by rapid changes in surface orientation of the object (i.e. the geometry of the object). Healey and Slater [8] and Finlayson *at al.*

[2] use illumination-invariant moments of color distributions for object retrieval. These methods are sensitive to object occlusion and cluttering as the moments are defined as an integral property on the object as one. In global methods in general, occluded parts will disturb retrieval. Slater and Healey [15] circumvent this problem by computing the color features from small object regions instead of the entire object.

In this paper, our aim is to propose new illumination invariant color models to be used for the purpose of image retrieval invariant to a substantial change in illumination. Further, our goal is to provide an image representation at different scales. To this end, a multi-scale representation is produced by applying Gaussian derivative filters, also called fuzzy derivative operators, at various scales on the proposed illumination-invariant color models. The uniqueness of the Gaussian kernel for scale-space filtering has been pointed out by Koenderink and van Doorn [9], where at a fixed level in scale-space, i.e. for a fixed scale parameter, it is possible to study the structure at that scale. Gaussian derivative operators are the natural operators in scale-space. This multi-scale illumination invariant index is then used to retrieve images on the basis of query by example and used to arrange images in the database as a binary tree for the purpose of image navigation.

The paper is organized as follows. In Section 2, assuming white illumination and dichromatic reflectance, we extend the theory that we have recently proposed on color invariant models [6]. Further, in Section 3, a change in spectral power distribution (SPD) of the illumination is considered to propose a new color constant color model. In Section 4, a multi-scale invariant image representation is proposed. In Section 5, two strategies are presented for the purpose of image retrieval and experimental results are discussed. Finally, in Section 6, conclusions are drawn.

2 Reflectance under White Illumination

Consider an image of an infinitesimal surface patch. Using the red, green and blue sensors with spectral sensitivities given by $f_R(\lambda)$, $f_G(\lambda)$ and $f_B(\lambda)$ respectively, to obtain an image of the surface patch illuminated by a SPD of the incident light denoted by $e(\lambda)$, the measured sensor values will be given by Shafer [14]:

$$C = m_b(\mathbf{n},\mathbf{s}) \int_\lambda f_C(\lambda)e(\lambda)c_b(\lambda)d\lambda + m_s(\mathbf{n},\mathbf{s},\mathbf{v}) \int_\lambda f_C(\lambda)e(\lambda)c_s(\lambda)d\lambda \qquad (1)$$

for $C = \{R,G,B\}$ giving the Cth sensor response. Further, $c_b(\lambda)$ and $c_s(\lambda)$ are the albedo and Fresnel reflectance respectively. λ denotes the wavelength, $\mathbf{n}$ is the surface patch normal, $\mathbf{s}$ is the direction of the illumination source, and $\mathbf{v}$ is the direction of the viewer. Geometric terms m_b and m_s denote the geometric dependencies on the body and surface reflection respectively.

Considering the neutral interface reflection (NIR) model (assuming that $c_s(\lambda)$ has a constant value independent of the wavelength) and white illumination (equal energy density for all wavelengths within the visible spectrum), then $e(\lambda) = e$ and $c_s(\lambda) = c_s$, and hence being constants. Then, we put forward that the measured sensor values are given by:

$$C_w = em_b(\mathbf{n}, \mathbf{s})k_C + em_s(\mathbf{n}, \mathbf{s}, \mathbf{v})c_s \int_\lambda f_C(\lambda)d\lambda \tag{2}$$

for $C_w \in \{R_w, G_w, B_w\}$ giving the red, green and blue sensor response under the assumption of a white light source. Further,

$$k_C = \int_\lambda f_C(\lambda)c_b(\lambda)d\lambda \tag{3}$$

is the compact formulation depending on the sensors and the surface albedo only.

If the integrated white condition holds (i.e. the areas under the three curves are equal):

$$\int_\lambda f_R(\lambda)d\lambda = \int_\lambda f_G(\lambda)d\lambda = \int_\lambda f_B(\lambda)d\lambda = f \tag{4}$$

we propose that the reflection from inhomogeneous dielectric materials under white illumination is given by:

$$C_w = em_b(\mathbf{n}, \mathbf{s})k_C + em_s(\mathbf{n}, \mathbf{s}, \mathbf{v})c_s f \tag{5}$$

2.1 Illumination Invariant Color Models under White Illumination

For a given point on a shiny surface, the contribution of the body reflection component $C_b = em_b(\mathbf{n}, \mathbf{s})k_C$ and surface reflection component $C_s = em_s(\mathbf{n}, \mathbf{s}, \mathbf{v})c_s f$ are added together $C_w = C_s + C_b$. Hence, the observed colors of the surface must be inside the triangular color cluster in the *RGB*-space formed by the two reflection components.

Hence, any expression defining colors on the same triangle, formed by the two reflection components, in *RGB*-space are photometric color invariants for the dichromatic reflection model with white illumination. To that end, we propose the following set of color models:

$$L^p = \frac{\sum_i a_i(C_i^1 - C_i^2)^p}{\sum_j a_j(C_j^3 - C_j^4)^p} \tag{6}$$

where $C_i^1 \neq C_i^2, C_j^3 \neq C_j^4 \in \{R, G, B\}, i, j, p \geq 1, a \in \mathcal{R}$.

Lemma 1. *Assuming dichromatic reflection and white illumination, L^p is independent of the viewpoint, surface orientation, illumination direction, illumination intensity, and highlights.*

Proof. By substituting eq. (5) in eq. (6) we have:

$$\frac{\sum_i a_i((em_b(\mathbf{n},\mathbf{s})k_{C_i^1}) - (em_b(\mathbf{n},\mathbf{s})k_{C_i^2}))^p}{\sum_j a_j((em_b(\mathbf{n},\mathbf{s})k_{C_j^3}) - (em_b(\mathbf{n},\mathbf{s})k_{C_j^4}))^p} =$$

$$\frac{\sum_i a_i(em_b(\mathbf{n},\mathbf{s}))^p(k_{C_i^1} - k_{C_i^2})^p}{\sum_j a_j(em_b(\mathbf{n},\mathbf{s}))^p(k_{C_j^3} - k_{C_j^4})^p} =$$

$$\frac{(em_b(\mathbf{n},\mathbf{s}))^p \sum_i a_i(k_{C_i^1} - k_{C_i^2})^p}{(em_b(\mathbf{n},\mathbf{s}))^p \sum_j a_j(k_{C_j^3} - k_{C_j^4})^p} =$$

$$\frac{\sum_i a_i(k_{C_i^1} - k_{C_i^2})^p}{\sum_j a_j(k_{C_j^3} - k_{C_j^4})^p}$$

only dependent on the sensors and the material's albedo, where $C_i^1 \neq C_i^2, C_j^3 \neq C_j^4 \in \{R,G,B\}, i,j,p \geq 1, a \in \mathcal{R}$.

For instance, for $p = 1$, we have the set: $\{\frac{(R-G)}{(R-B)}, \frac{(B-G)}{(R-B)}, \frac{(R-G)+(B-G)}{(R-B)}, \frac{(R-G)+3(B-G)}{(R-B)+2(R-G)}, \ldots\}$ and for $p = 2$: $\{\frac{(R-G)^2}{(R-B)^2}, \frac{(B-G)^2}{(R-B)^2}, \frac{(R-G)^2+(B-G)^2}{(R-B)^2}, \frac{(R-G)^2+3(B-G)^2}{(R-B)^2+2(R-G)^2}, \ldots\}$ where all elements are photometric color invariants for objects with dichromatic reflectance under white illumination.

Although any other instantiation of L^p could be taken, in this paper, we concentrate on the photometric color invariant model $l_1 l_2 l_3$ uniquely determining the direction of the linear triangular color cluster:

$$l_1 = \frac{(R-G)^2}{(R-G)^2 + (R-B)^2 + (G-B)^2} \tag{7}$$

$$l_2 = \frac{(R-B)^2}{(R-G)^2 + (R-B)^2 + (G-B)^2} \tag{8}$$

$$l_3 = \frac{(G-B)^2}{(R-G)^2 + (R-B)^2 + (G-B)^2} \tag{9}$$

the set of normalized color differences which is, similar to H, a photometric color invariant for matte as well as for shiny surfaces which follows from substituting eq. (5) in eq. (7) - (9), which for l_1 results in:

$$l_1(R_w, G_w, G_w) = \frac{(R_w - G_w)^2}{(R_w - G_w)^2 + (R_w - B_w)^2 + (G_w - B_w)^2} =$$

$$\frac{(em_b(\mathbf{n},\mathbf{s})(k_R - k_G))^2}{(em_b(\mathbf{n},\mathbf{s})(k_R - k_G))^2 + (em_b(\mathbf{n},\mathbf{s})(k_R - k_B))^2 + (em_b(\mathbf{n},\mathbf{s})(k_G - k_B))^2} =$$

$$\frac{(k_R - k_G)^2}{(k_R - k_G)^2 + (k_R - k_B)^2 + (k_G - k_B)^2} \tag{10}$$

only dependent on the sensors and the surface albedo.
Equal arguments hold for l_2 and l_3.

3 Reflectance under Colored Illumination

We consider the body reflection term of the dichromatic reflection model:

$$C_c = m_b(\mathbf{n}, \mathbf{s}) \int_\lambda f_C(\lambda) e(\lambda) c_b(\lambda) d\lambda \tag{11}$$

for $C = \{R, G, B\}$, where $C_c = \{R_c, G_c, B_c\}$ gives the red, green and blue sensor response of a matte infinitesimal surface patch of an inhomogeneous dielectric object under unknown spectral power distribution of the illumination.

Suppose that the sensor sensitivities of the color camera are narrow-band with spectral response be approximated by delta functions $f_C(\lambda) = \delta(\lambda - \lambda_C)$, then the measured sensor values are:

$$C_c = m_b(\mathbf{n}, \mathbf{s}) e(\lambda_C) c_b(\lambda_C) \tag{12}$$

It can easily be seen that $l_1 l_2 l_3$ change with a variation in color of the illumination. To this end, a new color constant color model is proposed in the next section.

3.1 Illumination Invariant Color Models under Colored Illumination

Existing color constancy methods require specific a priori information about the observed scene (e.g. the placement of calibration patches of known spectral reflectance in the scene) which will not be feasible in practical situations, [3], [4], [10] for example. To circumvent these problems, Funt and Finlayson [5] propose simple and effective illumination-independent color ratios for the purpose of image indexing. However, it is assumed that the neighboring points, from which the color ratios are computed, have the same surface normal. Therefore, the method depends on varying surface orientation of the object (i.e. the geometry of the objects) affecting negatively the retrieval performance. To this end, we propose a new color constant color ratio not only independent of the illumination color but also discounting the object's geometry:

$$m(C_1^{\mathbf{x}_1}, C_1^{\mathbf{x}_2}, C_2^{\mathbf{x}_1}, C_2^{\mathbf{x}_2}) = \frac{C_1^{\mathbf{x}_1} C_2^{\mathbf{x}_2}}{C_1^{\mathbf{x}_2} C_2^{\mathbf{x}_1}}, C_1 \neq C_2 \tag{13}$$

expressing the color ratio between two neighboring image locations, for $C_1, C_2 \in \{R, G, B\}$ where $\mathbf{x}_1$ and $\mathbf{x}_2$ denote the image locations of the two neighboring pixels. Note that the set $\{R, G, B\}$ must be colors from narrow-band sensor filters and that they are used in defining the color ratio because they are immediately available from a color camera, but any other set of narrow-band colors derived from the visible spectrum will do as well.

If we assume that the color of the illumination is locally constant (at least over the two neighboring locations from which the ratio is computed i.e. $e^{\mathbf{x}_1}(\lambda) = e^{\mathbf{x}_2}(\lambda)$), the color ratio is independent of the illumination intensity and color, and also to a change in viewpoint, object geometry, and illumination direction as shown by substituting eq. (12) in eq. (13):

$$\frac{(m_b^{\mathbf{x}_1}(\mathbf{n},\mathbf{s})e^{\mathbf{x}_1}(\lambda_{C_1})c_b^{\mathbf{x}_1}(\lambda_{C_1}))(m_b^{\mathbf{x}_2}(\mathbf{n},\mathbf{s})e^{\mathbf{x}_2}(\lambda_{C_2})c_b^{\mathbf{x}_2}(\lambda_{C_2}))}{(m_b^{\mathbf{x}_2}(\mathbf{n},\mathbf{s})e^{\mathbf{x}_2}(\lambda_{C_1})c_b^{\mathbf{x}_2}(\lambda_{C_1}))(m_b^{\mathbf{x}_1}(\mathbf{n},\mathbf{s})e^{\mathbf{x}_1}(\lambda_{C_2})c_b^{\mathbf{x}_1}(\lambda_{C_2}))} = \frac{c_b^{\mathbf{x}_1}(\lambda_{C_1})c_b^{\mathbf{x}_2}(\lambda_{C_2})}{c_b^{\mathbf{x}_2}(\lambda_{C_1})c_b^{\mathbf{x}_1}(\lambda_{C_2})} \tag{14}$$

factoring out dependencies on object geometry and illumination direction $m_b^{\mathbf{x}_1}(\mathbf{n},\mathbf{s})$ and $m_b^{\mathbf{x}_2}(\mathbf{n},\mathbf{s})$, and illumination $e^{\mathbf{x}_1}$ and $e^{\mathbf{x}_2}$ as $e^{\mathbf{x}_1}(\lambda_{C_1}) = e^{\mathbf{x}_2}(\lambda_{C_1})$ and $e^{\mathbf{x}_1}(\lambda_{C_2}) = e^{\mathbf{x}_2}(\lambda_{C_2})$, and hence only dependent on the ratio of surface albedos, where $\mathbf{x}_1$ and $\mathbf{x}_2$ are two neighboring locations on the object's surface not necessarily of the same orientation.

Note that the color ratio does not require any specific a priori information about the observed scene, as the color model is an illumination- invariant surface descriptor based on the ratio of surface albedos rather then the recovering of the actual surface albedo itself. Also the intensity and spectral power distribution of the illumination is allowed to vary across the scene (e.g. multiple light sources with different SPD's), and a certain amount of object occlusion and cluttering is tolerated due to the local computation of the color ratio. The color model is not restricted to Mondrian worlds where the scenes are flat, but any 3-D real world scene is suited as the color model can cope with varying surface orientation of objects. Further note that the color ratio is insensitive to a change in surface orientation, illumination direction and intensity for matte objects under white light, but without the constraint of narrow-band filters, as follows from substituting the body reflection component (i.e. $em_b(\mathbf{n},\mathbf{s})k_C$) in eq. (13):

$$\frac{(e^{\mathbf{x}_1}m_b^{\mathbf{x}_1}(\mathbf{n},\mathbf{s})k_{C_1}^{\mathbf{x}_1})(e^{\mathbf{x}_2}m_b^{\mathbf{x}_2}(\mathbf{n},\mathbf{s})k_{C_2}^{\mathbf{x}_2})}{(e^{\mathbf{x}_2}m_b^{\mathbf{x}_2}(\mathbf{n},\mathbf{s})k_{C_1}^{\mathbf{x}_2})(e^{\mathbf{x}_1}m_b^{\mathbf{x}_1}(\mathbf{n},\mathbf{s})k_{C_2}^{\mathbf{x}_1})} = \frac{k_{C_1}^{\mathbf{x}_1}k_{C_2}^{\mathbf{x}_2}}{k_{C_1}^{\mathbf{x}_2}k_{C_2}^{\mathbf{x}_1}} \tag{15}$$

only dependent on the sensors and the surface albedo.

Having three color components of two locations, color ratios obtained from a *RGB*-color image are:

$$m_1 = \frac{R^{\mathbf{x}_1}G^{\mathbf{x}_2}}{R^{\mathbf{x}_2}G^{\mathbf{x}_1}} \tag{16}$$

$$m_2 = \frac{R^{\mathbf{x}_1}B^{\mathbf{x}_2}}{R^{\mathbf{x}_2}B^{\mathbf{x}_1}} \tag{17}$$

$$m_3 = \frac{G^{\mathbf{x}_1}B^{\mathbf{x}_2}}{G^{\mathbf{x}_2}B^{\mathbf{x}_1}} \tag{18}$$

For the ease of exposition, we concentrate on m_1 based on the *RG*-color bands in the following discussion. Without loss of generality, all results derived for m_1 will also hold for m_2 and m_3.

Taking the natural logarithm of both sides of eq. (13) results for m_1 in:

$$\ln m_1(R^{\mathbf{x}_1}, R^{\mathbf{x}_2}, G^{\mathbf{x}_1}, G^{\mathbf{x}_2}) = \ln(\frac{R^{\mathbf{x}_1}G^{\mathbf{x}_2}}{R^{\mathbf{x}_2}G^{\mathbf{x}_1}}) =$$

$$\ln R^{\mathbf{x}_1} + \ln G^{\mathbf{x}_2} - \ln R^{\mathbf{x}_2} - \ln G^{\mathbf{x}_1} = \ln(\frac{R^{\mathbf{x}_1}}{G^{\mathbf{x}_1}}) - \ln(\frac{R^{\mathbf{x}_2}}{G^{\mathbf{x}_2}}) \tag{19}$$

Hence, the color ratios can be seen as differences at two neighboring locations $\mathbf{x}_1$ and $\mathbf{x}_2$ in the image domain of the logarithm of R/G:

$$d_{m_1}(\mathbf{x}_1, \mathbf{x}_2) = (\ln(\frac{R}{G}))^{\mathbf{x}_1} - (\ln(\frac{R}{G}))^{\mathbf{x}_2} \tag{20}$$

By taking these differences in a particular direction between neighboring pixels, the finite-difference differentiation is obtained of the logarithm of image R/G which is independent of the illumination color, and also a change in viewpoint, the object geometry, and illumination intensity. For pixels on a uniformly colored region (i.e. with fixed surface albedo), in theory, the color ratio will be zero and non-zero for pixels on locations where two regions of distinct surface albedo meet.

4 Multi-scale Illumination Invariant Indexing

In this section, a multi-scale representation is produced by applying Gaussian derivative filters, also called fuzzy derivative operators, at various scales on the proposed illumination-invariant color models $l_1 l_2 l_3$ (under white light) and $m_1 m_2 m_3$ (under colored light). The uniqueness of the Gaussian kernel for scale-space filtering has been pointed out by Koenderink and van Doorn [9], where at a fixed level in scale-space, i.e. for a fixed scale parameter, it is possible to study the structure at that scale. Gaussian derivative operators are the natural operators in scale-space.

4.1 $l_1 l_2 l_3$

For the ease of exposition we concentrate on l_1 in the following discussion. Without loss of generality, all results derived for l_1 will also hold for l_2 and l_3.

We use the partial derivatives of the Gaussian up to order 2 at scale σ for image l_1. Let $\{l_{1_x}, l_{1_y}, l_{1_{xx}}, l_{1_{yy}}, l_{1_{xy}}\}_\sigma$ denote the set of the first five partial Gaussian derivatives at scale σ. From these partial derivatives, we have computed the following set of invariants $\{l_1, l_{1_x}^2 + l_{1_y}^2, l_{1_{xx}} + l_{1_{yy}}, l_{1_{xx}} l_{1_{yy}} - l_{1_{xy}}^2\}_\sigma$, corresponding to image l_1, the gradient magnitude, the Laplacian, and the Hessian of image l_1. This set is invariant under the group of rotations in l_1. Note that Laplacian operator finds its roots in the modeling of certain psychophysical processes in mammalian vision and hence is suited to retrieve 'similar appearing' images. Further note that the Laplacian operator gives significant responses along lines and edges and hence is not particularly suitable as a cornerness operator. On the other hand, the Hessian operator does not respond to lines and edges but gives significant signals near corners and hence forms a cornerness measure. In this way, a set of invariants is obtained robust to a change in illumination (assuming white light) and independent of the choice of coordinate frame (up to rotations).

Having the 4 invariants computed for each image in the database by filtering it with the first five partial Gaussian derivatives at three different scales $\sigma = 1$, $\sigma = 3$ and $\sigma = 5$, a 12 dimensional multi-scale invariant vector is computed.

4.2 $m_1m_2m_3$

For the ease of exposition we concentrate on $m_1 = \ln(R/G)$ in the following discussion. Without loss of generality, all results derived for m_1 will also hold for m_2 and m_3.

We use the partial derivatives of the Gaussian up to order 2 at scale σ for image $\ln(R/G)$. Let $\{m_{1_x}, m_{1_y}, m_{1_{xx}}, m_{1_{yy}}, m_{1_{xy}}\}_\sigma$ denote the set of the first five partial Gaussian derivatives at scale σ. The set is computed for each location $\mathbf{x}$ in $\ln(R/G)$. From these partial derivatives, we have computed the following invariant set $\{m_{1_x}^2 + m_{1_y}^2, m_{1_{xx}} + m_{1_{yy}}, m_{1_{xx}} m_{1_{yy}} - m_{1_{xy}}^2\}_\sigma$, corresponding the gradient magnitude, the Laplacian, and the Hessian of image $\ln(R/G)$. In this way, a set of invariants is obtained robust to a change in spectral power distribution in the illumination and independent of the choice of coordinate frame (up to rotations).

Having the 3 invariants computed for each location in each image in the database by filtering it with the first five partial Gaussian derivatives at three different scales $\sigma = 1$, $\sigma = 3$ and $\sigma = 5$, a 9 dimensional multi-scale illumination invariant vector is obtained for each image location.

5 Image Retrieval

In this section, the multi-scale invariant image representations are used for the purpose of image retrieval by query example in Section 5.1 and for arranging the images in the database as a binary tree in Section 5.2.

5.1 Image Retrieval Based on Query by Example

We focus on image retrieval by histogram matching for comparison reasons in the literature. Obviously, transforming RGB to one of the invariant color models can be performed as a preprocessing step by other matching techniques.

Histograms are constructed on the basis of different color features representing the distribution of discrete invariant feature values in a n-dimensional color feature space, where $n = 12$ for each component of $l_1l_2l_3$ and $n = 9$ for each component of $m_1m_2m_3$. Histogram axes are partitioned uniformly with fixed intervals. The resolution on the axes follows from the amount of noise and computational efficiency considerations. We determined the appropriate bin size for our application empirically. This has been achieved by varying the same number of bins on the axes over $q \in \{2, 4, 8, 16, 32, 64, 128, 256\}$ and chose the smallest q for which the number of bins is kept small for computational efficiency and large for retrieval accuracy. The results show (not presented here) that the number of bins was of little influence on the retrieval accuracy when the number of bins ranges from $q = 8$ to $q = 256$ for all color spaces. Therefore, the histogram bin size used during histogram formation is $q = 8$ in the following. Due to the low number of bins, histogram matching becomes feasible for high-dimensional histograms. Then, for each test and reference image, a 12- and 9-dimensional histogram is created for each component of l_1l_2l and $m_1m_2m_3$ denoted by $\mathcal{H}_{l_1l_2l_3}$

and $\mathcal{H}_{m_1 m_2 m_3}$ respectively. For comparison reasons in the literature, in this paper, the histogram similarity function is expressed by histogram intersection [16].

The Experimental Results The image database is used developed at our laboratory. The database consists of $N_1 = 500$ reference images of multicolored 3-D domestic objects, tools, toys, etc.. Objects were recorded in isolation (one per image) with the aid of the SONY XC-003P CCD color camera (3 chips) and the Matrox Magic Color frame grabber. Objects were recorded against a white cardboard background. Two light sources of average day-light color are used to illuminate the objects in the scene. A second, independent set (the test set) of recordings was made of randomly chosen objects already in the database. These objects, $N_2 = 70$ in number, were recorded again one per image with a new, arbitrary position and orientation with respect to the camera, some recorded upside down, some rotated, some at different distances.

Fig. 1. *Left: 16 images which are included in the image database of 500 images. The images are representative for the images in the database. Right: Corresponding images from the query set.*

In Figure 1, 16 images from the image database of 500 images are shown on the left. Corresponding images coming from the query set are shown on the right. In Figure 3.a ranking statistics are shown differentiated for $l_1 l_2 l_3$ and $m_1 m_2 m_3$ by matching the set of 70 images on the database of 500 images. From the results of Figure 3.a we can observe that the discriminative power of $l_1 l_2 l_3$ and $m_1 m_2 m_3$ is very high achieving a probability of respectively 92 and 89 perfect matches out of 100.

To test retrieval accuracy for images under varying illumination intensity, an independent test set of recordings was made by randomly chosen 10 objects already in the database of 500 images. These objects were recorded again with the same pose but with spatially varying illumination intensity, see Figure 2. Then these 10 images were matched against the database of 500 images. From Figure 3.b is can be observed that the discriminative power of $l_1l_2l_3$ and $m_1m_2m_3$ (with 9 perfect matches out of 10) shows very high matching accuracy.

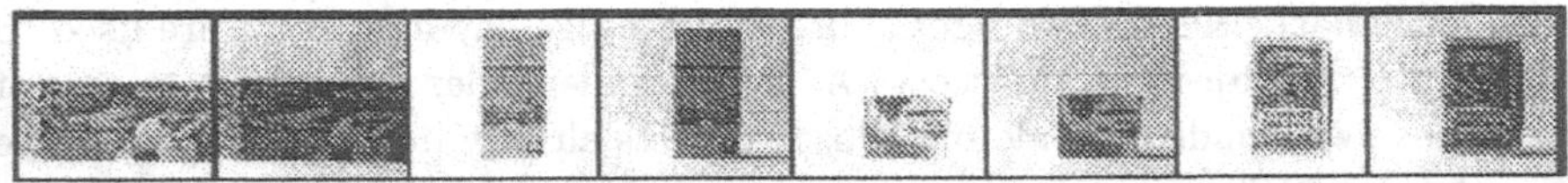

Fig. 2. *4 of the 10 objects with spatially varying illumination intensity.*

	$l_1l_2l_3$	$m_1m_2m_3$
rank 1	92%	89%
rank 2	1%	1%
rank 3	1%	2%
rank>3	6%	8%

a: Illumination constant

	$l_1l_2l_3$	$m_1m_2m_3$
rank 1	8	7
rank 2	2	1
rank 3	0	0
rank>3	0	2

b: Varying intensity

	$l_1l_2l_3$	$m_1m_2m_3$
rank 1	0	7
rank 2	0	1
rank 3	0	1
rank>3	10	1

c: Varying color

Fig. 3. *Ranking statistics a. matching the 70 images against the database of 500 images. b. matching the 10 images with spatially varying illumination against the database of 500 images. c. matching the 10 images with spatially varying color against the database of 500 images.*

To test retrieval accuracy for images under varying illumination color, another independent test set of recordings was made by randomly chosen 10 objects already in the database of 500 images. These objects were recorded again with the same pose but with spatially varying spectral power distribution (towards green illumination), see Figure 4. From the ranking statistics shown in Figure 3.c, we can observe that only the color ratio $m_1m_2m_3$ is robust to a change in illumination color.

5.2 Image Navigation: Arranging an Image db as a Decision Tree

In this section, we present a method to arrange an image database as a binary tree. The nodes in the decision or binary tree are image clusters, where similarity

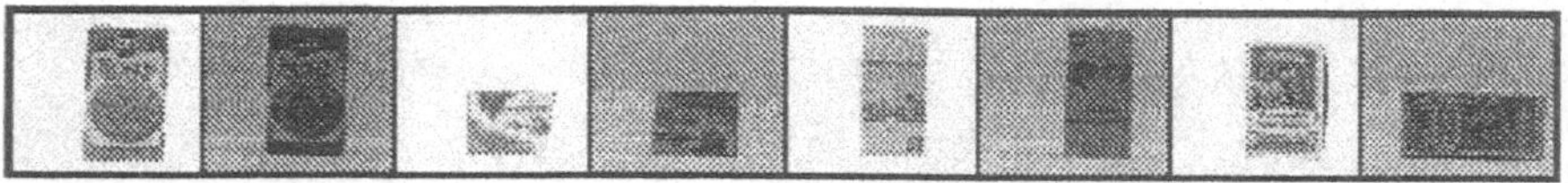

Fig. 4. *4 of the 10 objects with spatially varying illumination color.*

between images from the children of a node is higher than those from the parent node. Having build the decision or binary tree, the user is allowed to travel through the tree finding clusters of similar images at various scales along its path.

The Agglomerative Algorithm The decision tree inference method is similar to classic machine learning techniques [7], for example. The first step is to create a similarity matrix. The similarity matrix is a table of all the pair-wise distances between images. We have used histogram intersection distance between two images based on their multi-scale invariant histogram. Next, the smallest distance is found in the similarity matrix identifying the two clusters that are most similar to one another. These two clusters are then merged and the similarity matrix is updated by replacing the two rows that described the parent cluster with a new row describing the distance between the merged cluster and the remaining clusters. There are now $N-1$ clusters and $N-1$ rows in the similarity matrix. The merge step is repeated $N-1$ times resulting in one large cluster. At each iteration information is stored about which clusters were merged and how far apart they were. The distance between two clusters is given by the distance between the closest members. In this way, clusters are produced having the property that every image of a cluster is more closely related to at least one image of its cluster than to any image outside it. The agglomeration method creates hierarchical clusters and are represented as a binary or so-called decision tree.

The Experimental Results Because the size of large databases is impractical to show the images in the tree, experiments have been conducted to show very primarily results as the database of images only consists of 12 images. The input of the agglomerative algorithm are 12 images composed of 4 different objects but recorded three times under varying color and intensity. We have used the multi-scale illumination-invariant representation based on $m_1m_2m_3$ as input of the agglomeration method where the similarity measure is histogram intersection. The method creates the binary tree as shown in Figure 5. An examination of the tree shows that images of the same object recorded under varying illumination and pose are merged first. Then visual similar clusters are merged to obtain a hierarchical structure corresponding to our perception.

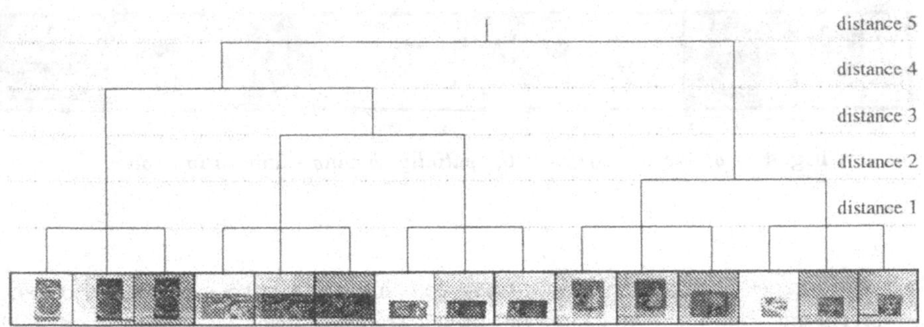

Fig. 5. *The binary tree generated for 12 images.*

6 Conclusion

The purpose is to arrive at image retrieval invariant to a substantial change in illumination.

We have extended the theory that we have recently proposed on illumination invariant color models [6]. Then, a multi-scale image representation has been proposed obtained by applying Gaussian derivatives at different scale levels on illumination invariant color models. In this way, an image index is obtained at various scales which is illumination-invariant and invariant under the group of rotations in the image domain. The multi-scale image representation has been taken as input for image retrieval by query by example (i.e. an example is given by the user) and image retrieval by arranging the image database as a binary tree (i.e. no example is given is available).

From the experimental results it can be observed that image retrieval by multi-scale invariant indexing provides high retrieval accuracy even under spatially and spectrally varying illumination.

References

1. Finlayson, G. D., Drew M. S., and Funt B. V.: Spectral Sharpening: Sensor Transformations for improved Color Constancy. J. Opt. Soc. Am. 11(5) (1994) 1553–1563
2. Finlayson, G. D., Chatterjee S. S., and Funt B. V.: Color Angular Indexing. ECCV96 II (1996) 16–27
3. Forsyth, D.: A Novel Algorithm for Color Constancy. International Journal of Computer Vision Vol. 5 1990 5–36
4. Funt, B. V. and Drew, M. S.: Color Constancy Computation in Near-Mondrian Scenes. In Proceedings of the CVPR IEEE Computer Society Press 1988 544–549
5. Funt, B. V. and Finlayson, G. D.: Color Constant Color Indexing. IEEE PAMI 17(5) 1995 522–529
6. Gevers, T. and Smeulders, A. W. M.: Image Indexing using Composite Color and Shape Invariant Features. ICCV Bombay India (1998)

7. Hartigan, J. A.: Clustering Algorithms. John Wiley and Sons U.S.A (1975)
8. Healey, G. and Slater D.: Global Color Constancy: Recognition of Objects by Use of Illumination Invariant Properties of Color Distributions. J. Opt. Soc. Am. A Vol. 11 No. 11 (1995) 3003–3010
9. Koenderink, J. J. and van Doorn A. J.: Representation of Local Geometry in the Visual System. Biological Cybernetics No. 55 (1987) 367–375
10. Land, E. H. and McCann, J. J.: Lightness and Retinex Theory. J. Opt. Soc. Am. Vol. 61 (1971) 1–11
11. Lee H.-C., Breneman E, J. and Schulte C. P.: Modeling Light Reflection for Computer Color Vision. IEEE Transactions on Pattern Analysis and Machine Intelligence Vol. 12 No. 3 (1990) 402–409
12. Levkowitz, H. and Herman G. T.: GLHS: A Generalized Lightness, Hue, and Saturation Color Model. CVGIP: Graphical Models and Image Processing Vol. 55 No. 4 (1993) 271–285
13. Nayar, S. K. and Bolle, R. M.: Reflectance Based Object Recognition. International Journal of Computer Vision Vol. 17 No. 3 1996 219–240
14. Shafer, S. A.: Using Color to Separate Reflection Components. COLOR Res. Appl. 10(4) (1985) 210–218
15. D. Slater and G. Healey: The Illumination-invariant Recognition of 3D Objects Using Local Color Invariants. IEEE Trans. PAMI 18(2) (1996)
16. Swain, M. J. and Ballard, D. H.: Color Indexing. International Journal of Computer Vision Vol. 7 No. 1 1991 11–32

Finding Pictures in Context

Rohini K. Srihari and Zhongfei Zhang

Center for Document Analysis and Recognition (CEDAR)
UB Commons, 520 Lee Entrance- Suite 202
State University of New York at Buffalo
Buffalo, NY 14228-2567
Phone: (716)-645-6164 ext. {102, 109}
Fax: (716)-645-6176
e-mail: {rohini, zhongfei}@cedar.buffalo.edu

Abstract. This research explores the interaction of textual and photographic information in multimodal documents. The WWW may be viewed as the ultimate, large-scale, dynamically changing, multimedia database. Finding useful information from the WWW without encountering numerous false positives (the current case) poses a challenge to multimedia information retrieval systems (MMIR). We exploit the fact that images do not appear in isolation, but rather with accompanying, *collateral* text. Taken independently, existing techniques for picture retrieval using (i) collateral text-based and (ii) image-based methods have several limitations. Text-based methods, while very powerful in matching context, do not have access to image content. Image-based methods compute general similarity between images and provide limited semantics. Our research focuses on improving *precision* and *recall* in a MMIR system by interactively combining text processing with image processing (IP) in both the indexing and retrieval phases. A picture search engine is demonstrated as an application.

1. Introduction

This research explores the interaction of textual and photographic information in multimodal documents. The World Wide Web (WWW) may be viewed as the ultimate, large-scale, dynamically changing, multimedia database. Finding useful information from the WWW poses a challenge in the area of multimodal information indexing and retrieval. The word "indexing" is used here to denote the extraction and representation of semantic content. Our research focuses on improving precision and recall in a multimodal information retrieval system by interactively combining text processing with image processing. We exploit the fact that images do not appear in isolation, but rather with accompanying text which we refer to as *collateral text.* Figure 1 illustrates such a case. The interaction of text and image content takes place in *both* the indexing and retrieval phases. An application of this research, namely a picture search engine which permits a user to retrieve pictures of people in various contexts is presented.

Taken independently, existing techniques for text and image retrieval have several limitations. Text-based methods, while very powerful in matching context

[13], do not have access to image content. There has been a flurry of interest in using textual captions to retrieve images [12]. Searching captions for keywords and names will not necessarily yield the correct information, as objects mentioned in the caption are not always in the picture. This results in a large number of false positives which need to be eliminated or reduced. In a recent test, we posed a query to a search engine to find pictures of Clinton and Gore resulting in 941 images. After applying our own filters to eliminate graphics and spurious images (e.g., white space), we were left with 547 potential pictures which satisfied the above query. A manual inspection revealed that only 76 of the 547 pictures contained pictures of Clinton or Gore! This illustrates the tremendous need to (i) employ image-level verification, and (ii) the need to use text more intelligently.

Typical image-based methods compute general similarity between images based on statistical image properties [9]. Examples of such properties are texture and color [17]. While these methods are robust and efficient, they provide very limited semantic indexing capabilities. There are some techniques which perform object identification; however these techniques are computationally expensive and not sufficiently robust for use in a content-based retrieval system. This is due to a need to balance processing efficiency with indexing capabilities. If object recognition is performed in isolation, this is probably true. More recently, other attempts to extract semantic properties of images based on spatial distribution of colour and texture properties have also been attempted [14]. Such techniques have drawbacks, primarily due to their weak disambiguation; these are discussed later. [19] describes an attempt to utilize both image and text content in a picture search engine. However, text understanding is limited to processing of HTML tags; no attempt to extract descriptions of the picture is made. More important, it does not address the interaction of text and image processing in deriving semantic descriptions of a picture.

In this paper we describe a system for finding pictures in context. A sample query would be *find pictures of victims of natural disasters*. Specifically, we have conducted experiments to effectively combine text content with image content in the **retrieval** stage. Text indexing is accomplished through standard statistical text indexing techniques and is used to satisfy the general context that the user specifies. Image processing consists of face detection and recognition; this is used to present the resulting set of pictures based on various visual criteria (e.g., the prominence of faces). We have experimented with two different scenarios for this task; results from both are presented. We also present some preliminary work in the intelligent use of collateral text in determining ***pictorial attributes***. Such techniques can be used independently, or combined with image processing techniques to provide visual verification. Thus, this represents the integration of text and image processing techniques in the **indexing** stage.

2. Important Attributes for Picture Searches

Before we describe techniques for extracting picture properties from text and images, it is useful to examine typical queries used in retrieving pictures. [7]

describes experimental work in the relative importance of picture attributes to users. Nine high-level attributes including *literal object, people, human attributes, art historical information, visual elements, color, location, description, abstract, content/story, viewer response and external relationship* were measured. It is interesting to note that *literal object* accounted for up to 31% of the responses. Human form and other human characteristics accounted for approximately 15%. Color, texture, etc. ranked much lower compared to the first two categories. The role of content/story varied widely, from insignificant to highly important. In other words, users dynamically combine image content and context in their queries.

Romer in [10] describes a wish list for image archive managers, specifically, the types of data descriptions necessary for practical retrieval. The heavy reliance on text-based descriptions is questioned; furthermore, the adaptation of such techniques to multimodal content is required. The need for visual thesuari [16, 3] is also stressed, since these provide a natural way of cataloguing pictures, an important task. An ontology of picture types would be desirable. Finally, Romer describes the need for "a precise definition of image elements and their proximal relationship to one another". This would permit queries such as *find a man sitting in a carriage in front of Niagara Falls.*

Based on the above analysis, it is clear that object recognition is a highly desirable component of picture description. Although object recognition in general is not possible, for specific classes of objects, and with feedback from text processing, object recognition may be attempted. It is also necessary to extract further semantic attributes of a picture by mapping low-level image features such as color, texture into semantic primitives. Efforts in this area [14] are a start, but suffer from weak disambiguation and hence can be applied in select databases; our work aims to improve this. We present improved text-based techniques for predicting image elements and their structural relationships.

3. Webpic: A Multimodal Picture Retrieval System

To demonstrate the effectiveness of combining text and image content, we have developed a robust, efficient and sophisticated picture search engine; specifically, our system will selectively retrieve pictures of people in various contexts. A sample query could be *find outdoor pictures of Bill Clinton with Hillary talking to reporters on Martha's Vineyard.* This should generate pictures where (i) Bill and Hillary Clinton actually appear in the picture (verified by face detection/recognition) and (ii) the collateral text supports the additional contextual requirements. By robust, we mean the ability to perform under various data conditions; potential problems could be lack of or limited accompanying text/HTML, complex document layout etc. The system should degrade gracefully under such conditions. Efficiency refers primarily to the time required for retrievals which are performed online. Since image indexing operations are time-consuming, they are performed offline. Finally, sophistication refers to the specificity of the query/response. In order to provide adequate responses to specific queries , it is necessary to perform more complex indexing on the data.

Figure 2 depicts the overall structure of our system. It consists of three phases. Phase 1 is the data acquisition phase- we download multimodal documents from WWW news sites (e.g., MSNBC, CNN, USA Today), In order to control the quality of data that is initially downloaded, we have implemented our own web crawler in Java which does more extensive filtering of both text and images. The inputs to the system are (i) a set of name keys (names of people), (ii) an initial set of URL's to initiate the search. We employ some pre-processing tools during this phase. One such tool is an image-based ***photograph vs. graphic*** filter. This filter is designed and implemented based on histogram analysis. Presumably a photograph histogram has much wider spectrum than that of a graphic image.

A woman adds to the floral tribute to
Princess Diana outside the gates of
Kensington Palace.
David Brauchli / AP
At Kensington Palace on Monday night, the crowd stood 10-deep
around an expanding carpet of floral tributes, while many waited
through the evening to sign condolence books, which remain available
24 hours a day.

Fig. 1. An example of an web image with collateral text of explicit and implicit caption. The explicit caption to the left of the photograph reads *A woman adds to the floral tribute to Princess Diana outside the gates of Kensington Palace.* Taken from www.msnbc.com

We also employ a *collateral text extractor* whose task is to determine the scope of text relevant to a given picture. Caption text appears in a wide variety of styles. News sites such as CNN and MSNBC use *explicit* captions for pictures. These are indicated through the use of special fonts and careful placement using HTML commands as illustrated in Figure 1. In other web pages, captions are not set off explicitly, but rather, are *implicit* by virtue of their proximity to the picture.

Explicit captions are detected based on the presence of strong HTML clues, as well as the usage of key phrases such as "left, foreground, rear" etc. These can be used to predict picture contents. General collateral text is detected based on the presence of words from the "ALT" tag, caption words, spatial proximity to picture etc. Such text, while not a powerful predictor of the contents of a

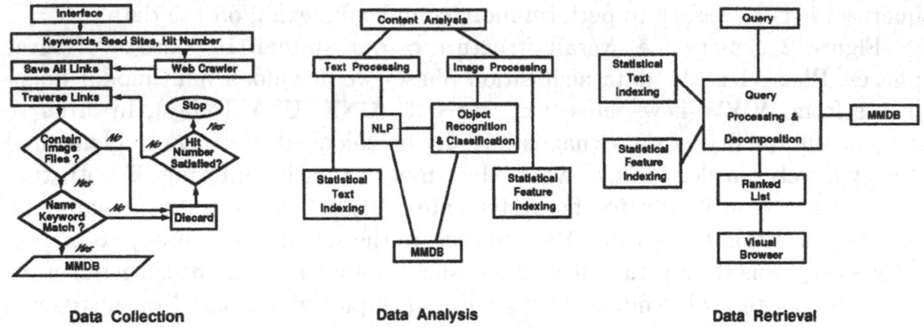

Fig. 2. Overall control structure of proposed system

picture, establishes the context of a picture. We also have access to an image-based caption extractor that extracts ASCII text which has been embedded in images (a common practice among news oriented sites).

Phase 2 is the content analysis or indexing phase (performed off-line). Phase 2 illustrates that both NLP and IP result in factual assertions to the database. This represents a more semantic analysis of the data than general text and image indexing based on statistical features. This is discussed in later sections.

Phase 3, retrieval, demonstrates the need to decompose the query into its constituent parts. A Web-based graphical user interface (GUI) has been developed for this. As Figure 3 illustrates, our system permits users to view the results of a match based on different visual criteria. This is especially useful in cases where the user knows the general context of the picture, but would like to interactively browse and select pictures containing his/her desired visual attributes. The interface also illustrates that further query refinement using techniques such as image similarity are possible. Finally, although the example illustrates a primary context query, it is possible for the original query to be based on pure image matching techniques. We have built the basic database infrastucture for a multimodal database using Illustra. This is used for data storage, as well as representing factual (exact) information. The ability to define new data types and associated indexing and matching functions is useful for this project.

4. Text Indexing

There are two main objectives to research in text processing for this effort: (i) statistical text indexing to capture general context, and (ii) advanced NLP techniques to extract picture attributes of a picture based on collateral text. This phase assumes that a previous process has extracted collateral text.

4.1 Statistical Text Indexing

The goal here is to capture the general context represented by collateral text. Though not useful in deriving exact picture descriptions, statistical text indexing

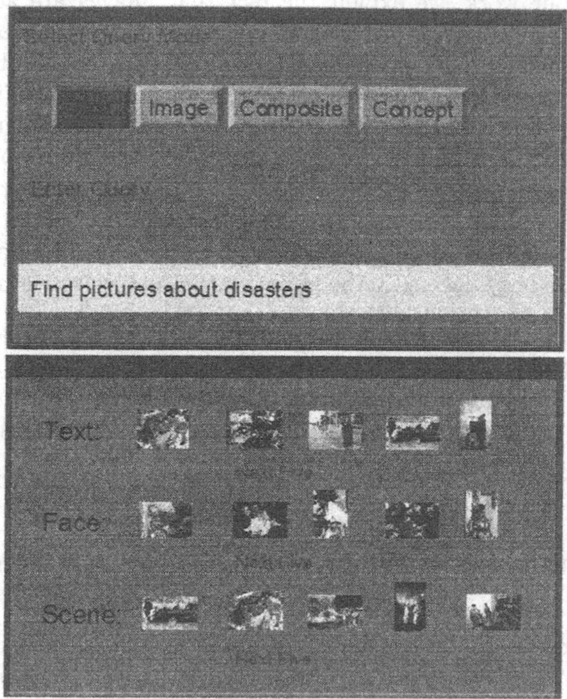

Fig. 3. The multimodal GUI used in retrieval

plays a key role in a robust multimodal IR system. The SMART [13] vector-space text indexing system is employed in this work; this is currently being interfaced with the Illustra DBMS. The problem being faced here differs from traditional document matching since the text being indexed, viz, collateral and caption text is frequently very sparse. For this reason, we have been experimenting with the use of NLP pre-processing in conjunction with statistical indexing. NLP pre-processing refers to methods such as Named Entity (NE) tagging [2] which classify groups of words as person name, location, etc. For example in the phrase, *Tiger Woods at the River Oaks Club*, *River Oaks Club* would be classified as a location. Applying NE tagging to captions and collateral text reduces errors typically associated with words having multiple uses. A query to "find pictures of woods containing oaks" should not give a high rank to the above caption. NE tagging of queries and captions leads to improved precision and recall. NE tagging is also a useful pre-processing step to parsing, described below.

4.2 Extracting Picture Attributes through NLP Techniques

The goal of image indexing is to automatically produce a semantic representation of the picture contents. Based on the current state of the art in image processing and computer vision techniques, only modest progress has been made towards this goal. In our work, we exploit the fact that images are accompanied by *collateral* text. By applying sophisticated natural language processing (NLP) techniques to such text, it is possible to derive vital information about a picture's content. Some organizations such as Kodak are manually annotating picture and video clip databases to permit flexible retrieval. Annotation consists of adding logical assertions regarding important entities and relationships in a picture. These are then used in an expert system for retrieval. [1] describes a system for image retrieval based on matching manually entered entities and attributes of pictures. Our goal is to *automatically* derive the following information which photo archivists have deemed to be important in picture retrieval:

- Determining which **objects** and **people** are present in the scene; the **location** and **time** are also of importance as is the **focus** of the picture. In the caption of Figure 1, Princess Diana should not be predicted as appearing in the scene. [15] discusses syntactic and semantic rules for this task.
- Preserving **event** (or activity) as well as **spatial** relationships which are mentioned in the text.
- Determining further **attributes** of the picture such as indoor vs. outdoor, mood, etc.

The goal here is to fill out *picture description templates* which represent image characteristics. Templates such as this are used by photo repository systems such as the Kodak Picture Exchange [11]. The templates carry information about people, objects, relationships, location, as well as other image properties. These properties include (i) indoor vs. outdoor setting, (ii) active vs. passive scene, i.e. an action shot vs. a posed photo, (iii) individual vs. crowd scene, (iv) daytime vs. night-time and (v) mood. As an example, consider Figure 4.2 which shows the output template from processing the caption and collateral text in Figure 1. Information about the mood is provided by the collateral text (the word "condolence").

We are currently employing a left corner, bottom up chart parser that uses detailed lexical entries supplying information such as subcategorization and agreement. The lexical information is shared with higher nodes of the parse tree by means of unification, as in theories such as HPSG implementations like the Tomita Parser [18]. The lexicon is derived from the online Oxford Advanced Learner's Dictionary (OALD) which has been reformatted into LISP-like syntax.

5. Image Indexing

In multimodal information retrieval, imagery is considered as the most frequently encountered modality next to text. Thus, it is important that techniques in effective and efficient image retrieval be available.

In this paper, we mainly focus our research on image retrieval of people or scene in a general context. This requires capabilities of face detection and/or

```
people: female(PER1)
objects: flowers
activity: pay_tribute(PER1, Princess Diana)
location: Kensington Palace, "outdoor"
event date: Monday, Sept 2, 1997
focus: PER1
```

Fig. 4. Picture Description Template

recognition in general imagegy domain. By a *general image domain*, it is meant that the appearances of the objects in question (e.g. faces) in different images may vary in size, pose, orientation, expression, background, as well as contrast. Since color images are very popular in use and very easy to obtain, we restrict our research in general color imagery.

Face detection and/or recognition have received focused attention in literature of computer vision and pattern recognition for years. A good survey in this topic may be found in [4]. Typically, face detection and recognition are treated separately in literature, and the solutions proposed are normally independent with each other. In this task, we propose a *streamlined solution* to both face detection and face recognition. By a streamlined solution, it is meant that both detection and recognition are conducted in the same color feature space, and the output of the detection stage is directly fed into the input of the recognition stage. Another major difference of this research from those existing in literature is that our system is a self-learning system, meaning that the face library used in face recognition is obtained through face detection and text understanding, using our earlier research system PICTION [15]. This allows the stage of face data collection for construction of the face library as an automatic part of data mining, as opposed to interactive, manual data collection usually conducted for face recognition in literature. Note that in many situations, it is impossible to do the manual data collection for certain individuals, such as Bill Clinton. For those people, their face samples can only be obtained through web, newspapers, etc. Thus, automatic data collection is not only efficient, but also necessary.

Face detection is approached as pattern classification in a color feature space. The detection process is accomplished in two major steps: **Feature Classification** and **Candidate Generation**. In the feature classification stage, each pixel is classified as face or non-face based on a standard Bayesian rule [6]. The classification is conducted based on pre-tuned regions for human face in a color feature space. The color features used in this approach are hue and chrominance. The pre-tuning of the classification region in the color feature space is conducted by sampling over 100 faces of different races from different web sites. In the candidate generation stage, first a morphological operation is applied to remove the noise, and then a connected component search is used to collect all the "clusters" which indicate the existence of human faces. Since the pre-tuned color feature region may also classify other parts of human body as candidates,

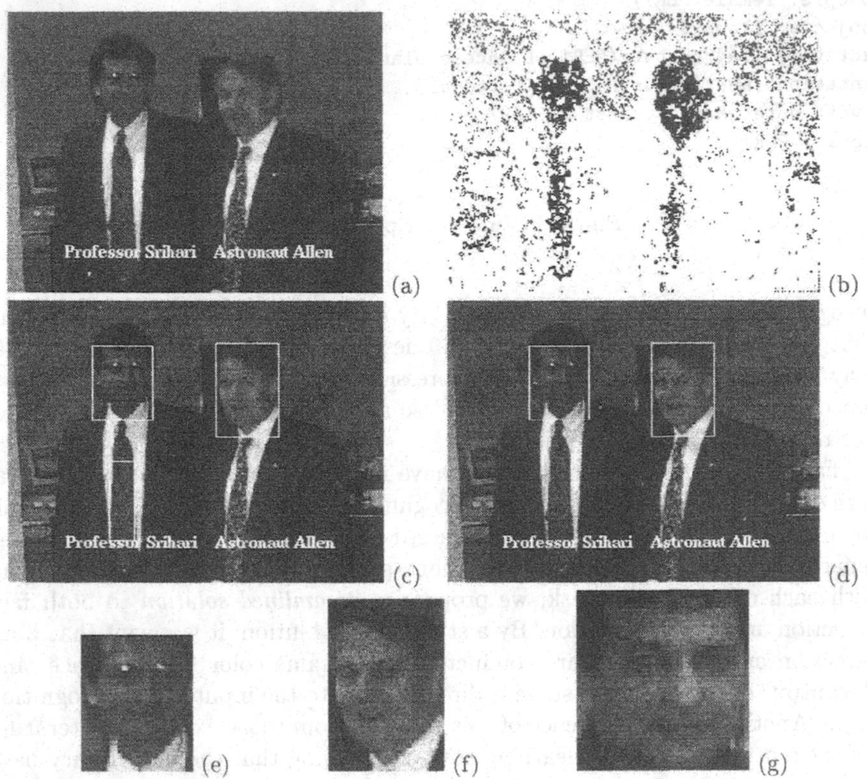

Fig. 5. An example of the result of automatic face detection and face recognition. (a) an original color image from Internet Web. (b) the binary image after pixel-wise classification. (c) result after morphological operations and connected component search. (d) final detection result after applying heuristics to reject false positives. (e) the first face candidate returned for face recognition. (f) the second face candidate returned for face recognition. (g) another face image of the same individual as in (e).

let alone certain other objects may also happen to be within the region in the color feature space, a simple heuristic checking is used to verify the shape of the returned bounding box to see if it is confirmed to the "golden ratio" law[1]. Fig. 5 shows the whole process of face detection and recognition for a web image. Note that each detected face is automatically saved into the face library if it has strong textual indication that who this person is (self-learning to build up the face library), or the face image is searched in the face library to find who the person is, if the query asks to retrieve images of this individual (query stage).

In the stage of face recognition, there are two modes of operation. In the

[1] It is believed that for a typical human face, the ratio of the width to the hight of the face is always around the magic value of $2/(1+\sqrt{5})$, which is called the golden ratio [5].

mode of face library construction, we assume that each face image has its collateral textual information to indicate identities of the people in the image. Face detection is first applied to detect all the faces. Based on the collateral information [15], the identities for each faces may be found and thus they may be saved into the library automatically. In the mode of query, on the other hand, the detected face needs to be searched in the library to find out the identity of this individual.

In order to find the semantic matching between two faces which may have undergone significant variation of pose, orientation, size, expression, and background. Fig. 5(e) and (g) are two face images of the same individual. Thus, the problem is reduced to finding *semantic* similarity between two face images in general image domain. At the time of writing up this paper, we have not been able to include the face recognition module into the system yet. Therefore, at this point we can only query general people (by face detection), or query individuals through face detection and finding collateral supporting text.

6. Matching

Processing queries requires the use of: (i) information generated from statistical text indexing, (ii) information generated from NLP of text, and (iii) information generated from image indexing, in this case, face detection and recognition. A *matching agent* has been implemented using C and Prolog. The output of the NLP module are assertions that are currently being stored separately.

A Prolog-based reasoning system has been designed; the purpose of this is to match queries to assertions in the database. We adopt an approach similar to that of [1] to perform inexact matching. For example, a query to find *Clinton shaking hands with Dole* should generate some degree of match with the caption *Clinton greeting Dole.* The use of WordNet [8] is critical in determining hierarchical relationships between entity classes and event classes. The matching algorithm is being extended to incorporate information from statistical text indexing as well as face detection/recognition.

It is the task of the matching agent to decide on how information should be combined. For example, for an unambiguous query such as *find pictures of Bill Clinton*, the face detection and recognition results will be automatically applied to produce a single ranking of images satisfying the query. However, for a more subjective query such as *find pictures of victims of natural disasters*, the general context will be first applied. The results will then be sorted based on various visual criteria, thus allowing the user to browse and make a selection. The next section discusses these scenarios. Currently, only results from statistical text indexing and face detection have been evaluated.

7. Retrieval Experiments

There were two experiments conducted in retrieval of pictures from multimodal documents. Each reflected a different strategy of combining information obtained by text indexing and image indexing. Both of these are now described.

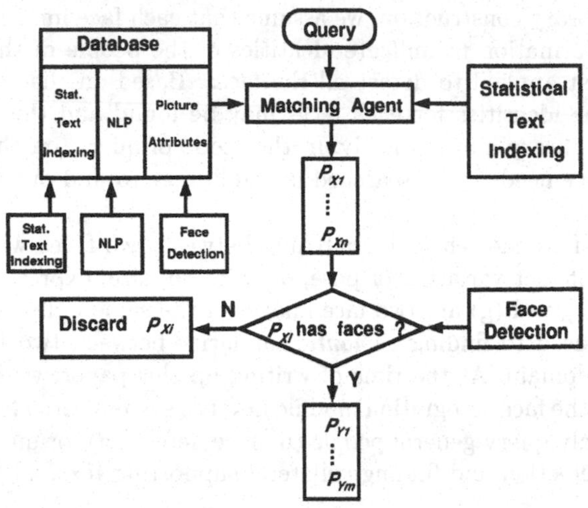

Fig. 6. Single-ranking strategy for combining text and image information in retrieval

7.1 Single Ranking Method

In this experiment, the queries are first processed using text indexing methods. This produces a ranking $P_{x_1} \ldots P_{x_n}$ as indicated in Figure 6. We then eliminate those pictures p_{x_i} which do not contain faces; this information is obtained by our automatic face detection module.

Fig 7 presents the results of an experiment conducted on 198 images that were downloaded from various news web sites. The original dataset consisted of 941 images; of these, 117 were empty (white space) and 277 were discarded as being graphics. From these, we chose a subset of 198 images for this experiment. There were 10 queries, each involving the search for pictures of named individual(s); some specified contexts also, such as *find pictures of Hillary Clinton at the Democratic Convention.* Due to the demands of truthing, we report the results for one query; more comprehensive evaluation is currently underway. The 3 columns in Figure 7 indicate precision rates using various criteria for content verification: (i) using text indexing (SMART) alone, (ii) using text indexing and manual visual inspection, and (iii) using text indexing and automatic face identification. As the table indicates, using text alone can be misleading- when inspected, many of these pictures do not contain the specified face. By applying face detection to the result of text indexing, we discard photographs that do not have a high likelihood of containing faces. The last column indicates that this strategy is effective in increasing precision rates. The number in parentheses indicates the number of images in the qiven quantile that were discarded due to failure to detect faces.

	Text Only	Text + Manual Insp	Text + Face Det
At 5 docs	1.0	1.0	1.0 (3)
At 10 docs	1.0	.70	1.0 (3)
At 15 doc	.80	.75	1.0 (2)
At 30 docs	.77	.67	NA

Fig. 7. Results of single-ranking strategy of combining text and image content. The last column indicates the result of text indexing combined with face detection. The number in parentheses indicates that this many images in the qiven quantile were discarded due to failure to detect faces.

Retrieval Results: Using Text Only

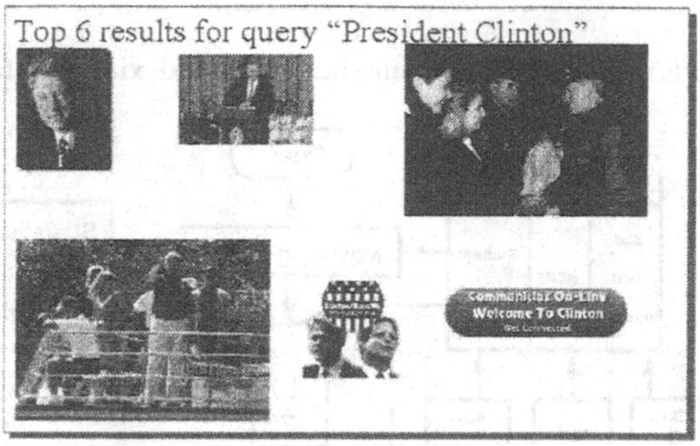

Top 6 results for query "President Clinton"

Fig. 8. Top 6 images based on text indexing alone

Sample output is shown in figures 8 and 9. Figure 8 illustrates the output based on text indexing alone. The last picture illustrates that text alone can be misleading! Figure 9 illustrates the re-ranked output based on results from face detection. This has the desired result that the top n images all are relevant. However, a careful examination reveals that due to the face detector's occasional failure to detect faces in images, relevant images are inadvertently being discarded. Thus this technique increases precision, but lowers recall. However, if the source of images is the WWW, this may not be of concern. We continue to improve the face detector to make it more robust to varied lighting conditions.

7.2 Multiple Ranking Method

In this experiment, we use a multiple ranking method for presenting candidate images to the user. This strategy is depicted in Figure 10. The context is first verified using statistical text indexing. We then *sort* these candidate images based on various visual properties. The first property is the presence of faces, the second represents the absence of faces (reflecting an emphasis on general scene context, rather than individuals). This reflects the assumption that the user doesn't know a priori exactly what kind of pictorial attributes he is looking

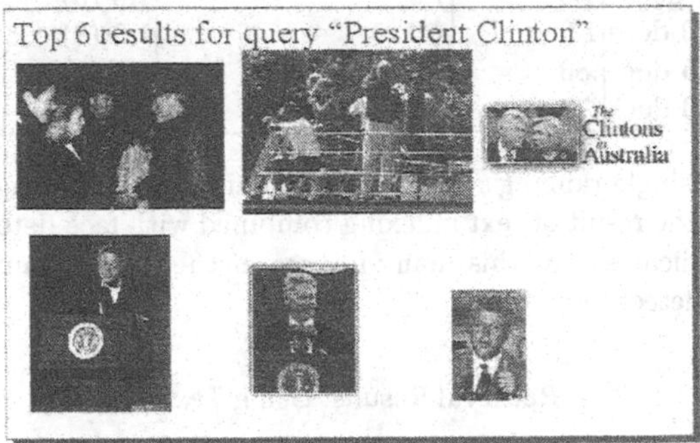

Fig. 9. Top 6 images based on combining text indexing and face detection

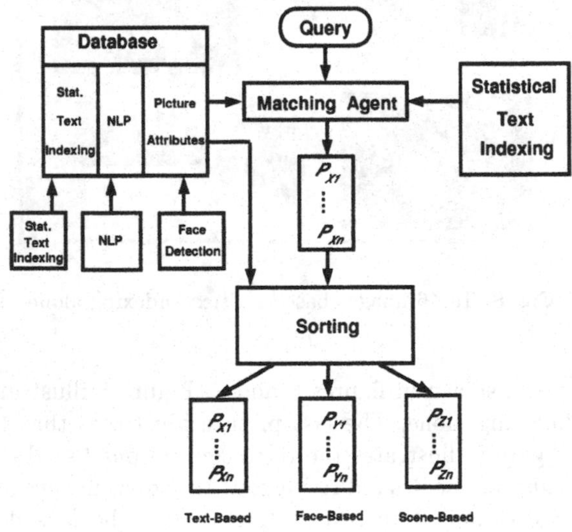

Fig. 10. Multiple-ranking strategy for combining text and image information in retrieval

for- he would like to browse.

Figure 11 depicts the top ranked images for the query *victims of disasters.* Space did not permit us to show the text accompanying these images; many of these refer to the recent aircrash in Indonesia, partially blamed to heavy smoke from forest fires. Some images depict victims, some depict politicians discussing the situation. Based on a threshold imposed, only the top 10 images returned by text retrieval were considered. As the results show, this produces a different ranking of images, where the lower row clearly emphasizes people. Had a lower threshold for text retrieval been used, the difference would have been more dramatic.

Evaluating precision and recall for such a technique is challenging. The precision rate for a given sorting criteria is based on both the text relevance and the presence of the required pictorial attributes (e.g., presence of faces). The text retrieval precision for the top 10 images is 90%. However, when "presence of faces" is used as a sorting criteria, the precision in the top 10 images drops to 40%. This is primarily due to the presence of very small faces in the image which are detected by the face detector. Since the manual annotators were instructed to disregard faces below a certain size, these are judged to be erroneous (e.g., the last picture in the second row of Figure 11. Thus, assigning relevance judgments based on pictorial attributes must be reinvestigated.

Finally, our future work includes the addition of other sorting criteria (e.g. natural scenery) as well as query refinement using image similarity and relevance feedback.

8. Conclusion

This paper has presented a system for searching multimodal documents for pictures in context. Specifically, two different techniques for combining information from text processing and image processing have been presented. This work represents efforts towards satisfying users' needs to efficiently browse for pictures. It is also one of the first efforts to automatically derive semantic attributes of a picture, and to subsequently use this in content-based retrieval. Our design for a multimodal indexing and retrieval system supports several types of text indexing (statistical, NLP-based) and image indexing (object detection, similarity-based). In this paper, we have only discussed the use of two of these indexing methods. We are currently engaged in incorporating NLP indexing as well; these results should be available soon. Finally, more comprehensive testing and evaluation of these techniques is required and is being conducted.

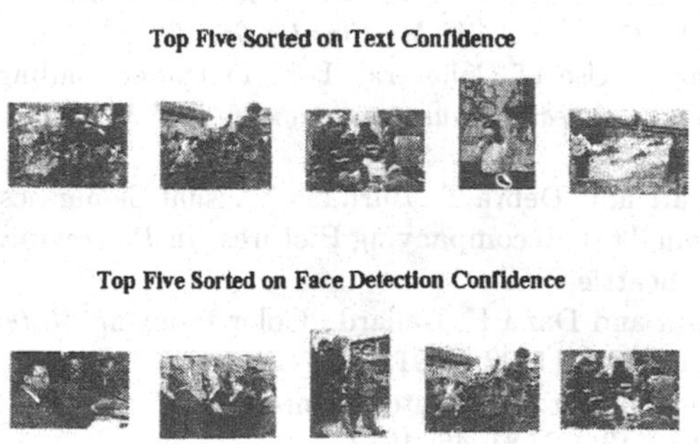

Fig. 11. Top 5 images based on combining text indexing and face detection

References

1. Y. Alp Aslandogan, Chuck Thier, Clement T. Yu, Jon Zou, and Naphtali Rishe. Using Semantic contents and WordNet in Image Retrieval. In *Proceedings of the 20th Annual International ACM SIGIR Conference on Research and Development in Information Retrieval*, pages 286–295. ACM, 1997.
2. Daniel M. Bikel, Scott Miller, Richard Schwartz, and Ralph Weischedel. Nymble: a High-Performance Learning Name-finder. pages 194–201, 1997.
3. C. C. Chang and S. Y. Lee. Retrieval of Similar Pictures on Pictorial Databases. *Pattern Recognition*, 24(7):675–680, 1991.
4. R. Chellappa, C.L. Wilson, and S. Sirohey. Human and machine recognition of faces: A survey. *Proc. of the IEEE*, 83(5), 1995.
5. L.G. Farkas and I.R. Munro. *Anthropometric Facial Proportions in Medicine.* Charles C. Thomas, 1987.
6. K. Fukunaga. *Introduction to Statistical Pattern Recognition, 2nd Ed.* Academic Press, 1990.
7. Corinne Jorgensen. An Investigation of Pictorial Image Attributes in Descriptive Tasks. In Bernice E. Rogowitz and Jan P. Allenbach, editors, *Proceedings of SPIE Vol. 2657, Human Vision and Electronic Imaging.* SPIE Press.
8. G.A. Miller, R. Beckwith, C. Fellbaum, D. Gross, and K. Miller. Introduction to WordNet: An On-line Lexical Database. *International Journal of Lexicography*, 3(4):235–244, 1990.
9. W. Niblack, R. Barber, W. Equitz, M. Flickner, E. Glasman, D. Petkovic, P. Yanker, C. Faloutsos, and G. Taubin. The qbic project: Querying images by content using color, texture, and shape. In *Storage and Retrieval for Image and Video Databases.* SPIE, 1993.
10. Donna M. Romer. Research agenda for cultural heritage on information networks: Image and multimedia retrieval.
11. Donna M. Romer. A Keyword is Worth 1,000 Images. Kodak internal technical report, Kodak, 1993.
12. N. Rowe and E. Guglielmo. Exploiting Captions in Retrieval of Multimedia Data. *Information Processing and Management*, 29(4):453–461, 1993.
13. Gerard Salton. *Automatic Text Processing.* Addison-Wesley, 1989.
14. J.R. Smith and S.F. Chang. Visualseek: a fully automated content-based image query system. In *Proc. of ACM Multimedia 96*, 1996.
15. Rohini K. Srihari. Use of Collateral Text in Understanding Photos. *Artificial Intelligence Review (special issue on integration of NLP and Vision)*, 8:409–430, 1995.
16. Rohini K. Srihari and Debra T. Burhans. Visual Semantics: Extracting Visual Information from Text Accompanying Pictures. In *Proceedings of AAAI-94*, pages 793–798, 1994. Seattle, WA.
17. Michael J. Swain and Dana H. Ballard. Color Indexing. *International Journal of Computer Vision*, 7(1):11–32, 1991.
18. M. Tomita. An Efficient Augmented-Context-Free Parsing Algorithm. *Computational Linguistics*, 13(1-2):31–46, 1987.
19. Webseer. http://webseer.cs.uchicago.edu.

Image Enhancement and Improvement of Both Color and Brightness Contrast Based on Lateral Inhibition Method

Takashi Sakamoto Toshikazu Kato

1704, Intelligent Systems Division, Electrotechnical Laboratory,
1-1-4 Umezono, Tsukuba 305-8568, Japan
sakamoto@etl.go.jp, kato@etl.go.jp

Abstract. We propose a new lateral inhibition method for image enhancement which improves both color and brightness contrast. Our method deserves attention for the following reasons: (1) it can adapt itself to the objective image automatically, (2) physiological and psychological behavior of early visual system has been considered and (3) it can affect locally and parallelly both the light region and dark region of the objective image. This method has been derived on the basis of our early vision system and modelized by a simple mathematical function which forms reverse-S shaped stimulus-response curve, and additionally, it can simulate lateral inhibition effects. Our model explains how the lateral inhibition mechanism with local and adaptive processing system realizes the robustness for various input images and detects objects from wide varieties of visual stimuli. The proposed method can make the maximum use of the lateral inhibition effects, perform mild and powerful image enhancement and improve image quality very naturally both at the light and dark regions.

1. Introduction

From the view point of image processing, we have developed computational physiology of visual mechanism and its application to human-centered multimedia technology (or Human Media [1]) which provides information processing facilities in a human friendly manner.

Human vision mechanism is composed of huge neural networks: retinal cells to sense light stimulus, neural pathways and visual cortex for high level visual cognition [2]. These components can be classified into a hierarchical structure on the basis of their information processing role at different levels, that is physical level, physiological level, psychological level and cognitive level. In every level, many rational and well-ordered mechanisms can be found that should be applied in information processing. Such neural computing and generic algorithm are good examples of the application using biological mechanisms.

In this paper, a new method for enhancement and improvement of image contrast is shown as a computational model of lateral inhibition mechanism which is a physiological level phenomenon of early vision system. Different kinds of methods to enhance and improve image contrast are used such as gray or color scale transformation (liner, gamma, logarithm, etc. [3-7]), histogram modification [8-11] and some kinds of filtering or masking techniques [12-16]. These methods, however, bring some unsatisfactory results which arise for the following reasons: (1) we have to choose ad hoc one of them according to the needs of the moment, because they do not adapt themselves into an objective image automatically; or / and (2) these methods are based on using numerical or statistical features of the image, but they lack a consideration of physiological and psychological behavior of our visual systems.

The proposed method in this paper gets over both of the above mentioned problems (1, 2) and can show the rationality of human vision mechanism to realize the robustness for various input images and detect objects from wide varieties of visual stimuli. The lateral inhibition model used in this method is on the basis of some physiological and psychological knowledge. This model can be written as a simple mathematical function which has a reverse-S shaped nonlinear response curve. The proposed method makes the maximum use of the lateral inhibition effects and realizes the robustness and adaptivity for input images.

2. Modeling of Visual System

2.1 Lateral Inhibition

Lateral inhibition in retinal cells is a phenomenon that has the effect of making the cells compare the amount of light at one position with the amount at neighboring regions [17]. Lateral inhibition makes concentric arrangement of retinal regions that respond in the opposite sense. These regions, called receptive fields, are classified into ON-center type, and OFF-center type [18]. Receptive fields are very common in the retinae of many species, and the same arrangement is found in second and higher order neurons. We can regard the total response by the linear summation of the responses from each receptive field.

2.2 Mathematical Formulation

Let ν be the intensity of the input stimulus, V be response to the stimulus, ν_0 and ν_1 be minimum and maximum intensity of the input stimulus ν (also to avoid zero value). We will consider ON-center type and OFF-center type neural channels. These channels respond to positive images (ν - ν_0) and negative images (ν_1 - ν), respectively. The positive channel is to detect brightness, while the negative one is to detect darkness.

According to Weber's law, the difference of the intensity $\Delta\nu$ in both channels would be normalized by the power of the background at each cell;

$$\Delta V \propto \frac{\Delta\nu}{\nu - \nu_0}, \qquad \text{(Eq.1a)}$$

$$\Delta V \propto \frac{\Delta\nu}{\nu_1 - \nu}. \qquad \text{(Eq.1b)}$$

Such normalization is assumed to perform scale invariance in responding to a wide range of intensity stimulus. According to Fechner's law, the response to the intensity is proportional to the integrated forms of (Eq.1a, Eq.1b) as follows;

$$V \propto \log(\nu - \nu_0), \qquad \text{(Eq.2a)}$$

$$V \propto -\log(\nu_1 - \nu). \qquad \text{(Eq.2b)}$$

Finally, we get the total response by a linear summation of both channels. It is proportional to the algorithm of the ratio of positive channel to negative channel as follows;

$$V \propto \log(\nu - \nu_0) - \log(\nu_1 - \nu) = \log\frac{\nu - \nu_0}{\nu_1 - \nu} \triangleq K(\nu). \qquad \text{(Eq.3)}$$

2.3 Numerical Behavior

Let us observe the behavior of the function $K(v)$. The overall figure of the non-linear function $K(v)$ shows a reverse-S form. We can control the curvature of $K(v)$ by the constant parameters v_0 and v_1 as shown in Fig.1. When $|v_0| = |v_1|$, $K(v)$ shows symmetrical reverse-S, and when $|v_0| = |v_1| \gg 1$, $K(v)$ approaches linear. When $|v_0| \gg |v_1|$, $K(v)$ approaches an exponential curve or gamma-curve cv^γ ($\gamma < 1$), while when $|v_0| \ll |v_1|$, $K(v)$ approaches a logarithmic curve or gamma-curve cv^γ ($\gamma > 1$).

A light stimulus-response of a retinal cell forms an S-shape non-linear curve and this curve shifts to the right or left according to the intensity of background light [18]. In order to obtain linear response on the overall wide stimulus range using the above stimulus response behavior, it is necessary to adjust its non-linear response to a quasi-linear one by using such a reverse-S-shape function $K(v)$ (see Fig.2). Our mathematical model can be described as a simple formula, nevertheless it can explain several visual phenomena and the numerical and physiological necessity of both ON and OFF type channels. Some visual phenomena were simulated and the results are shown in the section 4.

3. Improvement of Image Contrast

It is assumed that both ON type and OFF type responses in the early vision system are the mechanisms to enhance the contrast of a point that we are looking at and its neighborhood under the various background intensities (of light or color). Let us discuss active control method of the shape of reverse-S function $K(v)$, according to the level of background light or color of every point and its neighborhood. Some kinds of ON type and OFF type cells exist to detect light and shade, other kinds to detect colors and their physiological complementary colors. So, we have to formulate both light/dark type for gray images and color type for color images.

3.1 The Proposed Method for Gray Images

For the enhancement of gray images, (Eq.3) is modified as follows;

$$K(v) = \log \frac{v - v_L + \delta_{posi}}{v_U - v + \delta_{nega}} . \qquad \text{(Eq.4)}$$

Let v_U and v_L be a maximum value and a minimum value of gray level v, δ_{posi} and δ_{nega} be parameters to control the shape of the reverse-S function $K(v)$. To every point and its neighboring area (e.g. 3×3 neighborhood) in a gray image, these parameters are defined as follows;

$$\delta_{posi} = \alpha_{posi} \cdot (v_{\min} - v_L) + \beta_{posi}, \qquad \text{(Eq.5a)}$$

$$\delta_{nega} = \alpha_{nega} \cdot (v_U - v_{\max}) + \beta_{nega} . \qquad \text{(Eq.5b)}$$

where $v_{\max}$ and $v_{\min}$ are a maximum and a minimum gray level in the neighbor area. α_{posi} and α_{nega} decide the degrees of enhancement. Ratio of α_{posi} to β_{posi} and α_{nega} to β_{nega} indicate dependency on local gray level distribution (also β_{nega} avoids zero division). To enhance images using (Eq.4), these parameters are decided beforehand to according to the user's preferences.

3.2 The Proposed Method for Color Images

Any color image is composed of more than three pieces of color information such as R, G, B. It is assumed that an objective image has RGB signals and their intensities are r, g and b respectively. For the enhancement of color images, we need three kinds of (Eq.4) as follows;

$$K_R(r) = \log\frac{r - r_L + \xi_{posi}}{r_U - r + \xi_{nega}}, \quad \text{(Eq.6a)}$$

$$K_G(g) = \log\frac{g - g_L + \eta_{posi}}{g_U - g + \eta_{nega}}, \quad \text{(Eq.6b)}$$

$$K_B(b) = \log\frac{b - b_L + \zeta_{posi}}{b_U - b + \zeta_{nega}}. \quad \text{(Eq.6c)}$$

Every part in (Eq.6) corresponds to those in (Eq.4). For the sake of brevity, let δ represent any one of ξ, η and ζ. Now they are defined in d form by

$$\delta_{posi} = \alpha_{posi}\cdot(l_{\min} - l_L) + \beta_{posi}\cdot(c_{\min} - c_L) + \gamma_{posi}, \quad \text{(Eq.7a)}$$

$$\delta_{nega} = \alpha_{nega}\cdot(l_U - l_{\max}) + \beta_{nega}\cdot(c_U - c_{\max}) + \gamma_{nega}. \quad \text{(Eq.7b)}$$

In (Eq.7), l and c represent lightness (or intensity) and colorfulness (or saturation) respectively. When we deal with the color image on a display, we can use luminance for NTSC-RGB system as l (coefficient numbers are simplified):

$$l = f_{\text{luminance}}(r, g, b) = 0.3r + 0.6g + 0.1b. \quad \text{(Eq.8)}$$

Colorfulness c can be defined variously for using computer vision [15-17] and its general description is

$$c = f_{\text{colorfulness}}(r, g, b). \quad \text{(Eq.9)}$$

It is adequate for our processing that c indicates color balance at every point of image and has ranges between any color and its complementary color. Thus we can define c as follows;

$$c = \begin{cases} r/(r + 0.5g + 0.5b) & \text{for both } \xi_{posi} \text{ and } \xi_{nega}, \\ g/(0.5r + g + 0.5b) & \text{for both } \eta_{posi} \text{ and } \eta_{nega}, \\ b/(0.5r + 0.5g + b) & \text{for both } \zeta_{posi} \text{ and } \zeta_{nega}. \end{cases} \quad \text{(Eq.10)}$$

where $0 \le c \le 1$ and when achromatic ($r = g = b$), $c = 0.5$.

3.3 Image Enhancement Results

In this subsection, our new method for enhancement and improvement of both gray and color contrast has been evaluated by comparison with gamma correction and histogram equalization [3][4]. These experimental results are shown in Fig.3 and Fig.4.

Gamma correction method is a kind of gray or color scale transformation, thus it is comparable with $K(v)$ transformation. The gamma method does not have local adaptivity, so that it can enhance either the light or dark region alternatively. By contrast, the reverse-S method can improve image contrast both at the light and dark regions, because it is scale invariant and stable for a wide variety of input levels.

Histogram equalization is one of the statistical methods and it can considerably enhance image contrast. However, the results of this method are so extreme that its visual quality is unnatural, especially when it contains any natural object. Moreover, if a histogram of the original image is flat, this method does not affect the image. By contrast, the reverse-S method performs a mild and powerful image enhancement and improves image quality very naturally.

4. Simulation of Visual Phenomena

Several visual phenomena are well known as they originate from the lateral inhibition mechanism of early vision system. The proposed computational model written in (Eq.3) is simple, but this model can simulate and give a good explanation of some visual phenomena, such as Mach bands and after-image phenomenon, for both gray and color images. These experimental results for color vision are shown in Fig.5 and Fig.6.

The response of our visual system to the abrupt changes in luminance or color (edges or contours) displays overshoots known as "Mach bands" [17]. This phenomenon leads to the effect of enhancing or deblurring edges, as shown in the former experiment (Fig.3). We can assume two type of edge enhancement effects: between light and dark and between a color and its complementary color. Fig.5 shows the later type and the luminance of the left sided image (original image) had been equalized to avoid effects of the lightness. This is the reason why yellow looks dark and blue looks bright in the original image. Several overshoots can be felt "psychologically" in the left sided image and observed "numerically" in the right sided.

An after-image can be observed when we fix our eyes on a black or colored square for a little while and then suddenly look at a white paper [19]. Then, for a moment, we can observe a "positive" after-image which has the same color but is pale. For a moment later, we can observe a "negative" after-image which is tinged with a complementary color and that will fade out in a white background. In order to simulate these phenomena, assume that an image effect $v_{t-\Delta t}$ observed just before is exponentially decreasing and an image effect v_t observed just on time is decided by the weighted average between the former image effect and a new input image we are now seeing:

$$v_t = v_{t-\Delta t} \cdot e^{-\lambda \Delta t} + v_{\text{new}} \cdot (1 - e^{-\mu \Delta t}). \qquad \text{(Eq.11)}$$

Fig.6 shows the simulation results on the assumption that we fix our eyes on a red square in a black background and move our eyes abruptly to a white background. In this experiment, an appropriate time scale is used for simulation so that the displayed time is not consistent with that of real phenomena.

5. Concluding Remarks

We have proposed a new method for enhancement and improvement of both gray and color contrast of the image. Our method is on the basis of a visual system and it can simulate lateral inhibition effects. This method can perform a mild and powerful image enhancement and improve image quality very naturally both at the light and dark regions. The proposed model shows that the lateral inhibition mechanism with local and adaptive processing function has the rationality to realize the robustness for various input images and detect objects from wide varieties of visual stimulus.

A further problem is to consider and to implement a high-order visual system, such as LGN (lateral geniculate nucleus) and visual cortex into our model. These modifications enable us to enhance psychological color.

The proposed method is based on local and parallel processing mechanism which exist in early visual system so that it can be developed into parallel processing hardware. In the near future, we can integrate this model into various devices such as cameras, displays, and so on.

References

[1] T. Kato: "Multimedia Database with Visual Interaction Facilities", in Proc. of IFCS-96, Data Science, Classification and Related Methods, 46-49 (1996).

[2] J.E.Dowling & B.B.Boycott: "Organization of the primate retina: Electron microscopy", Proceedings of the Royal Society Series B, 166, 80-111 (1966).

[3] W.K.Pratt: Digital Image Processing (2nd. Edition), John Wiley & Sons (1991).

[4] J. C. Russ: The Image Processing (2nd. Edition), CRC Press (1995).

[5] T.K.De & B.N.Chatterji: "The concept of de-enhancement in digital image processing",Pattern Recognition Letters, vol.2, 329-332 (1984).

[6] B. Chanda, B. B. Chaudhuri & D. D.Majumder: "Some algorithms for image enhancement incorporating human visual response" Pattern Recognition, vol.17, 4, 423-428 (1984).

[7] A.Toet: "Adaptive Multi-Scale Contrast Enhancement Through Non-Linear Pyramid Recombination", Pattern Recognition Letters, vol.11, 735-742 (1990).

[8] E. L. Hall: "Almost uniform distribution for image enhancement" IEEE Trans. vol.C-23, 2, 207-208 (1974).

[9] W.Frei: "Image Enhancement by Histogram Hyperbolization", Computer Graphics and Image Processing, vol.6, 3, 286-294 (1977).

[10] R.A.Hummel: "Image Enhancement by Histogram Transformation", Computer Graphics and Image Processing, vol.6, 2, 184-195 (1977).

[11] S.M.Pizer, et al.: "Adaptive Histogram Equalization and Its Variations", Computer Vision, Graphics, and Image Processing, vol.39, 3, 355-368 (1987).

[12] W.F.Schreiber: "Wirephoto Quality Improvement by Unsharp Masking", J. Pattern Recognition, vol.2, 111-121 (1970)

[13] J. Granrath & B. R. Hunt: "A two-channel model of image processing in the human retina", Proc. of SPIE, vol.199, 126-133 (1979).

[14] J. D. Fahnestock & B. R. Hunt: "The maintenance of sharpness in magnified digital images" , Computer Vision, Graphics, and Image Processing, vol.27, 32-45 (1984).

[15] S. Hahn & E.E. Mendoza: "Simple enhancement techniques in digital image processing", Computer Vision, Graphics, and Image Processing, vol.26, 233-241 (1984).

[16] G.Ramponi, et al.: "Nonlinear Unsharp Masking Methods for Image-Contrast Enhancement", Journal of Electronic Imaging, vol.5, 3, 353-366 (1996).

[17] F. Ratliff: Mach Bands: Quantitative Studies of Neural Networks in the Retina, Holden-Day (1965).

[18] L. Spillmann & J. S. Werner: Visual Perception, The neurophysiological Foundations, Academic Press (1990).

[19] G.S. Brindley, "Afterimages", Scientific American, 10 (1963).

[20] O.D.Faugeras, "Digital Color Image Processing Within the Framework of a Human Visual Model", IEEE Trans. Acoustics, Speech, and Signal Processing, ASSP-27, 4, 380-393 (1979).

Figures

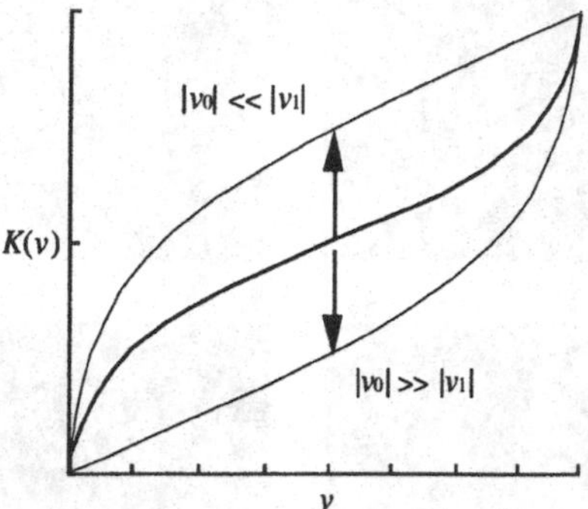

Fig.1 The Curvature of $K(v)$ can be controlled by the constant parameters v_0 and v_1.

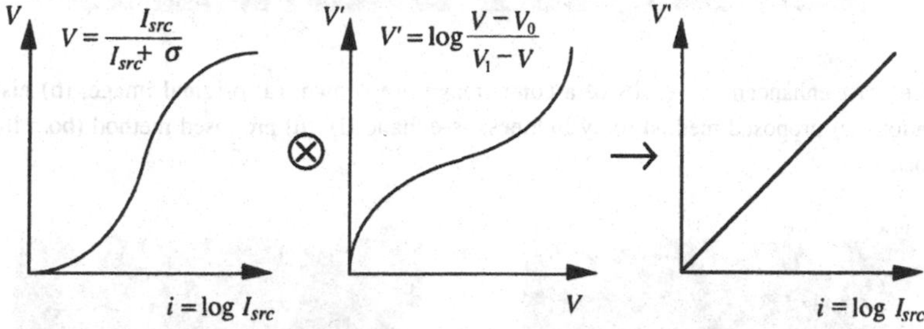

Fig.2 For non-linear stimulus response curve, the quasi-linear response can be obtained by using reverse-S-shape function.

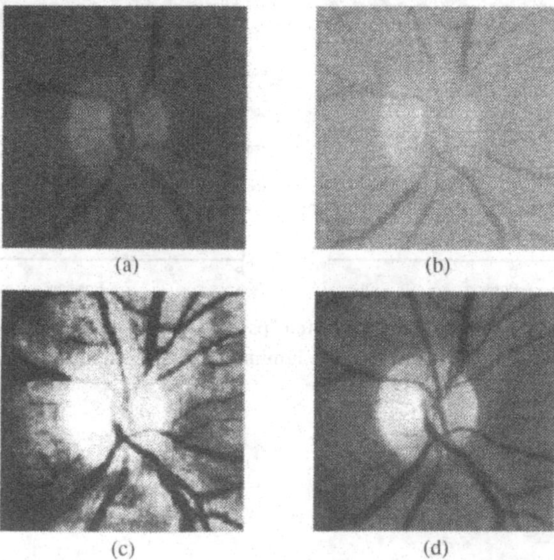

Fig.3 Contrast enhancement results of gray images are shown. Our method is compared with gamma correction ($\gamma = 0.4$) and histogram equalization: (a) original image, (b) gamma correction ($\gamma = 0.4$), (c) histogram equalization, and (d) proposed method.

Fig.4 Contrast enhancement results of a color image are shown: (a) original image, (b) histogram equalization, (c) proposed method (only lightness is enhanced), (d) proposed method (both lightness and color)

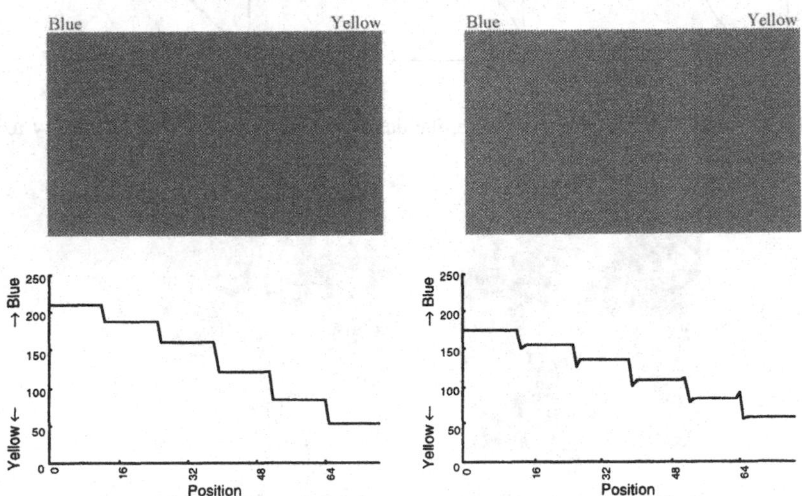

Fig.5 Several overshoots (Mach bands) can be felt "psychologically" in the left sided color image and observed "numerically" in the right sided. The luminance of the left sided image was equalized to avoid the effects of the lightness.

Fig.6 The simulation results are based on the assumption that we fix our eyes on a red square in a black background (t=0) and move our eyes abruptly to a white background. Appropriate time scale is used for this experiment.

An Area-Based Shape Representation for Affine Invariant Content-Based Retrieval

Horace H S Ip, Dinggang Shen, Wai-him Wong, Ken C K Law

Image Computing Group, Department of Computer Science
City University of Hong Kong
Tat Chee Avenue, Kowloon, Hong Kong
Email: cship@cityu.edu.hk, edgshen@ntu.edu.sg
Fax: (852)2788-8614

Abstract. In this paper, we present an area-based affine invariant representation of shapes and a point-correspondence algorithm which supports multi-scale similarity matching for shape-based retrieval. This method is affine invariant and stable against noise and shape deformations. Since the area-based representation scheme simultaneously captures both local and global affine invariant features of shapes, it makes the solution to the correspondence problem very robust.

1 Introduction

Works on image retrieval are frequently based upon iconic features, such as colour indexing [eg. 1,2], texture [eg. 3,4], and sketch or shape [eg. 5,6]. The trade-off between adopting a global or local features for matching purpose are that global features are typically relatively stable to noise, more sensitive to occlusion and not effect for shapes which are similar. Local features, on the other hand, such as geometric invariants [7], corners [8] and curvature [9] are insensitive to occlusion, sensitive to noise and the amount of local information available is usually insufficient for robust matching. Recently, Sclaroff and Pentland [10] described objects in terms of generalised symmetries, which were defined in terms of the object's eigenmodes. Since modes provide a global-to-local ordering of deformation, modal matching allowed for selecting the appropriate types of deformations that were used in object alignment and comparison. Since the technique requires computing and interpolating eigenmodes, it is computationally expensive and therefore not particularly suited to interactive image retrieval applications.

This paper presents a multi-scale similarity matching method for shaped-based retrieval from an image database. Specifically we have proposed a new affine-invariant representation scheme which is relatively insensitive to shape deformation and simultaneously capture both local and global shape information.

We will first show analytically that the local area of a region, defined in terms of curvature, is an affine invariant feature. Based on this observation, we propose a new area-based shape representation scheme which captures both local and global affine invariant features. A similarity matching error for shape-based image retrieval will be defined and a multi-scale approach of similarity matching will be given. Finally, the feasibility of the approach is demonstrated experimentally.

2 From Curvature to Area Representation of Curves

Let $r(t) = [x(t), y(t)]$, where t is the parameter, represent a shape (or curve) C in Cartesian coordinate system. The curvature $k(t)$ of the shape C is hence defined as

$$k(t) = \frac{(\dot{x}(t)\ddot{y}(t) - \ddot{x}(t)\dot{y}(t)) \cdot (dt)^3}{(ds)^2} = \frac{\begin{vmatrix} \dot{x}(t) & \ddot{x}(t) \\ \dot{y}(t) & \ddot{y}(t) \end{vmatrix} \cdot (dt)^3}{(ds)^2} \tag{1}$$

where

$$(ds)^2 = (dx)^2 + (dy)^2$$

$$\dot{x}(t) = \frac{dx(t)}{dt},\ \ddot{x}(t) = \frac{d^2x(t)}{dt^2},\ \dot{y}(t) = \frac{dy(t)}{dt},\ \text{and}\ \ddot{y}(t) = \frac{d^2y(t)}{dt^2}.$$

and s is the arc length of C.

We can interpret (1) as $k(t) = \frac{ang}{dist}$, where

$$ang = \begin{vmatrix} \dot{x}(t) & \ddot{x}(t) \\ \dot{y}(t) & \ddot{y}(t) \end{vmatrix} \cdot (dt)^3 \text{ and } dist = (ds)^2, \tag{2}$$

Under this interpretation, we can show that *ang* is an affine invariant quantity when properly normalised. We approximate the first and the second derivatives of $x(t)$ and $y(t)$ with respect to t as follows.

$$\dot{x}(t) = \frac{dx(t)}{dt} \doteq \frac{x(t) - x(t - \Delta t)}{\Delta t}$$

$$\ddot{x}(t) = \frac{d^2x(t)}{dt^2} \doteq \frac{x(t) - 2\cdot x(t-\Delta t) + x(t-2\cdot\Delta t)}{\Delta t\cdot\Delta t}$$

$$\dot{y}(t) = \frac{dy(t)}{dt} \doteq \frac{y(t) - y(t-\Delta t)}{\Delta t}$$

$$\ddot{y}(t) = \frac{d^2y(t)}{dt^2} \doteq \frac{y(t) - 2\cdot y(t-\Delta t) + y(t-2\cdot\Delta t)}{\Delta t\cdot\Delta t} \tag{3}$$

Combining (2) and (3) gives:

$$\frac{1}{2}ang = \frac{1}{2}\begin{vmatrix} \dot{x}(t) & \ddot{x}(t) \\ \dot{y}(t) & \ddot{y}(t) \end{vmatrix}\cdot(dt)^3 = \frac{1}{2}\begin{vmatrix} dx(t) & d^2x(t) \\ dy(t) & d^2y(t) \end{vmatrix}$$

$$\doteq \frac{1}{2}\begin{vmatrix} x(t)-x(t-\Delta t) & x(t)-2\cdot x(t-\Delta t)+x(t-2\cdot\Delta t) \\ y(t)-y(t-\Delta t) & y(t)-2\cdot y(t-\Delta t)+y(t-2\cdot\Delta t) \end{vmatrix}$$

$$= \frac{1}{2}\begin{vmatrix} x(t)-x(t-\Delta t) & x(t-2\cdot\Delta t)-x(t-\Delta t) \\ y(t)-y(t-\Delta t) & y(t-2\cdot\Delta t)-y(t-\Delta t) \end{vmatrix}$$

$$= \frac{1}{2}\begin{vmatrix} x(t-2\cdot\Delta t) & x(t-\Delta t) & x(t) \\ x(t-2\cdot\Delta t) & y(t-\Delta t) & y(t) \\ 1 & 1 & 1 \end{vmatrix}$$

The above formula shows that the value of $\frac{1}{2}ang$ is equal to the area of a triangle, whose three vertices are respectively $(x(t-2\cdot\Delta t), y(t-2\cdot\Delta t)), (x(t-\Delta t), y(t-\Delta t))$ and $(x(t), y(t))$. Moreover, it can be shown that the area of any triangle is also relatively affine invariant. Let $area_{t_1,t_2,t_3}$ be the area of a triangle, whose three vertices are $(x(t_1), y(t_1))$, $(x(t_2), y(t_2))$ and $(x(t_3), y(t_3))$. That is

$$area_{t_1,t_2,t_3} = \frac{1}{2}\begin{vmatrix} x(t_1) & x(t_2) & x(t_3) \\ y(t_1) & y(t_2) & y(t_3) \\ 1 & 1 & 1 \end{vmatrix}$$

Let $C'=[u(\tau), v(\tau)]$ be the affine-transformed version of the shape C. Here τ is also a parameter, corresponding to t. Mathematically, the relationship between two shapes C and C' under affine transformation can be expressed as follows.

$$\begin{bmatrix} u(\tau) \\ v(\tau) \\ 1 \end{bmatrix} = \begin{bmatrix} a_{11} & a_{12} & b_1 \\ a_{21} & a_{22} & b_2 \\ 0 & 0 & 1 \end{bmatrix}\begin{bmatrix} x(t) \\ y(t) \\ 1 \end{bmatrix} \tag{4}$$

If we use the same form as equation (4) to express the affine transformation between two shapes C and C', then the relationship between the areas of the same triangle before and after any affine transformation can be obtained as follows:

$$\begin{bmatrix} u(\tau_1) & u(\tau_2) & u(\tau_3) \\ v(\tau_1) & v(\tau_2) & v(\tau_3) \\ 1 & 1 & 1 \end{bmatrix} = \begin{bmatrix} a_{11} & a_{12} & b_1 \\ a_{21} & a_{22} & b_2 \\ 0 & 0 & 1 \end{bmatrix} \begin{bmatrix} x(t_1) & x(t_2) & x(t_3) \\ y(t_1) & y(t_2) & y(t_3) \\ 1 & 1 & 1 \end{bmatrix}$$

which leads to

$$area^{affine}_{t_1,t_2,t_3} = \frac{1}{2}\begin{vmatrix} u(\tau_1) & u(\tau_2) & u(\tau_3) \\ v(\tau_1) & v(\tau_2) & v(\tau_3) \\ 1 & 1 & 1 \end{vmatrix} = \frac{1}{2}\begin{vmatrix} a_{11} & a_{12} & b_1 \\ a_{21} & a_{22} & b_2 \\ 0 & 0 & 1 \end{vmatrix}\begin{vmatrix} x(t_1) & x(t_2) & x(t_3) \\ y(t_1) & y(t_2) & y(t_3) \\ 1 & 1 & 1 \end{vmatrix}$$

$$= (a_{11}a_{22} - a_{12}a_{21})\cdot area_{t_1,t_2,t_3}$$

Under affine transformation, the area value is just linearly increased by a scale factor of $(a_{11}a_{22} - a_{12}a_{21})$. This implies that the area of any triangle when suitably normalised is an affine-invariant quantity.

3 Area-Based Affine Invariant Shape Representation

In the following, we present a feature vector, which simultaneously captures both local and global affine-invariant information, for every feature point on the given shape boundary.

Let $\{(x_1, y_1), (x_2, y_2), \ldots, (x_N, y_N)\}$ be the set of feature points representing the given shape C in a particular scale. We assume N to be an odd number. These feature points can be extracted, for example, simply by taking sample points evenly along the shape boundary. To make feature points robust to noise, we collect coordinates of neighboring points at a candidate feature point and average them out to obtain a final feature point. This setup guarantees complete coverage of all the points defining the shape with overlapping regions of support between adjacent feature points. For the i-th feature point (x_i, y_i), its corresponding feature vector is defined as

$$F_i = [f_{1i} \;\; f_{2i} \ldots f_{Mi}]^T$$

where $f_{ji} = \frac{1}{2}\begin{vmatrix} x_{(i-j)} & x_i & x_{(i+j)} \\ y_{(i-j)} & y_i & y_{(i+j)} \\ 1 & 1 & 1 \end{vmatrix}$ is the area of a triangle formed by (x_{i-j}, y_{i-j}), (x_i, y_i), and (x_{i+j}, y_{i+j}). Notice that $x_i = x_{(i+2N)\%N}$ and $y_i = y_{(i+2N)\%N}$ in the expression of f_{ji}, particularly for $i,j<0$ or $i,j>N$. It is easy to verify that there are at most $\frac{N-1}{2}$ measures for each feature point, ie. $M=\frac{N-1}{2}$, and the size of matrix F is $\frac{N-1}{2}\times N$. It can be seen that f_{ji} represents the local feature of the shape when j is small, and f_{ji} gradually becomes a global feature of the shape when j becomes large and when j is close to M, f_{ji} serves to represent local feature again. In other words, the elements in the feature vector F_i can be roughly grouped into three parts, local features, global features and local features such that

$$F_i=\left[\underbrace{f_{1i}\ \cdots\ f_{ki}}_{local\ features}\ \underbrace{f_{(k+1)i}\cdots f_{li}}_{global\ features}\ \underbrace{f_{(l+1)i}\cdots f_{Mi}}_{local\ features}\right]^T$$

In the above section, we have shown that an affine transformation transforms a feature vector F_i to $(a_{11}a_{22}-a_{12}a_{21})F_i$, $i=1,2,\ldots,N$. To make such feature vectors affine invariant, we normalize the feature vector as follow:

$$\hat{F}_i=\frac{F_i}{\sum_{i=1}^{N}\sum_{j=1}^{M}\left|f_{ji}\right|}$$

This defines our affine invariant area-based feature for a point on an arbitrary shaped curve. It should be noted that taking different number of sample points along the curve, ie. using different values of N, gives raise to representations of the same curve at different scales.

4 Establishing Point Correspondence

Based on this affine invariant description of a shape, we can define a process of determining point correspondence between two shapes by searching for an optimal matching of the two corresponding feature vectors. Let the shape C' be the affine-transformed version of the shape C. Two feature vectors, $\{\hat{F}_i,\ i=1,2,\ldots,N\}$ and $\{\hat{F}'_i,\ i=1,2,\ldots,N\}$, correspond to the shape C and the shape C', respectively. The feature vector $\hat{F}'_{(i+s)}$ is expected to be identical to the feature vector $\hat{F}_i$, if the *(i+s)*-th feature point of the shape C' corresponds to the i-th feature point of the shape C and the requirement of match becomes:

$$E_{i,s}=\left(\hat{F}'_{(i+s)}-\hat{F}_i\right)^T\cdot G\cdot\left(\hat{F}'_{(i+s)}-\hat{F}_i\right)=0$$

where s is an integer and G is an $M \times M$ diagonal matrix defined by a Gaussian function, i.e.

$$G = [g_{mm}]_{M \times M} \text{ and } g_{mm} = \frac{1}{\sqrt{2\pi}\sigma} e^{-\frac{\left(\frac{m}{M}-\frac{1}{2}\right)^2}{2\sigma^2}}.$$

G is a weighting matrix to express the notion that the global features should be more important than the local features for establishing point correspondence and to reduce the sensitivity to noise and small shape deformation.

In practice, due to the presence of noise and/or possible shape deformations, we seek to minimise the error measure (E) instead in correspondence matching. That is, we look for the minimal error from the summed errors along the diagonals of the feature matrix F . The mathematical form is as follows.

$$E = \min_{s \in \{0,1,2,\ldots,N-1\}} E_s$$

$$E_s = \sum_{i=1}^{N} E_{i,s} = \sum_{i=1}^{N} \left(\hat{F}'_{(i+s)} - \hat{F}_i\right)^T \cdot G \cdot \left(\hat{F}'_{(i+s)} - \hat{F}_i\right)$$

where s is the shift in the starting feature indices. When two shapes are mirror images of each other, the point matching order will be reversed. Accordingly, the method of searching for the minimal error (E) of correspondence matching should be as follows.

$$E = \min_{s \in \{0,1,2,\ldots,N-1\}} E_s$$

$$E_s = \sum_{i=1}^{N} E_{i,s} = \sum_{i=1}^{N} \left(\hat{F}'_{(N-i+s)} - \hat{F}_i\right)^T \cdot G \cdot \left(\hat{F}'_{(N-i+s)} - \hat{F}_i\right)$$

4.1 Similarity matching for content-based retrieval

The correspondence matching process described in section 3 gives us a globally optimal solution between two sets of feature points. However, it does not tell us whether the two shapes are really similar, or not. Thus, a means of verifying whether the two shapes are similar is required. We integrates two sources of error information, i.e. error in correspondence matching and error in the consistency among individual feature points, to obtain a similarity measure.

The first source of information is the value of the minimal summed absolute error of the correspondence matching, ie. the error measure E described in section 3. For similar shapes, E should be small. If E is large, the two shapes differ a lot. The second source is consistency among individual feature points. For each feature point, we find for it a mapping to the nearest neighbour from the set of feature points in the other shape such that the measured error is at its local minimal. Let's define this optimal local mapping as viewed from each feature point (x_i, y_i) as a local shift s_j. If the global scheme is indeed a good match, we should expect a lot of these local optimal shifts s_j to agree with the global optimal shift s. Let n denote the number of local optimal mapping equivalent to the global optimal scheme. The ratio n:N (N is the number of feature points) provides us with a measure on the goodness of the matching results. The optimal ratio is the case when n=N, in which all feature points voting the global scheme are also their local optimal, giving us a high confidence to the matching result. The worst case is n=0, which is rare but still possible. Definitely, with n=0, we can confidently reject the hypothesis that the two shapes are similar.

To integrate the two sources of error information, we define the error measure as:

$$E_{Total} = 2^{h(N/n)} E$$

where $h(x) = \begin{cases} x, & x < UP \\ UP, & x \geq UP \end{cases}$. UP is an upper bound for the value $\frac{N}{n}$. In our experiments, we set this bound at 5 to cap the error term and to give meaningful comparison while at the same time avoid the divided-by-zero error. For content-based retrieval, this function serves as a similarity measure between two shapes. The higher the value, the larger is the error and the lower the possibility that the two shapes are visually similar.

5 Illustration of the Approach

5.1 Evaluating the affine transformation parameters

Once the correspondence between two shapes is established, the estimation of the affine-transformation relating these two shapes can be easily determined. This estimation is useful if we want to use additional information such as internal textures or grey level distribution for comparing the two shapes. Affine-transform matrix A relating shape C and C' can be calculated by

$$A = (Q' \cdot Q^T)(Q \cdot Q^T)^{-1}$$

where

$$A=\begin{bmatrix} a_{11} & a_{12} & b_1 \\ a_{21} & a_{22} & b_2 \\ 0 & 0 & 1 \end{bmatrix},\ Q=\begin{bmatrix} x_1 & x_2 & \dots & \dots & x_N \\ y_1 & y_2 & \dots & \dots & y_N \\ 1 & 1 & \dots & \dots & 1 \end{bmatrix},\ \text{and}\ Q'=\begin{bmatrix} u_{1+j_0} & u_{2+j_0} & \dots & \dots & u_{N+j_0} \\ v_{1+j_0} & v_{2+j_0} & \dots & \dots & v_{N+j_0} \\ 1 & 1 & \dots & \dots & 1 \end{bmatrix}.$$

For ease of discussion and without loss of generality, we have assumed that the $(i+j_0)$-th feature point (u_{i+j_0}, v_{i+j_0}) of the shape C' corresponds to the i-th feature point (x_i, y_i) of the shape C. Since $Q \cdot Q^T$ is a 3×3 matrix, its inverse matrix can be easily calculated. That means, affine transform A can be obtained immediately.

5.2 Similarity matching

Some examples on similarity matching are shown in figures 1-3. In each of these illustrations, the lines show the correspondence established between the feature points of the two shapes in consideration, with error measure shown at the bottom. From our experience, for very dissimilar shapes, the error measures typically larger than 0.5. The diagrams show matching results for different number of sample points along the curves.

Figure 1 shows the matching results of two similar fish. Our method gives good correspondence and consistent values (around 0.02) for the error measure across scales. Figure 2 shows two hammers that look very much alike, with one of them having a deformed handle. The deformation as well as the changes due to viewing parameters do not affect correct matching. Figure 3 shows that the error measure increases quickly as the degree of difference between the two shapes being compared increases. Some of the test input data used in the experiment were obtained from the author of [10].

6 Conclusion

In this paper, we have presented an area-based affine invariant shape representation scheme for content-based image retrieval. We also develop a point correspondence matching algorithm and a similarity measure for shape retrieval. Point correspondence of two shapes was formulated as an error minimization process of matching two feature vectors. Since our feature vectors capture both local and global affine invariant information of shapes is very robust to noise and shape deformation. Theoretical analysis and experimental results demonstrated the feasibility of our techniques.

References

1. B. M. Mehtre, M. S. Kankanhalli, A. D. Narasimhalu and G. C. Man, Color matching for image retrieval, *Pattern Recognition Letters*, 16(3), 325-331, 1995.
2. M. S. Kankanhalli, B. M. Mehtre, and J. K. Wu, Cluster-based color matching for image retrieval, *Pattern Recognition*, 29(4), 701-708, 1996.
3. G. L. Gimelfarb and A. K. Jain, On retrieving textured images from an image database, *Pattern Recognition*, 29(9), 1461-1483, 1996.
4. F. Liu and R. W. Picard, Periodicity, directionality, and randomness - Wold features for image modeling and retrieval, *IEEE Trans. on PAMI*, 18(7), 722-733, 1996.
5. R. Mehrotra and J. Gary, Similar-shape retrieval in shape data management, *Computer*, Sept. 1995.
6. K. Hirata and T. Kato, Rough sketch-based image information retrieval, *NEC Research & Development*, 34(2), 263-273, Apr. 1993.
7. E. Rivlin, and I. Weiss, Local invariants for recognition, *IEEE Trans. on PAMI*, 17(3), 226-238, March 1995.
8. A. Rattarangsi, and R. T. Chin, Scale-based detection of corners of planar curves, *IEEE Trans. on PAMI*, 14(4), 430-449, 1992.
9. F. Mokhtarian, Silhouette-Based isolated object recognition through curvature-scale space, *IEEE Trans. on PAMI*, 17(5), 539-544, May 1995.
10. A. Sclaroff, and A. P. Pentland, Modal matching for correspondence and recognition, *IEEE Trans. on PAMI*, 17(6), 545-561, June 1995.

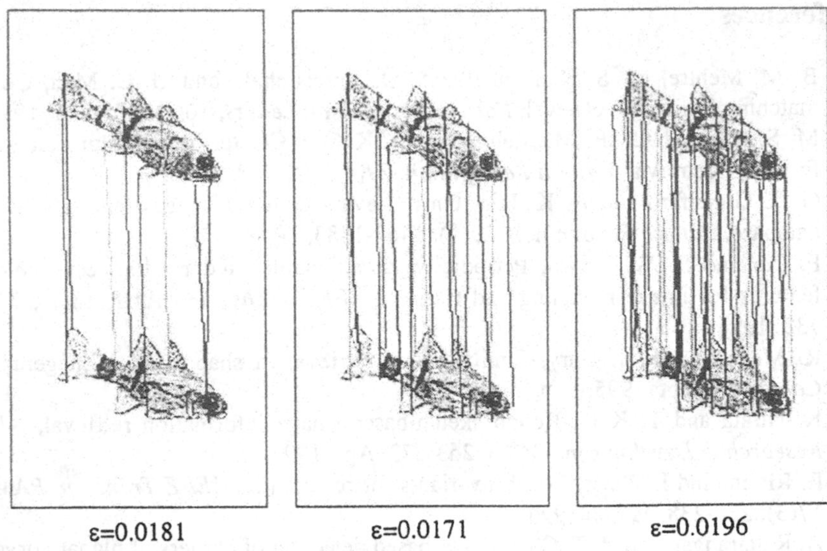

ε=0.0181 ε=0.0171 ε=0.0196

Figure 1 Comparing two similar fish.
The lines show correspondence while the figures show the error measures. Note also the consistency in error measures across scales (ie. for different values of *N*)

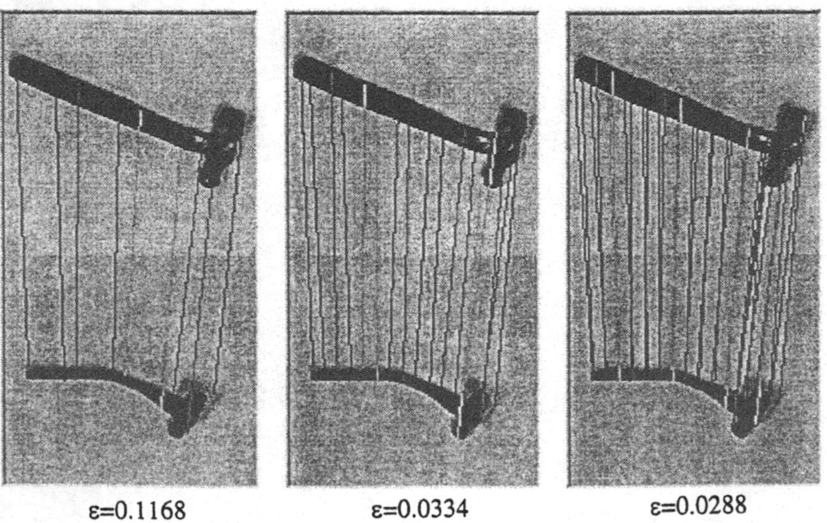

ε=0.1168 ε=0.0334 ε=0.0288

Figure 2 Comparing two similar hammers.
The lines show correspondence while the figures show the error measures. Note that the deformation does not affect correct matching.

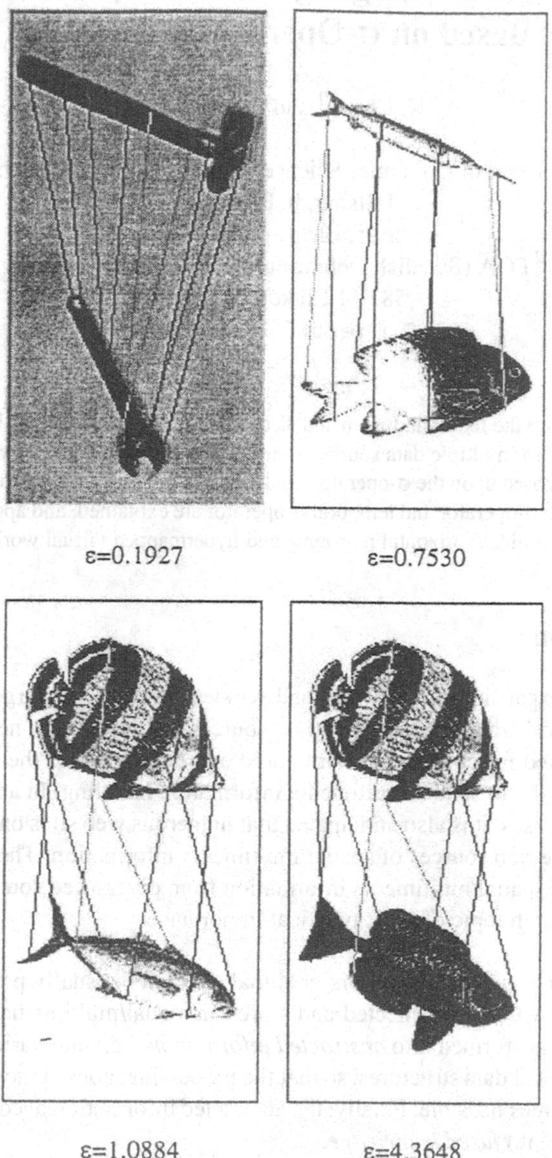

Figure 3 Comparing error measures on unlike shapes.
The error term grows as the shapes under comparison are more complex. This agrees with the general notion that shapes with more distinct features should be compared with greater confidence, even at lower resolutions.

A Spatial Query Language for Multiple Data Sources Based on σ-Operator Sequences

S.-K. Chang[1] and Erland Jungert[2]

[1] Dept. of Computer Science, University of Pittsburgh
Pittsburgh, PA 15 260
Chang@cs.pitt.edu
[2] FOA (Swedish Defence research Est.), Box 1165,
S-581 11 Linköping, Sweden
Jungert@lin.foa.se

Abstract: To support the retrieval, fusion and discovery of visual/multimedia information, a spatial query language for multiple data sources is needed. In this paper we describe a spatial query language which is based upon the σ-operator sequence and in practice expressible in an SQL-like syntax. The general σ-operator and temporal σ-operator are explained, and applications of the σ-query language to vertical/horizontal reasoning and hypermapped virtual world are explored.

1 Introduction

With the rapid expansion of the wired and wireless networks, a large number of soft real-time, hard real-time and non-real-time sources of information need to be quickly processed, checked for consistency, structured and distributed to the various agencies and people involved in spatial/multimedia information handling. In addition to spatial/multimedia databases, it is also anticipated that numerous web sites on the World Wide Web will become rich sources of spatial/multimedia information. The retrieval, fusion and discovery of spatial/multimedia information from diversified sources emerges as a challenging research topic of great practical importance.

Data sources such as camera, sensors or signal generators usually provide continuous streams of data, which are collected and stored in visual/multimedia databases. Such data need to be transformed into *abstracted information*, i.e., into various forms of spatial/temporal/logical data structures, so that the processing, consistency analysis and fusion of data becomes possible. Finally, the abstracted information needs to be integrated and transformed into *fused knowledge*.

Figure 1 illustrates the relationships among data sources, data, abstracted information and fused knowledge, with emphasis on the diversity of data sources and multiplicity of abstracted representations. For example a video camera is a data source that generates video data. Such video data can be transformed into various forms of abstracted representations including: text, keyword, assertions, time sequences of frames, qualitative spatial description of shapes, frame strings, and projection strings [1]. For example, to describe a frame C_{t_1} containing two objects 'a' and 'b', the text is "a is north-west of b", the keywords are {a, b}, and the assertion is (a northwest b). The x-directional projection string is (u: a < b). The time sequence of three frames C_{t_1} i, C_{t_2} i and C_{t_3} i is (t: $C_{t_1} < C_{t_2} < C_{t_3}$). Finally, the qualitative spatial description QSD of a rectangular object 'a' is (0,+,1)[=,<,2:3](0,+,1)[=,<,3:2](0,+,1)[=,>,2:3](0,+,1) [=,<,3:2] [2]. Some of these transformations will be explained later.

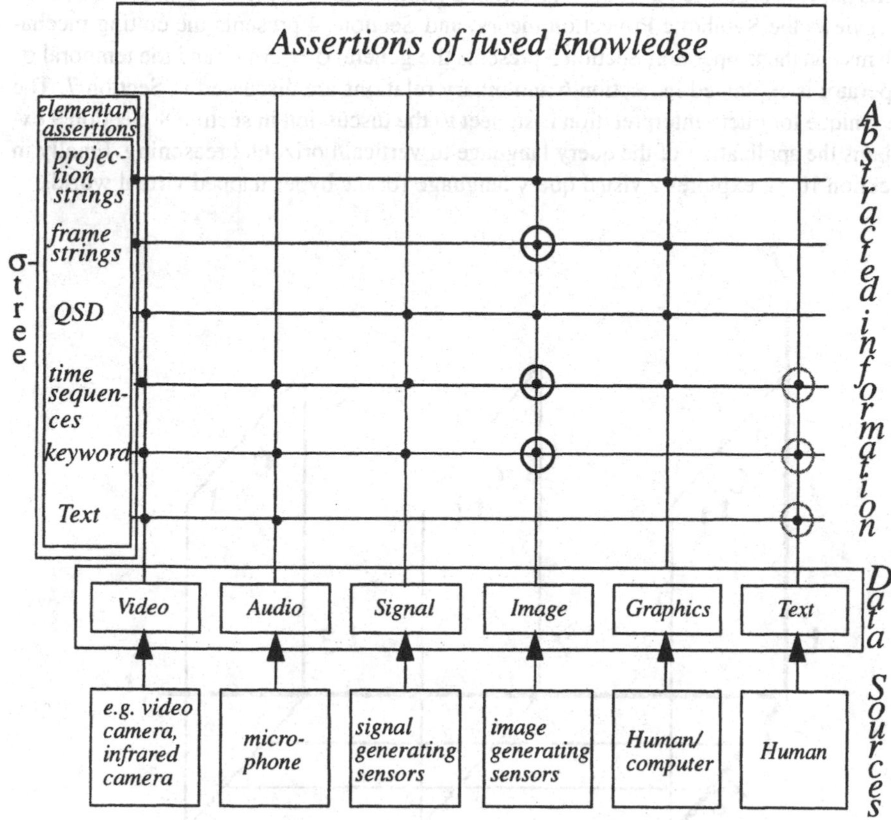

Fig. 1. A framework for information retrieval and fusion.

In Figure 1 a potentially viable transformation from data to abstracted representation is indicated by a black dot. Thus, from video it is possible to transform into almost all kinds of abstractions. A supported transformation is indicated by a large circle. Thus the image data will be transformed into keywords, assertions, qualitative spatial description and projection strings. The information sources may include hard real-time sources (such as the signals captured by sensors), soft real-time sources (such as pre-stored video), and non-real-time sources (such as text, images and graphics from a visual database or a web site).

To retrieve spatial/multimedia information and to discover relevant associations among media objects, a powerful spatial query language is needed. In this paper a spatial query language for multiple data sources is described. This spatial query language is based upon the σ-operator sequence. In practice the queries can be expressed in an easy-to-understand SQL-like syntax. It can be extended to a visual query language in the hypermapped virtual world environment.

This paper is organized as follows. Section 2 explains the concept of σ-queries. Section 3 reviews the Symbolic Projection theory, and Sections 4 presents the cutting mechanisms and the κ-operator. Section 5 presents the general σ-operator and the temporal σ-operator is explained in Section 6 and binary relations are discussed in Section 7. The technique for query interpretation is subject to the discussion in section 8. Section 9 explains the application of the query language to vertical/horizontal reasoning. Finally in Section 10 we explores a visual query language for the hypermapped virtual world.

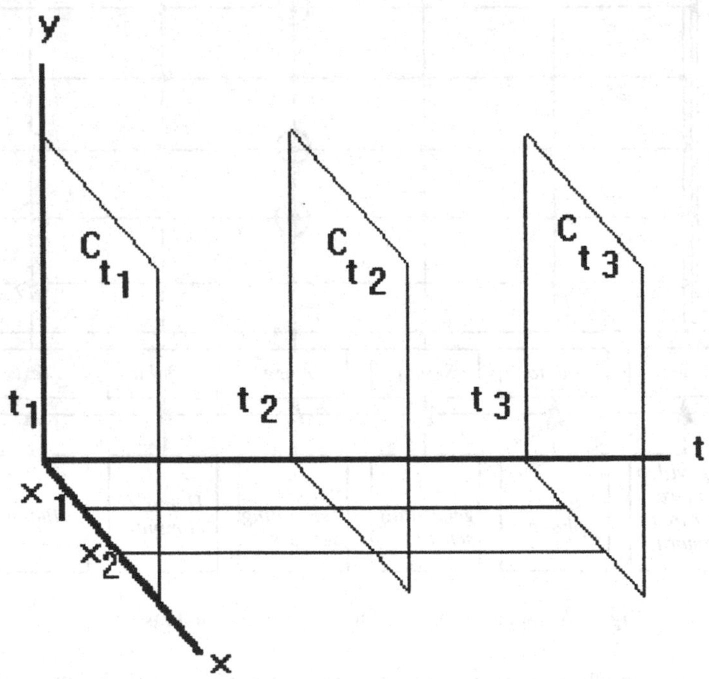

Fig. 2. Example of a video source consisting of three time slices (frames).

2 The σ-query concept

As mentioned in Section 1, the σ-query language is a spatial query language for information retrieval from multiple sources. Its strength is its simplicity: the query language is based upon a single operator - the σ-operator. Yet the concept is natural and can easily be mapped into an SQL-like query language. The σ-query language is useful in theoretical investigation, while the SQL-like query language is easy to implement and is a step towards a user-friendly visual interface.

A simple example is illustrated in Figure 2. The source R, sometimes also called a universe, consists of time slices of 2D frames. To extract three pre-determined time slices from the source R, the query in mathematical notation is:

$$\sigma_t(t_1, t_2, t_3)\ R$$

The meaning of the σ-operator in the above query is *select*, i.e. we want to select the time axis and three slices along this axis. The subscript t in σ_t indicates the selection of the time axis. In SQL-like language the query can be expressed as:

```
SELECT    t
CLUSTER   t1, t2, t3
FROM      R
```

A new keyword "CLUSTER" is introduced, so that the parameters for the σ-operator can be listed, such as t_1, t_2 and t_3. The word CLUSTER indicates that a cluster may have a sub-structure which can be subject to another (recursive) query. A cluster is always contained in a *frame*. The mechanism for clustering will be discussed later in Section 4. Each cluster can be *open* (objects inside visible) or *closed* (objects inside not visible). The notation is t_2^o for an open cluster and t_2^c or simply no superscript for a closed cluster. In the SQL-like language the keyword "OPEN" is used:

```
SELECT    t
CLUSTER   t1, OPEN t2, t3
FROM      R
```

With the above described notation, it is quite easy to express a complex, recursive query. For example, to find the spatial relationship of two objects 'a' and 'b' from the three time slices of a source R, as illustrated in Figure 3, the query in mathematical notation is:

$$\sigma_x(x_1^o, x_2^o)(\sigma_t(t_1, t_2, t_3)\ R)$$

In SQL-like language the query can be expressed as:

```
SELECT    x
CLUSTER   OPEN x1, OPEN x2
FROM SELECT   t
CLUSTER   t1, t2, t3
FROM      R
```

The extracted information is the spatial relationship of the objects 'a' and 'b'. How to express this spatial relationship depends upon the spatial data structure used. In the next section we explain Symbolic Projection as a means to express this spatial relationship.

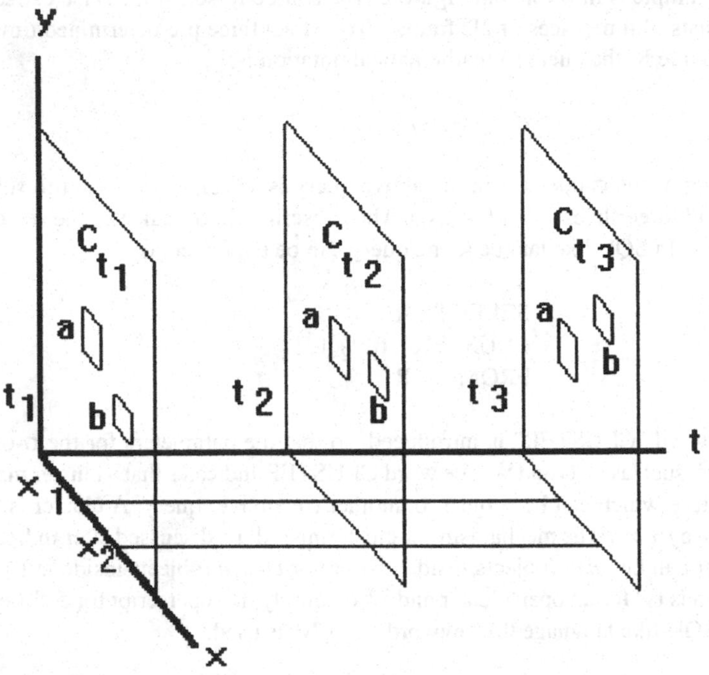

Fig. 3. Example showing details in the time slices (frames) of the video source.

3 Symbolic Projection for spatial queries

Symbolic projection, originally proposed by Chang et al. [3], is a formalism where space is represented as a set of strings. Each string is a formal description of space, including all existing objects, and their relative positions viewed along the corresponding coordinate axis of the string in a symbolic form.This representation is qualitative since it corresponds to sequences of projected objects and their relative relations. A simple example of the *Fundamental Symbolic Projection* is given in figure 4a, where the U-string corresponds to the projections along the x-axis and the V-string corresponds to the projections along the y-axis. Figure 4b illustrates the *Interval Projection* method [1].

We can use Symbolic Projection as a means for expressing the spatial relations extracted by a spatial query. In the σ-query, the operator σ indicates that the selection is made according to the information-loseless default clustering mechanism generated by the canonical cutting (see Section 4). To select the x-axis, we define $\sigma_x = \sigma_x(x_1, ..., x_n)$. We will also use the notation * to indicate default clustering: $\sigma_x = \sigma_x(*)$. With this notation, therefore, (σ_x, σ_y) is exactly the same as the pair of symbolic projections (u,v).

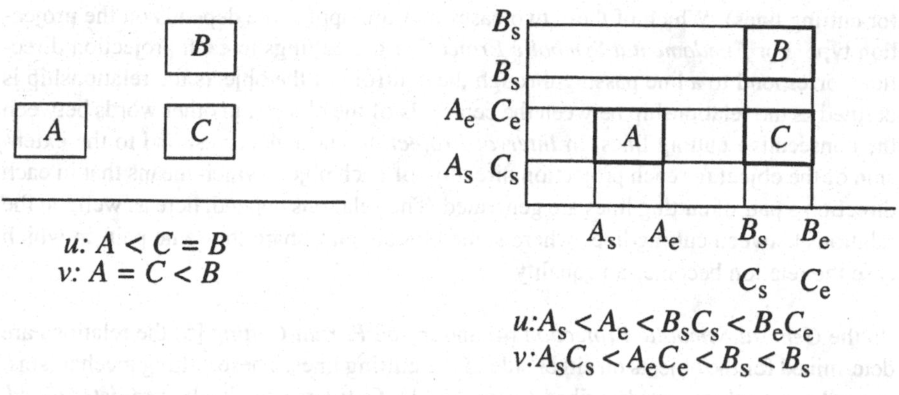

Fig 4. The original approach to symbolic projection including the resulting projection strings(a) and the same scene applied to interval projection (b).

Continuing the example illustrated by Figure 3, for time slice C_{t_1} its x-projection using the Fundamental Symbolic Projection is:

$$\sigma_x C_{t_1} = \sigma_x(x_1, x_2) C_{t_1} = (U: \mathrm{a} < \mathrm{b})$$

and its y-projection is:

$$\sigma_y C_{t_1} = \sigma_y(y_1, y_2) C_{t_1} = (V: \mathrm{b} < \mathrm{a})$$

The query $\sigma_t(t_1, t_2, t_3)$ R yields the following result:

$$(\mathrm{t}: C_{t_1} < C_{t_2} < C_{t_3})$$

Similarly, the query $\sigma_x\,(x_1{}^o, x_2{}^o)\sigma_t\,(t_1, t_2, t_3)$ R yields the following result:

$$(\mathrm{t}: (\mathrm{u}: \mathrm{a} < \mathrm{b}) < (\mathrm{u}: \mathrm{a} < \mathrm{b}) < (\mathrm{u}: \mathrm{a} < \mathrm{b}))$$

The query result depends upon the clustering mechanism, which can be specified by the cuttings, as described in the next section.

4 Cuttings

Cuttings are an important part of Symbolic Projection since a cutting determines both how to project and also the relationships at each cutting, which generally means the relationships between the objects or partial objects to the left and to the right of the cutting line. In other cases the relationships are determined between two consecutive cuttings (or cutting lines). Which of these two cases that are applicable depends on the projection type. For *Fundamental Symbolic Projection* the cuttings in each projection direction correspond to a line passing through the centroid of the objects the relationship is defined as the relationship between the centroids of the objects, in other words between the consecutive cutting lines. In *Interval Projection* cuttings correspond to the extension of the object for each projection direction of each object which means that in each direction a pair of cutting lines are generated. The relations depend, here as well, on the relations between cutting lines where some objects may share the same pair, in which case the relation becomes an equality.

In the *General Symbolic Projection* [4] and in the *Partial Cutting* [5] the relations are determined for the objects on either side of the cutting lines. These cutting mechanisms, as well some others, are described further in [1]. Cuttings can also be *pre-determined*, which is the case in most of the examples given in this paper; in that case the cuttings are given just as ordered lists where the cuttings, for simplicity, correspond to the Fundamental Symbolic Projection. In this perspective it is easy to see the need to *explicitly* determining which particular cutting mechanism that should be applied in a particular σ-query. This is the purpose of the κ-operator.

The κ-operator indicates the specific cutting mechanism and generates the corresponding cuttings according to the given cutting mechanism. Thus, the κ-operator should be applied to the source in focus. Cuttings along an arbitrary coordinate axis, here called the s-axis, of the source R thus becomes:

$$\kappa_s(\Xi)R = (\delta_0, \delta_1, \ldots, \delta_n)R = \Delta_s R$$

where $\Delta_s = (\delta_0, \delta_1, \ldots, \delta_n)$ and Ξ is the cutting mechanism. The above formula indicates a formal substitution of $\kappa_s(\Xi)$. by $(\delta_0, \delta_1, \ldots, \delta_n)$ in the context of the source R.

The κ-operators indicating cuttings along multiple coordinate axes are allowed as well, e.g. κ_{xy} indicates cuttings along both the x and y axes. However, the following relations between the different κ-operators must be observed:

$$\kappa_{s1s2}(\Xi)R = (\Delta_{s1}; \Delta_{s2})R$$

The canonical cutting mechanism is defined as follows:

$$\Delta s = (*)$$

where * indicates the set of cuttings used in the Fundamental Symbolic Projection.

Here, as well, clusters can be either open or closed: $\kappa_s^{o}(\Xi)R$ refers to a set of open clusters while $\kappa_s(\Xi)R$ refers to closed clusters.

5 The general σ-operator

The general σ-operator is defined by the following expression where, in order to make different cutting mechanisms available, the κ-operator is included:

$$\sigma_{\{string-type_i\}_1^k} \kappa\left(\{\Xi_i\}_1^k\right) \langle cluster-string \rangle$$

$$= stl_1 : \langle string_1 \rangle ; stl_2 : \langle string_2 \rangle ; \ldots ; stl_k : \langle string_k \rangle$$

The general σ-operator operates on arbitrary types of cluster strings (source) which may be an ordinary data source, i.e. data from a specified sensor. From such cluster strings new strings of different types can be generated by further σ-operator whose types are indicated by the given index in the σ-operator. For each string type a set of cuttings must be invoked either by the κ-operator or by the canonical cutting mechanism. The generated strings correspond to the common types of the fundamental/temporal projection strings:

x | y | z | t | xy | xz | yz | xt | yt | zt | xyt | xzt | yzt

Some observations, regarding the characteristics of the resulting projection strings after the application of the σ-operator to some arbitrary projection directions, can now made:

For cuttings in 3D or 2D + time space plane clusters are generated.
For multiple cuttings (σ_{s1s2}) in 3D or 2D + time space line clusters are generated.
For pairs of cuttings in 3D or 2D + time line clusters are generated.

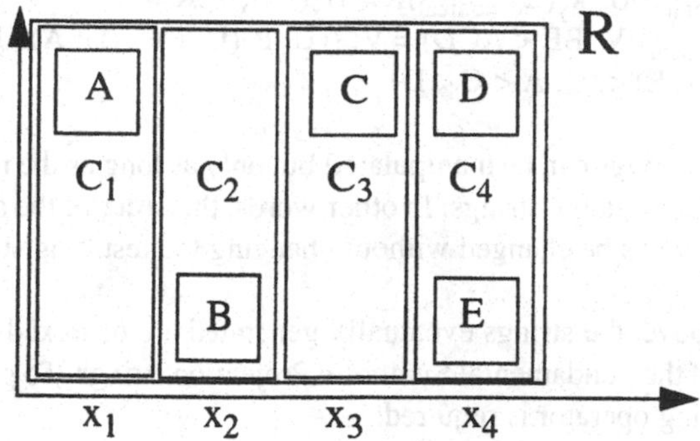

Fig. 5. A source, R, including five objects, A through E, and four clusters, C_1 through C_4, to which $\sigma_y\sigma_x$ is applied.

When applying σ_s to a 4D space (i.e. x,y,z,t) a 3D space is generated. In principle, when applying a σ-operator to a single coordinate axis of specified type, that axis is collapsed and a new sub-space perpendicular in all directions to that axis is created. This is controlled by κ and each sub-space either correspond to a cluster or can be handled as such a structure, in other words as a source. Thus the application of a σ-operator can be described in the following two steps:

Step1: Apply cutting operation (*unless pre-determined cuttings are applied*) to the source; thus generating cutting positions for new strings with open or closed clusters.

Step 2: Apply the operator σ_s to generate a string of type 's' where closed clusters are treated as non-decomposable objects while open clusters are treated as sets of objects.

An illustration of the usage of multiple σ-operators applied in sequence can be seen in Figure 5, where four cluster determines four cuttings in the x-direction whereas in the y-direction canonical cuttings are requested, i.e.:

$$\sigma_y\ \kappa_y(\Xi_{canonical})\sigma_x(x_1,x_2,x_3,x_4)R = \sigma_y\ \kappa_y(\Xi_{canonical})(U{:}\ C_1 < C_2 < C_3 < C_4)$$
$$= u{:}\ (v{:}A) < (v{:}B) < (v{:}C) < (v{:}\ D < E)$$

Here the clusters, C_1 through C_4, control the final string structure, but for certain cases there is no real difference; consider, for instance, the case where the canonical cuttings are applied, in other words:

$$\sigma_y\kappa_y(\Xi_{canonical})\sigma_x\kappa_x(\Xi_{canonical})R = \sigma_y(*)\sigma_x(*)R = \sigma_y(*)(U{:}\ A < B < C > DE)$$
$$= U{:}\ (V{:}\ A) < (V{:}\ B) < (V{:}C) < (V{:}\ D < E)$$

in which case the final result is the same. The opposite ordering of the operators, i.e. $\sigma_x\sigma_y$, gives the following projection strings:

$$\sigma_x\ \kappa_x(\Xi_{canonical})\sigma_y\ \kappa_y(\Xi_{canonical})R = \sigma_y(*)\sigma_x(*)R =$$
$$\sigma_x\kappa_x(\Xi_{canonical})(V{:}\ BE < ACD) = V{:}\ ((U{:}\ B)(U{:}\ E) < (U{:}\ A)(U{:}\ C)(U{:}\ D))$$
$$= V{:}((U{:}\ B < E) < (U{:}\ A < C < D)$$

To some extent the strings can be manipulated but only as long as the required information is present in the available strings. In other words, the order of the application of the operators cannot always be changed without changing the resulting strings.

In the examples above, the strings eventually generated are of mixed type and include just the x-strings of the Fundamental Symbolic Projection strings. To get *both* canonical strings, the following operator is required:

$$\sigma_{xy}\kappa(\Xi_x;\Xi_y)R = (U{:}\ A < B < C < DE)(V{:}\ BE < ACD)$$

Earlier variations of the σ-operator have been described in [6] and [1] and a more recent version close to the one here was described in [7].

6 The σ_t-operator

The σ_t-operator is used to generate *event projection strings* indicating possible changes in space over time where the events may take place between clusters. The σ_t-operator should primarily be applied to sequences of images and for this reason it may be used in temporal reasoning, i.e. reasoning about changes in space over time. The sequential sets of images may correspond to video sequences; the use of the σ_t-operator for this purpose can be illustrated as:

$$\sigma_{xy}\kappa_{xy}{}^{o}(\Xi_{canonical})\sigma_t(t_1,t_2,t_3)R$$
$$= (u: str_{xt1})(v: str_{yt1}) < (u: str_{xt2})(v: str_{yt2}) < (u: str_{xt3})(v: str_{yt3})$$

The events that may occur in the image sequences may be determined from the sequences of clusters, i.e. the sequences of (u,v) strings. Furthermore, the cuttings in this type of time sequences can either be determined by a κ-operator (for any specific cutting mechanism) by pre-determined cuttings or by the canonical cuttings. In video sequences, the latter alternative means all existing frames in a given time interval are selected.

7 Binary Relations as Assertions

String types corresponding to binary relations, such as, for instance, direction, distance or topological relations, are to a certain degree different compared to the strings of elementary type. Such binary relations can be represented as strings, as well, which will different compared to the traditional projection strings in that they basically are inferred from the latter by means of various, in many cases, simple inference rules. Consequently, although binary relations are assertions in the normal sense we can also regard them as fused knowledge since they correspond to local object relations inferred from global assertion structures like the Fundamental Symbolic Projection strings. Furthermore, the resulting relations from one query can be extended stepwise with further relations from later queries using the same information sources. This fits well with the blackboard architecture, described in [7]. Therefore a set of binary directional relations may look like:

$$\text{direction: } (a_1\ rel_1\ a_2)\ \&\ (a_3\ rel_2\ a_4)\ \&\ \ldots\ldots$$

where rel ={north, north-east, east, south-east, south, south-west, west, north-west, same_location} [8] and where & is just a delimiter

A σ-expression for the determination of directional relations could be:

$$\sigma_{find\text{-}direction}(\Delta_{direction})\sigma_{xy}\kappa(\Xi_x;\Xi_y)\sigma_t(\Delta_t)\sigma_{source}\kappa(\Delta_{media\_source})R$$

where the query is applied to the media source R to select temporal and then spatial projection strings, and finally extract directional strings. The $\Delta_{direction}$, Δ_t and Δ_{media_source}, for instance, could be the pre-determined cuttings, such as $\Delta_{direction}$={west, north}, $\Delta_t = (t_0, ..., t_n)$ and Δ_{source}= radar. An example of a returned direction string could be:

(A west F) & (C west F) & (D north G)

This result could be combined with assertions concluded from earlier queries.

The $\sigma_{find\text{-}direction}$-operator refers to the *find* query that was discussed in [7]. Similar references to the basic query types that were described in that work can now be applied by defining suitable σ-operators, that eventually will return fused knowledge corresponding to those query types which are demonstrated in Table 1 [7].

query type	σ-operator	meaning
Are-there-any	σ_{ata}	object classification
which	σ_{which}	attributes
where	σ_{where}	locations/positions
when	σ_{when}	events
how	σ_{how}	moving patterns
find relation	σ_{find}	object relations
orientation	$\sigma_{orientation}$	object orientation

With this open set of operators, the σ-operators can, as demonstrated by the $\sigma_{find\text{-}direction}$-operator, be extended further, e.g. $\sigma_{which\text{-}shape}$ will return the shape description of the current object.

8 The σ-query interpreter

The syntax of σ-query language is as follows:

```
<query>::= SELECT  <select_target>
CLUSTER <cluster>
FROM    <source>
WHERE   <condition>
<select_target>::= x | y | z | t | image_object | audio_object | video_object
                   | type
<cluster>::=  * | <x_cluster> | <y_cluster> | <z_cluster> | <t_cluster>
<x_cluster>::=  <x_val> | <x_val> <x_cluster>
<x_val>::= <val> | OPEN <val>
<source>::= <query> | <source_name>
<condition>::=  <string>
```

σ-queries can be interpreted and translated into Fundamental Symbolic Projection strings as explained below. In the query $\sigma_t(t_1, t_2,..., t_n)$ R, R is a cluster string representing a source, and t is not necessarily time but merely a formal parameter. The above query is interpreted as follows:

Procedure Query_Interpretor(σ_t $(t_1, t_2,..., t_n)$ R)
Input: (1) A multimedia static information structure MSS
(2) A σ-query

Output: The retrieval results
Step 1: Apply $\kappa_t(\Xi)$ to each existing (canonical) cluster in R to find its sub-clusters C_{t_1}, C_{t_2}, ... ,C_{t_n}.
Step 2: For each sub-cluster C_i, if t_i^c is closed it is treated as a single object. If t_i^o is open it is treated as a set of objects.
Step 3: Encode these clusters according to σ_t and return the result.

The above algorithm is precursive, i.e., each R may itself be the form $\sigma_w(w_1, w_2,..., w_n)Q$ and can be evaluated recursively.

The multimedia static information structure MSS consists of a list of nodes encoded as follows: (node_name; media_type; child_node_list; description).

Example 1:

Input MSS:

(R; video; R_1, R_2, R_3; (t: R_1<R_2<R_3))
(R_1; frame; null; (t:t_1) and (U:a<b) and (V:a<b))
(R_2; frame; null; (t:t_2) and (U:a<b) and (V:a=b))
(R_3; frame; null; (t:t_3) and (U:a<b) and (V:a>b))

Input Query:

SELECT t
CLUSTER t_1, t_2, t_3
FROM R

Output: $(t: R_1 < R_2 < R_3)$

Example 2:

Input MSS: Same as that of Example 1

Input Query:

SELECT t
CLUSTER t_1, OPEN t_2, t_3
FROM R

Output: $(t: R_1 < ((t{:}t_2) \text{ and } (u{:}a<b) \text{ and } (v{:}a=b)) < R_3)$

Example 3:

Input MSS: Same as that of Example 1
Input Query:

SELECT x
CLUSTER OPEN x_1, OPEN x_2
FROM SELECT t
CLUSTER t_1, t_2, t_3
FROM R

Output: $(t{:}\ (u{:}a < b) < (u{:}a < b) < (u{:}a < b))$

Example 4:

Input MSS:

(n_1; composite; n_2, n_3, n_4; null)
(n_2; video; v_1, v_2, ..., v_n; ($t{:}v_1<v_2<...<v_n$))
(n_3; audio; a_1, a_2, ..., a_n; ($t{:}a_1<a_2<...<a_n$))
(n_4; text; null; ($keyword_1$, $keyword_2$))

Input Query:

SELECT t
CLUSTER OPEN *
FROM n_1

Output: $(t{:}\ (v_1\ a_1) < (v_2\ a_2) < \ldots\ (v_n\ a_n))$

9 Horizontal and vertical reasoning

As illustrated in Figure 1, knowledge retrieval, discovery and fusion can be accomplished when there are several circles in the same horizontal row. For example if transformation of image and text into assertions and keywords is supported, consistency checking of assertions or keywords become feasible. Such reasoning is called the *horizontal reasoning* because it combines information along horizontal rows. Another type of reasoning works with different abstracted representations so that they can be combined and checked for consistency. Such reasoning is called the *vertical reasoning* because it combines information along vertical columns.

As an example patient information are abstracted from different media sources, including imaging devices, signal generators, instruments, etc. (*vertical reasoning*). Once abstracted and uniformly represented, the decision maker is invoked to make a tentative diagnosis *horizontal reasoning*. Using the σ-query, similar patient records are found (*vertical reasoning*). A retrieved patient record is compared with the target patient record *horizontal reasoning*. If similar records lead to similar diagnosis then the results are consistent and the patient record (with diagnosis) is accepted and integrated into the knowledge base. If the diagnosis is different then the results are inconsistent and the negative feedback can also help the decision maker learn. In the vertical reasoning phase, in addition to comparing patient data, we can also compare images to determine whether we have found similar patient records. This example illustrates the *alternating application of horizontal reasoning and vertical reasoning.*

Using the σ-query, a vertical reasoning example is presented below:

$$\sigma_x (x_1^o, x_2^o)\sigma_t(t_1, t_2, t_3) \text{ media\_source}$$

In vertical reasoning we first extract time slices from media_source and then extract the following spatial information: two open clusters along the x axis. The query can be expressed as:

```
SELECT    x
CLUSTER   OPEN x1, OPEN x2
FROM SELECT    t
CLUSTER   t1, t2, t3
FROM      media_source
```

Similarly, a horizontal reasoning example is as follows:

$$\sigma_{type}(c_1, c_2)\ (\sigma_t(t_1, t_2, t_3)\ (\sigma_{audio\text{-}object}(*)\ \text{audio\_source})\ \text{or}$$
$$(\sigma_t\ (t_1, t_2, t_3)\ (\sigma_{video\text{-}object}(*)\ \text{video\_source})$$

In horizontal reasoning we first extract time slices of objects from audio source and the corresponding time slices of objects from video source, combine such information, and then extract two types of objects. The query can be expressed as:

```
SELECT   type
CLUSTER c1, c2
FROM (SELECT    t
CLUSTER   t1, t2, t3
FROM    SELECT audio-object
CLUSTER *
FROM    audio_source
OR
SELECT    t
CLUSTER   t1, t2, t3
FROM    SELECT video-object
CLUSTER *
FROM    video_source)
```

In horizontal reasoning the fused information must be at the same level of abstraction. In the above example audio source and video source are different. After objects are extracted then they can be combined. Indeed, with the appropriate σ-operators, shape, orientation, size, etc. can all be adjusted so that objects become consistent and fusion is possible.

10 Conclusions

In this paper a spatial query language based on σ-operator sequences was presented. The σ-query language is a spatial query language for information retrieval from multiple sources. Its strength is its simplicity: the query language is based upon a single operator - the σ-operator. Yet the concept is natural and can easily be mapped into an SQL-like query language. The σ-query language is useful in theoretical investigation, while the SQL-like query language is easy to implement and is a step towards a user-friendly visual interface.

This new query language is specially suitable to the Hypermapped Virtual World (HVW) information model [9]. The σ-query may serve as the basis of a visual query language for the Hypermapped Virtual World. Hypermaps can be used advantageously as a metaphor for the representation of all the multimedia hyperbase elements. In Geo-Anchor [10] a map can be built dynamically as a view of the multimedia hyperbase. Each displayed geometry is an anchor to either a geographic node or to a related node. Hence, the map on the screen acts both as an index to the nodes and as a view to the multimedia hyperbase. In a Virtual Library a hypermap can also be used as a metaphor to link the most frequently accessed items such as reading rooms, book shelves, etc. to present different views to the end user. This combined metaphor of Hypermapped Virtual Library (which is a combination of the VR information space and the logical information hyperspace [11] may lead to efficient access of multimedia information in a digital library.

References

1. Chang, S.-K. and Jungert, E.: Symbolic Projection for image information retrieval and spatial reasoning. Academic Press (1996) 62-70
2. Jungert, E.: A qualitative approach to recognition of man-made objects in laser-radar images. Proceedings of the conf. on Spatial Data Handling, Vol II, Delft, The Netherlands, August 13-16 (1996) A15-A26
3. Chang, S.-K. , Shi, O. Y. and Yan, C. W.: Iconic indexing by 2D strings. IEEE transactions on Pattern Analysis and Machine Intelligence (PAMI) 9 (3) (1987) 413-428
4. Chang, S.-K. and Jungert E.: Pictorial data management based upon the theory of Symbolic Projection. Journal of Visual Languages and Computing, vol. 2, no 3 (1990) 195-215
5. Lee, S.-Y. and Hsu, F.-S.: Spatial reasoning and similarity retrieval of images using 2D C-string knowledge representation. Pattern Recognition, vol. 25 (1992) 305-318
6. Jungert, E. and Chang, S.-K.: The σ-tree - a symbolic spatial data model. Proceedings of the 11th IAPR International Conference on Pattern Recognition, The Hague, The Netherlands, Aug 30 - Sept. 3 (1992) 461-465

7. Chang, S.-K. and Jungert, E.: Human and system directed fusion of multimedia and multimodal information using the σ-tree data model. Proceedings of the 2nd International conference on Visual information systems, San Diego, CA, December 15-17 (1997) 21-28
8. Frank, A. U.: Qualitative spatial reasoning about cardinal directions. Autocarto 10, Baltimore, MD (1991) 148-167
9. Chang, S. K.: Media Enhanced Input/Output Metaphors For Content-Based Access to Multimedia Documents. Department of Computer Science, University of Pittsburgh, Technical Report, December (1997)
10. Caporal, A J., Viemont, A Y. A Y.: Maps as a Metaphor in a Geographical Hypermedia System. Journal of Visual Languages and Computing, vol. 8, No 1, February (1997) 3-25
11. Chang, S. K. and Costabile, M. F.: Visual Interface to Multimedia Databases. in Handbook of Multimedia Information Systems, W. I. Grosky, R. Jain and R. Mehrotra (eds.) Prentice Hall (1997) 167-187

Video Segmentation Using Color Difference Histogram

C.F. Lam and M.C. Lee

Department of Computer Science and Engineering
The Chinese University of Hong Kong
Shatin, N.T., Hong Kong
{cflam,mclee}@cse.cuhk.edu.hk

Abstract. This paper proposes a video segmentation algorithm based on a color difference histogram (*CDH*) which is insensitive to illuminations, object motions and camera movements. The relative high performance of the algorithm relies to some extent on the newly devised video scene detection method (*DD*) based on the analysis of the changes of video frame differences. We have identified characteristic patterns for the changes of frame difference values around the frame positions involving a scene break, or a flashlight. The paper demonstrates experimentally that the proposed algorithm out-performs other existing algorithms and that the *DD* method can identify flashlights besides detecting scene breaks.

1 Introduction

With the recent advances in digital storage and compression technologies, application such as video on demand, video databases, etc., which often involves a voluminous data size, has become more feasible and popular than ever before. These applications often support some form of content-based search and retrieval. To facilitate such content-based search and retrieval operations, a digital video stream should be indexed. The content-based indices are often based on one or more image features. Before indexing, a video stream should be first segmented into its constituent video scenes. Therefore, video segmentation has become an important research topic.

During the video segmentation process, certain feature values are first extracted from each video frame. Then inter-frame differences, commonly called frame differences, are computed by using the extracted feature values for consecutive frames. Finally, the frame differences are analyzed to determine scene change locations. The overall video segmentation process is as depicted in Fig. 1.

There are two ways to define a video scene, namely *logically* or *physically*, depending on whether or not camera movements are considered. If defined logically, a video scene will consist of a sequence of frames which do not have significant changes in object contents and in background of the scene, irrespective of the number of different camera shots it is composed of. On the other hand, if camera movements are considered, a video scene corresponds to one camera shot which

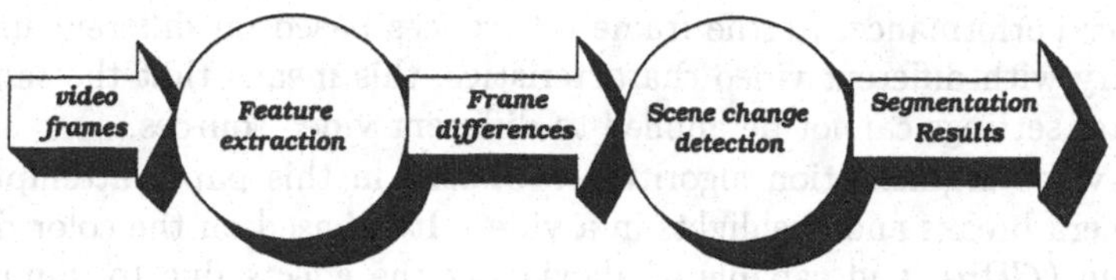

Fig. 1. Video segmentation process

may involve camera panning or zooming. In this paper, we adopt the later definition, where a video scene involves either camera panning from side to side, up and down or in any fixed direction, or a zoom in resulting in a close-up view, or a zoom out resulting in a distant view.

A transition from one scene to another is called a scene change. There are two kinds of scene changes, namely *camera break* and *gradual transition* such as fade-in, fade-out, dissolve or wipe. Typically, camera break is the most common type of scene change in any video, which involves an abrupt change in image feature contents between two consecutive frames. Gradual transitions, however, are edited effects involving usually a slow successive change in the image contents in a sequence of not less than 2 frames. In addition to the above two kinds of scene changes, sudden or gradual variations in brightness often exist in most video scenes. For example, a camera flash or any other light source can cause variations in intensities to the sequence of frames being taken. Any such sudden increase in intensities tends to make a video segmentation algorithm falsely detect a scene break. Of course, for a frame sequence being lighted by a camera flash or other light sources, a scene change should not be declared if the same frames, not subjected to the said camera flash or light sources, are all similar to each other. Otherwise, a scene change should be declared despite the introduction of the noise caused by the lighting effect.

In recent years, different video segmentation methods have been proposed. These methods can be classified into different categories based on the image features used to compute the frame differences. These categories include pixel-based comparison ([5, 7, 8]), histogram-based comparison ([5, 8]), DCT-based comparison ([3, 9]) and color ratio-feature based comparison ([1]). Ahanger and Little [2] present a survey of the existing video segmentation algorithms, and we have investigated some of the algorithms in depth, showing that there are still problems inherent in many of the currently available segmentation algorithms. The frame differences based on pixel comparison, histogram comparison, DCT comparison ([3, 5, 7–9]) are found to be sensitive to illumination changes, fast camera movements and object motions. Adjeroh, et al. [1] propose a segmentation method using the neighborhood color ratio feature to alleviate foregoing problems. However, the method involves a much computation overhead. Zhang, et al. [8] and Adjeroh, et al. [1] propose different methods to automate the detection of video scene cuts. However, they are not robust enough since they require human interaction to fine tune various parameters such as the global or local thresholds for better performance. As the frame differences based on different image features vary with different video characteristics, this means that the same set of parameter settings cannot be applied to different video sources.

The video segmentation algorithm proposed in this paper attempts to detect camera breaks and flashlights in a video. It is based on the color difference histogram (*CDH*), and capable of alleviating the effects due to illuminations, object motions and camera movements. The proposed frame difference analysis method, called *DD*, declares scene cuts and flashlights by identifying the respective characteristic patterns of changes of frame differences around the scene cut and flashlight locations.

This paper is organized as follows. First, the related work is given in Sect. 2. Then, the color difference histogram is defined in Sect. 3. Section 4 presents the detail of the *DD* scene change detector. Experimental results are presented in Sect. 5. Finally, conclusion and future work are given in Sect. 6 and Sect. 7 respectively.

2 Related Work

Ahanger and Little [2] present a survey on various existing segmentation methods, which can be grouped into different categories based on the image features used to compute the frame differences of a video. Such methods include pixel-based comparison, histogram-based comparison, DCT-based comparison, and color ratio based feature comparison. There are also different frame difference analysis methods for locating the scene change positions and identifying the types of scene changes. Despite the numerous number of proposals available, as outlined below, there are inherent problems in many of the current video segmentation methods.

(a) **Pixel-based Comparison:** In this category of segmentation methods, the pixels of two consecutive frames are compared pair-wise. For uncompressed video frames, Nagasaka and Tanaka [5] use the sum of pixel gray level or color differences to represent amount of content changes between two frames. Zhang, et al. [8] compare the pixels pair-wise with a user-defined threshold t to determine if there is a change for each pair of pixels from two frames. The proportion of changed pixels is then used to represent the content difference between the two frames. As for compressed images, Yeo and Liu [7] generate a DC-image sequence for an MPEG-I [4] compressed video source and apply pair-wise pixel comparison to two consecutive dc-images to obtain the frame difference. As these methods compare corresponding pixels across two frames, one major problem is that their results would be very sensitive to camera and object motions.

(b) **Histogram-based Comparison:** The video segmentation process of this category of methods involves the generation of gray level or color histograms, one for each frame, and a pair-wise comparison of the histogram bins ([5, 8]). Nagasaka and Tanaka [5] tested both gray level and color histograms and found that the regional χ^2 test on color histograms give the best segmentation results with high robustness to momentary noises. Notwithstanding the histogram-based methods are less sensitive to camera or object movements, they are quite sensitive to changes in brightness.

(c) **DCT-based Comparison:** Arman, et al. [3], Zhang, et al. [9] use the DCT coefficients extracted from motion JPEG or MPEG-I video to evaluate the content difference between two frames. In Arman's method [3], each frame is represented by a vector constructed from a subset of DCT coefficients of a subset of blocks. Since the DCT coefficients are mathematically related to the spatial domain, the correlation between two frames, represented by a dot product of the two vectors, is used to determine whether the contents of two frames are similar. Instead of using the dot product of DCT coefficients, Zhang, et al. [9] compute the frame differences by comparing pair-wise the DCT coefficients between two frames. However, the performances of these methods are also affected by illumination variations.

(d) **Color-ratio Based Comparison:** Adjeroh, et al. [1] propose a neighborhood color ratio feature to cope with both illumination changes and motions in a video. The method has also been extended to apply to transform domain images. When compared with pixel-based, histogram-based and DCT-based methods, color ratio based methods are more complicated and require higher computation overhead.

Concerning the detection of scene changes based on frame differences, Zhang, et al. [8] provide a statistical method for automatic selection of a global threshold. However, a fixed threshold may not be effective in detecting scene cuts since the noises introduced by camera and object movements can cause false alarms. To eliminate such noises, Otsuji and Tonomura [6] use a projection-detecting filter. But this filter is more computationally expensive than the simple global threshold method since, in order to get the filtered difference sequence, projection-detecting filter requires about 5 passes of computation and each stage needs different parameter settings. Another approach, known as sliding window, is proposed by Yeo and Liu [7] who view a scene change as a local activity. They also extended their method to detect gradual transitions and flashlights. However, this method also requires the tuning of several parameters for better performance.

3 Color Difference Histogram

One useful feature of color histograms is that they do not capture information about the spatial structures in an image. Thus, color histograms are relatively robust to object motions and camera movements. However, they are affected by intensity changes. Nagasaka and Tanaka [5] use a regional approach and ignore those regions with large histogram differences to eliminate the illumination effects. Nevertheless, based on our experiments, their approach results in more misses. Adjeroh, et al. [1] tackle the problems of illumination changes and motion in the video by using the neighborhood based color ratios. However, the relatively high complexity of their method makes it not a very good candidate for practical use. This section introduces a frame difference metric using a two-dimensional color difference histogram (*CDH*) for video segmentation. The proposed metric involves a relatively low time complexity. Moreover, it is insensitive to illumination changes, and invariant to translation, orientation and motion.

3.1 Color Differences and CDH

Unlike other color histogram based video segmentation methods which make use of the actual red (*R*), green (*G*) and blue (*B*) values of an image to generate color histograms, the proposed *CDH* method removes the brightness information from *R*, *G* and *B* components by making use of the two ***color differences***, namely *R-G* and *G-B*, to produce a two-dimensional ***color difference histogram*** (*CDH*) for each uncompressed video frame.

For a given 24-bit *RGB* color image *I* with $X \times Y$ pixels, its color difference histogram (CDH_I) can be generated by counting the number of pixels with a specific pair of color differences (*R-G*, *G-B*) as described below. For each pixel at (x, y), the two color differences are evaluated by the following formula

$$rg_I(x,y) = \left\lfloor \frac{R_I(x,y) - G_I(x,y) + 255}{Q} \right\rfloor . \tag{1}$$

$$gb_I(x,y) = \left\lfloor \frac{G_I(x,y) - B_I(x,y) + 255}{Q} \right\rfloor . \tag{2}$$

where $rg_I(x,y)$ and $gb_I(x,y)$ denote respectively the color differences between red and green and between green and blue components of the pixel (x,y) of the image *I*; $R_I(x,y)$, $G_I(x,y)$ and $B_I(x,y)$ represent the red, green and blue intensity values of the pixel (x,y) respectively; *Q* is a quantization factor to quantize the number of levels of color differences to a user-defined constant *L* such that $0 \leq rg_I(x,y) \leq L-1$ and $0 \leq gb_I(x,y) \leq L-1$. For instance, when $Q = 16$, the color differences will be quantized to have $L = 32$ discrete levels.

Then, we can generate the two-dimensional color difference histogram for the image *I*, $CDH_I(0..L-1, 0..L-1)$, as follows:

```
for r = 0 to L − 1 do
  for s = 0 to L − 1 do
    CDH_I(r, s) ← 0
  end for
end for
for x = 0 to X − 1 do
  for y = 0 to Y − 1 do
    increment the value of CDH_I(rg_I(x, y), gb_I(x, y)) by 1
  end for
end for
```

3.2 Inter-frame Similarity Measure

Traditionally, inter-frame differences $\{D_i\}$, $0 \leq i \leq N-1$, are formed by summing the pair-wise histogram bin differences for each pair of consecutive frames of a video with N frames. However, the *CDH* method uses the similarity-based approach; it determines the similarity S between two video frames by finding the intersection of two color difference histograms for two successive frames:

$$S_i(f_i, f_{i+\phi}) = \frac{\sum_{r=0}^{L-1} \sum_{s=0}^{L-1} \min\left(CDH_{f_i}[r,s], CDH_{f_{i+\phi}}[r,s]\right)}{X \times Y} . \qquad (3)$$

where f_i and $f_{i+\phi}$ denote respectively the ith and $(i+\phi)^{th}$ frames of a video and ϕ is a temporal skip factor. S_i will have high values, typically greater than 75% (with Q set to 16), for frames with similar contents. In general, frames within a scene usually have similar contents, thus a scene change may be indicated by an abrupt decrease in the values of S_i.

The capability of (3) in discriminating images and the amount of time needed to compute S_i depend on the size of the color difference histograms, $L \times L$, which in turn depends on the value of the quantization factor Q. With large values of Q (e.g. above 50), smaller sized histograms with less detail are generated. This results in faster computation of S_i. However, S_i will tend to be less sensitive to changes in image contents in the two images, giving relatively high values of S_i even for images with quite different contents. On the other hand, if Q is set to very low values such as 1, 2 or 4, the histograms will have larger number of bins and hence contain greater detail. However, it will require more time to obtain S_i. In practice, we may set the value of Q as 2^l, where l ranges from 3 to 5, so that the *CDH*'s will have a moderate size and capture adequate image detail, and S_i's can be computed in a moderate time. In addition, the division of color difference by Q can be efficiently implemented as a bit-shifting operation.

3.3 Orientation and Motion Invariant

When generating the color difference histograms, the spatial structures exhibited in an image are lost. Thus, S_i computed by using (3) will be less sensitive to object movements such as translation or rotation when compared with the frame differences based on ordinary pixel-based methods where two images are compared pixel by pixel.

Figure 2 shows four frames extracted from a camera panning scene (frames 444, 453, 462, 472 of a sample MPEG-I video *ad1.mpg*). During the panning, the camera view was gradually changing with some objects disappearing from the view while some showed up; for example, the motorcycle and the driver gradually showed up in frames 444–472. The background remained more or less the same. The similarities among these four frames using three settings of Q (8, 16, 32) are as shown in Table 1, which shows that the similarity value for any pair of the frames within the panning scene is quite high. It thus demonstrates that *CDH* is invariant to camera or object movement.

Fig. 2. Frames 444, 453, 462 and 472 of a camera panning sequence in *ad1.mpg*

Table 1. Similarities of frames for the camera panning scene in Fig. 2

		Similarity (%)					Similarity (%)		
Frame 1	**Frame 2**	$Q = 8$	$Q = 16$	$Q = 32$	**Frame 1**	**Frame 2**	$Q = 8$	$Q = 16$	$Q = 32$
444	453	89.7	91.8	97.4	453	462	88.7	91.5	95.9
444	462	82.9	86.5	94.5	453	472	83.4	85.5	90.6
444	472	77.4	79.5	89.0	462	472	89.7	90.5	93.7

3.4 Insensitivity to Illumination Changes

Since each color component contains brightness information, the frame differences computed by various existing segmentation methods such as pair-wise pixel comparison ([5, 8]), histogram comparison ([5, 8]) and DCT-based comparison ([3, 7, 9]) would be affected by intensity changes across successive frames. The subtraction performed in (1) and (2) to obtain color differences between two separate color intensities tends to cancel out the effects of intensity variation of pixels across different frames.

Figure 3 shows eight frames (frames 425–432 of the video *ad1.mpg*), one subsequence with increasing brightness and the other sub-sequence with decreasing brightness. There is a strong flashlight at frame 428 and also a scene change at frame 429. Some of the similarities among the eight frames are tabulated in Table 2. The relatively high frame similarities among frames 425–427 (with increasing intensity) and among frames 429–432 (with decreasing intensity) indicate that *CDH* is relatively illumination invariant. However, the low similarity between frames 427 and 428 shows that *CDH* cannot cope with a strong flash. Nevertheless, this problem can be resolved by the newly devised scene change detection method (*DD*) discussed in Sect. 4.

4 DD Scene Change Detector

In this section, we propose a relatively powerful scene-cut detection method called *DD*, based on the analysis of the changes of frame differences. The pros and cons of two existing methods to detect camera breaks are identified first before presenting the *DD* method.

Fig. 3. Frames 425–432 of the video *ad1.mpg*

Table 2. Similarities of frames with different brightness in Fig. 3

Frame 1	Frame 2	Similarity (%)			Frame 1	Frame 2	Similarity (%)		
		$Q = 8$	$Q = 16$	$Q = 32$			$Q = 8$	$Q = 16$	$Q = 32$
425	426	89.1	90.4	94.7	429	430	78.0	84.0	94.9
425	427	81.3	82.2	88.7	429	431	68.4	73.3	88.0
426	427	88.3	89.9	92.5	430	431	81.7	84.5	90.5
427	428	45.9	59.1	66.6	430	432	74.7	77.5	88.4
428	429	44.0	58.0	66.1	431	432	89.6	91.6	95.2

4.1 Global Threshold Method

With this method, a scene cut will be declared at frame f_{i+1} whenever the frame difference D_i is greater than a pre-set threshold T_b. This method assumes that all frames having a scene break exhibit a clear boundary from other frames in terms of the frame difference values D_i's; that is, the D_i's of the scene break frames are all larger than those of the other frames, meaning that scene break frames have their D_i's all larger than a certain threshold T_b.

However, the frame differences computed based on pixel comparison, histogram comparison or DCT coefficient comparison are so greatly affected by illumination changes, camera and object motions that they fail to have the foregoing property. Thus, a global threshold method may result in many false alarms if the threshold is set too low or many dismissals if the threshold is set too high. In addition, a different T_b should be set for each individual video since different videos tend to have different characteristics.

For example, Fig. 4 shows a plot of frame differences (a *FD*-plot) over a selected video sequence for the *CDH*-based method and three other tested algorithms (*U1, U2* and *U3*) introduced in Sect. 5. In a *FD*-plot, a peak can be interpreted as an occurrence of a scene cut while two consecutive peaks usually indicate an occurrence of a flashlight. For the particular case depicted in Fig. 4, *CDH* has the smallest frame difference values among other methods. So

T_b should be set to a small value, say 20, for scene cut detection. For *U1*, it has the highest frame difference values among other methods except for the segment D_{490}–D_{491}, thus a bigger T_b, say 70, should be employed. For *U2* and *U3*, T_b could be set to 35. For any carefully chosen fixed threshold, there will often be some false detections and/or misses. However, in general, it is not possible to find a threshold that can minimize both false detections and dismissals at the same time. In fact, when one is being improved, the other will become worse. Thus, despite its simplicity, the performance of this detector cannot be guaranteed.

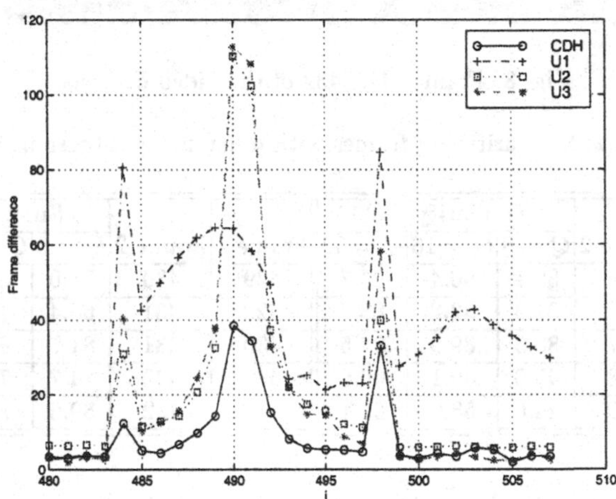

Fig. 4. Frame differences for frames 480–508 of *ad1.mpg*

4.2 Sliding Window Method

Proposed by Yeo and Liu [7], the sliding window method takes into account the changes occurring within a window of $2m - 1$ frame differences. This method compares the first two maximums in the window and declares a scene cut at the location of the first maximum if they differ by a factor of n. The parameter m determines the minimum scene distance while the parameter n depends on the overall rate of changes of scene contents.

As the value of m increases, the number of false detections will decrease, but at the same time, the number of dismissals will increase. The same effect occurs when the value of n is gradually increased. One reason is that, as m increases, more peaks will fall into the same window. Since at most one cut can be declared from one window, the number of dismissals therefore increases. Similarly, if n is increased, for any frame declared to be a cut, the underlying first maximum frame difference must be very big and stands out among others. Therefore, the number of misses will increase. For the same reason, the number of false detections decreases as m or n increases.

4.3 DD Scene Change Detection Method

This section presents the details of the proposed scene change detection method (*DD*), which uses the differences of the inter-frame similarities computed by using the color difference histograms (*CDH*'s) or the differences of frame differences computed by other segmentation methods.

Let the similarity sequence for consecutive frames be $\{S_i : 1 \leq i \leq N-1\}$, where N is the total of number of frames in the video. To make *CDH* compatible with other segmentation methods, we first convert the frame similarities into frame differences by the relation

$$D_i = 100 - S_i, \quad 1 \leq i \leq N-1 \ .$$

From this sequence $\{D_i\}$, a sequence of differences of frame differences $\{DD_i\}$ can be computed by

$$DD_i = D_{i+1} - D_i, \quad 1 \leq i \leq N-2 \ .$$

Figure 5 presents a plot of a sequence of differences of frame differences (a *DD*-plot) for frames 480 to 508 of the video ***ad1.mpg***.

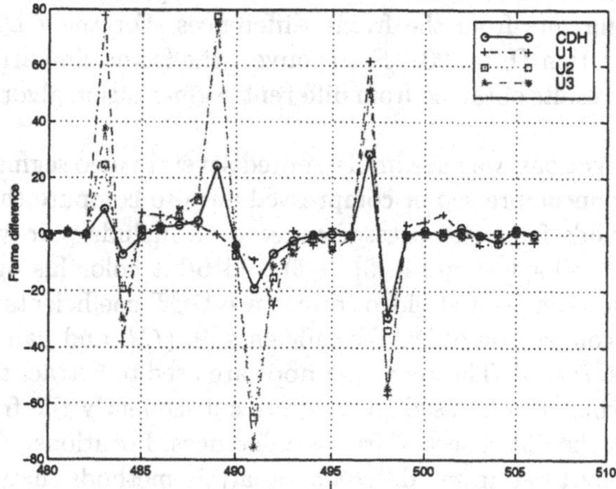

Fig. 5. Differences of frame differences for frames 480–508 of ***ad1.mpg***

Comparing the variations of differences of frame differences in Fig. 5 with the variations of frame differences in Fig. 4, we notice a characteristic pattern for the variations of differences of frame differences around the frame positions having a scene cut; that is, each peak indicating a scene cut in *FD*-plot is transformed to one positive peak followed immediately by one negative peak in *DD*-plot. It is concluded that a scene cut can be declared at frame f_{i+2} if

1. $DD_i > T_{\mathrm{d}}$ (C1)

2. $DD_{i+1} < -T_d$ (C2)

where T_d is a user-defined threshold which determines the degree of changes of differences of frame differences to indicate a scene cut.

Two consecutive peaks in a *FD*-plot usually indicate an occurrence of a flashlight. These two peaks are transformed to a positive peak, followed by a very small value, followed by a negative peak in a *DD*-plot. Thus, the presence of a flashlight frame can be detected by checking for the combined conditions:

1. $DD_i > T_f$ (F1)
2. $DD_{i+2} < -T_f$ (F2)

where T_f is a user-defined threshold for flashlight detection. If both the conditions (F1) and (F2) are satisfied, it can be declared that frame f_{i+2} contains flashlight.

5 Experiments

In order to study the proposed methods in depth, we have implemented the *CDH*-based method together with several other existing methods so that their performances can be compared by using the same set of sample video files. Two scene cut detection methods have also been implemented to determine the locations of scene cuts from the frame differences. For the *CDH* method, we have used the relation $D_i = 100 - S_i$ to convert the frame similarities into frame differences. The results obtained from different segmentation algorithms are then compared.

In our investigations, we have implemented several video segmentation methods using either uncompressed or compressed data to compute the frame differences. The methods for uncompressed video data include pair-wise pixel comparison [5] (*U1*), color histogram [5] (*U2*) and 6-bit color histogram [5] (*U3*); whereas, those for compressed video frames are DCT-coefficients correlation [3] (*C1*), pair-wise comparison of DCT-coefficients [9] (*C2*) and pair-wise comparison of dc-images [7] (*C3*). The above methods are used to extract their respective feature values which can be used to compute subsequently the frame difference values as well as the differences of frame differences. Locations of scene changes are then found by three frame difference analysis methods, namely the global threshold method, the sliding window method and the *DD* method.

To evaluate the performance of a video segmentation algorithm, we have employed the *precision* and *recall* metrics widely used in the areas of information retrieval. Let N_d be the number of correctly detected scene changes, N_f the number of falsely detected scene changes, N_m the number of missed scene changes. Then, we have

$$\text{recall} = \frac{N_d}{N_d + N_m} \qquad \text{precision} = \frac{N_d}{N_d + N_f} .$$

To facilitate the computation of the recall and precision for any segmentation run, all sample videos have been manually examined to prepare the actual scene change data. The locations of scene changes generated by any segmentation algorithm are matched against the manually prepared data automatically to produce the precision and recall statistics, and three data lists for correct detections, false detections and missing scene changes respectively.

We have implemented the algorithms in C based on the public domain MPEG-I software decoder obtained from the website of the U.C. Berkeley MPEG Research (http://bmrc.berkeley.edu/projects/mpeg/mpeg_play.html). The sample videos have been prepared through a Pentium PC by using a video capture card to convert several TV programs including news, advertisements, cartoons, music videos, movies and documentaries from analog sources in videotapes to digital files of MPEG-I formats.

Three sets of experiments are performed to evaluate the effectiveness of our method and to validate the illumination and motion invariant properties of the *CDH* method. The results are analyzed in the following sub-sections.

5.1 Performance of Color Difference Histogram

We have applied the *CDH*-based segmentation method (with Q set to 16) to a number of videos. Table 3 summarizes the results of applying each tested algorithm using different scene change detectors on the video ***ad1.mpg*** which is a TV advertisement with 1437 frames and 33 scene cuts. The video contains several scenes with camera panning, object motions, and flashlights. For each method, different threshold settings are tried and those which correspond to the best precision and recall values are chosen for comparison purpose.

Table 3. Performance statistics of video segmentation algorithms for ***ad1.mpg***

	DD		Global Threshold		Sliding Window	
Metric	Recall	Precision	Recall	Precision	Recall	Precision
CDH	0.94	0.89	0.94	0.65	0.88	0.97
U1	0.88	0.97	0.91	0.75	0.70	1.00
U2	0.88	0.71	0.70	0.48	0.82	1.00
U3	0.85	0.93	0.88	0.60	0.85	0.97
C1	0.88	1.00	0.79	0.54	0.85	0.90
C2	0.85	0.97	0.76	0.58	0.88	0.66
C3	0.91	0.97	0.94	0.84	0.76	0.96

From the table, we can see that the *CDH* method coupled with the *DD* detector compares favorably with other methods. Also, the *DD* detector gives better precision for similar values of recall when compared with the global threshold and the sliding window method.

As for flashlight detection, when T_f is set to 10, the *CDH* and *DD* combination can correctly detect all three flashlight frames (428, 439, 491) in the video without any false detection. Methods *U2, U3, C2* and *C3* can also detect two flashlights correctly when combined with the *DD* method. For *U1* and *C1*, they miss all the flashlights. In addition, for *U1, U2, U3, C1, C2, C3*, there are always some false detections and dismissals no matter how many different T_f's are tried.

We have also experimented with other types of videos. Table 4 shows the recall and precision statistics for a cartoon video *temple.mpg*, which consists of 960 video frames. This cartoon has 26 scenes and some of them contain fast object motions. Again, in general, the *CDH* and *DD* combination out-performs other methods.

Table 4. Performance statistics of video segmentation algorithms for *temple.mpg*

	DD		Global Threshold		Sliding Window	
Metric	Recall	Precision	Recall	Precision	Recall	Precision
CDH	1.00	1.00	0.96	0.86	1.00	1.00
U1	0.84	1.00	0.84	0.81	0.92	0.70
U2	0.92	1.00	0.88	0.79	1.00	0.96
U3	1.00	0.96	0.88	0.79	0.96	0.96
C1	0.28	0.88	0.56	0.31	0.68	0.85
C2	0.52	0.52	0.36	0.24	0.52	0.50
C3	0.92	0.62	0.80	0.45	0.80	0.83

5.2 Effect of Brightness Changes

We have chosen a specific video sequence having various lighting effects for testing the sensitivity of different methods to intensity variations across the video frames. Figure 6 shows the selected sequence containing frames 483 to 497 of the video *ad1.mpg*. From Fig. 4, which shows the plots of frame differences over the selected sequence for the methods *CDH, U1, U2* and *U3*, we can see that the use of color differences can reduce the sensitivity of *CDH* to intensity changes. *CDH* has smaller frame differences than others in those sequences with successive changes in intensity, for example, the difference sequences D_{485}–D_{489} and D_{492}–D_{497}. On the other hand, for *CDH*, the peaks that stand for scene changes are still obvious. Thus, *CDH* can help improve the precision of the video segmentation results by avoiding the false alarms caused by a gradual increase in light intensity.

5.3 Missing Scene Breaks

Similar to the simple color histogram, the *CDH* method computes the color distribution within a frame, thus making it invariant to any spatial changes of objects (e.g. translation and rotation). One disadvantage of the method is that it may not detect any difference if two frames involving in a camera break have similar objects and similar backgrounds. The above problem can be observed in frames 484 and 485 in Fig. 6 where there is a camera break at frame 485. However,

Fig. 6. Frames 483 to 497 of *ad1.mpg*

the similarity between the two frames is so high (87.6%) that *CDH* treats them as similar frames belonging to the same scene. As described in Sect. 1, there can be two different definitions for a video scene, namely logical scene and physical scene. *CDH* copes with only ***logical*** scenes. As the above mentioned two frames are quite similar and related logically; they should belong to the same logical scene. Thus, we need not consider the above as a missing scene break.

One problem with the *DD* scene change detection method is that the two conditions (C1) and (C2) for scene break detection will not hold simultaneously when there is a flashlight frame just before a scene change. This will cause the scene change to be missed by our detector. For instance, the frame 492 in Fig. 6 starts a new scene. But frame 491 contains flashlight, which makes it difficult to detect a scene cut between frames 490 and 492 by just using the information from two consecutive frames: either frames 490 and 491 or frames 491 and 492. To solve this problem, one way is to look forward one frame and measure the similarity between the frames before and after the flashlight to determine if there is a scene change immediately after the flashlight.

6 Conclusion

The main sources of errors in video segmentation are due to camera panning or zooming, object motion and illumination changes. In this paper, we have proposed a new video segmentation method based on color difference histogram (*CDH*), which is invariant to illumination changes, object or camera motions. We have also developed a new scene cut detection method (*DD*) by analyzing the patterns of the changes in frame differences at the locations of scene cuts and flashlights. The performance of the proposed video segmentation method has been compared with other existing algorithms on different videos containing flashlight, camera movements or object motions. Experimental results confirm that *CDH* is less sensitive to brightness changes than other histogram based methods, and less sensitive to camera movements and object motions than other pixel based methods. Experiments also show that the *CDH* method coupled with the *DD* detector out-performs other video segmentation methods. The *DD* detector gives higher precision for similar values of recall when compared with both the global threshold and the sliding window method. Flashlights can also be correctly detected by the *DD* method.

7 Future Work

As far as scene change detection is concerned, there are still some unsolved problems. For instance, a frame sequence involving fast scene changes, fast camera panning or zooming, fast object motion, dark or unclear images, irregular lighting is not easy to be segmented even with the help of human eyes. Fig. 7 shows an example of unclear images, where an object passes rapidly from right to left in front of the main objects in some of the frames. No detection methods discussed in this paper can handle this case.

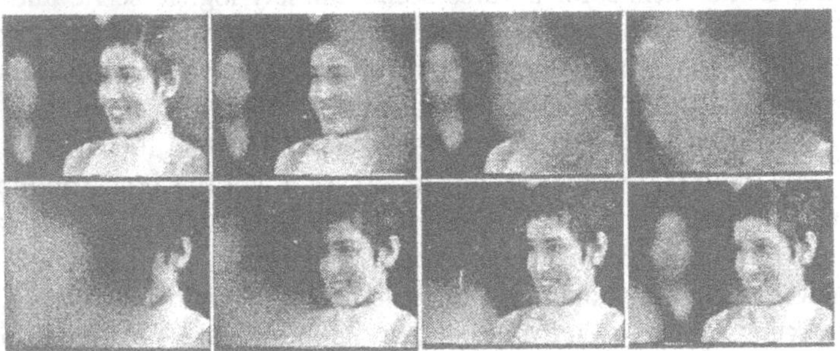

Fig. 7. Frame sequence containing a front moving object (frames 829–836 of *ad1.mpg*)

Concerning the *CDH* method itself, we have noticed that there is a very uneven distribution of counts within an array representing each color difference histogram. That is, the histograms concerned are sparse matrices. In practice, many of the non-zero counts are confined to the central region of the histogram, i.e. the probability of finding a non-zero $CDH(r, s)$, where r and s range from say 0 to 10 or 22 to 31, is very close to zero. To improve the speed of segmentation process, we can ignore those zero entries of $CDH(r, s)$ while computing the inter-frame similarities. On the other hand, the current definition of *CDH* is for uncompressed *RGB* images. It could be extended to transform domain images.

Finally, as for scene change detection methods, different optimal settings of the parameters such m, n, or T_{d} have to be prepared for various videos in order to obtain satisfactory results. Therefore, there is a need to devise new adaptive techniques to automate the optimal parameter setting process according to the individual video characteristics. Moreover, methods for detecting gradual transitions and for reducing errors from brightness changes, camera movements, and object motions, should be further investigated.

References

1. D.A. Adjeroh, M. C. Lee and Cyril U. Orji, "Techniques for Fast Partitioning of Compressed and Uncompressed Video", *Multimedia Tools and Applications*, **4**, pp. 225–243, 1997.
2. G. Ahanger and Thomas D. C. Little, "A Survey of Technologies for Parsing and Indexing Digital Video", *Journal of Visual Communication and Image Representation*, **7**, 1, pp. 28–43, 1996.
3. F. Arman, A. Hsu and M.Y. Chiu, "Image processing on encoded video sequences", *Multimedia Systems*, Springer Verlag, **1**, pp. 211–219, 1994.
4. D. Le Gall, "MPEG: A video compression standard for multimedia applications", *Communications of the ACM*, **34**, 4, pp. 46–58, 1991.
5. A. Nagasaka and Y. Tanaka, "Automatic Video Indexing and Full-Video Search for Object Appearances", *Visual Database Systems II*, Elsevier Science Publishers, pp. 113–127, 1992.
6. K. Otsuji and Y. Tonomura, "Projection-detecting filter for video cut detection", *Multimedia Systems*, Springer Verlag, **1**, pp. 205–210, 1994.
7. B.L. Yeo and B. Liu, "Rapid Scene Analysis on Compressed Video", *IEEE Transactions on Circuits and Systems For Video Technology*, **5**, 6, pp. 533–544, 1995.
8. H.J. Zhang, A. Kankanhalli and Stephan W. Smoliar, "Automatic partitioning of full-motion video", *Multimedia Systems*, **1**, pp. 10–28, 1993.
9. H.J. Zhang, C.Y. Low and Stephan W. Smoliar, "Video Parsing and Browsing Using Compressed Data", *Multimedia Tools and Applications*, **1**, pp.89–111, 1995.

A New Scene Breakpoint Detection Algorithm Using Slice of Video Stream

Kong Weixin, Ren Yao, Lu Hanqing

National Laboratory of Pattern Recognition
Institute of Auotomation, Chinese Academy of Sciences
Beijing 100080, P.R.China
Email: kongwx@prlsun1.ia.ac.cn

Abstract. Automatic scene breakpoint detection is the first step and also an important step for content-based parsing and indexing of video data. Several methods have been introduced to address this problem, e.g. pixel-by-pixel comparisons and histogram comparisons. Each has some advantages, but all of them are slow because they need process the data of entire image frames. Furthermore, none of these methods fully utilize the spatio-temporal attribute of video stream. In this paper, we propose a new scene breakpoint detection algorithm, which is based on the 2-D vertical and horizontal slice images of video stream. In this way, we largely reduce the amount of data to be processed. Experimental results show that our method is effective in detecting abrupt scene breakpoints.

1 Introduction

Image and video have become important elements in multimedia computing and communication. The applications such as video on demand (VOD) and digital librariy(DLI) are moving into pratical services. Such sevices rely on efficient management of image and video data. But most tradational databases are designed to handle only alphanumber data. The content-based method proposed recently is an efficient method for building image and video databases[1].

In this method, prior to storage in a database, a video sequence must first be segmented into shots. A shot is a sequence of video frames generated during a continuous camera operation and therefore represents a continuous action in time and space. It can serve as the elemental index unit in the database. Detection of shots requires algorithms to automatically determin scene breakpoint. Scene breakpoints mark the transition from one shot to another.

The detection of scene breakpoint is also important to other applications, such as compression. Motion based compression alogorithm like MPEG can obtain higher compressin rates without sacrificing quality when the scene breakpoints are known.

We begin with a survey of related work on detection of scene breakpoint in section 2. These methods most rely on color or luminance data in two entire consecutive frames. We then present our algorithm to this problem in section 3. We use vertical and horizontal slices of video stream. Exprimental results are discussed in section 4 and some concluding remarks are drawn in section 5.

2 Related Work

There are two types of shot transition: abrupt transition and gradual transition. Abrupt transition results from camera break, while gradual transition results from various video edit modes such as fade-in, fade-out, dissolve and wipe. When abrupt transition occurs, the difference in gray level and color infomation between two consecutive frames is large. However, those differences are not so large when gradual transition occurs. For the development of existing various scene breakpoint detection algorithms, an assumption is always made that the content of frames should not change too much within a video shot.

In general, scene breakpoint is detected by employing a difference metric to measure the changes between two frames. A scene change is declared if the difference between the two frames exceeds a certain threshold. Most algorithms can be grouped into two categories. In the first category, a difference metric is computed by pixel-by-pixel comparison[2], this method is very sensitive to motion and noise. In the second category, global attribute such as intensity/color histogram is employed[3,4]. This method reduces the effects of noise, but is prone to missing breakpoint when two scene have similar intensity/color distributions. The combination of the above methods has also been reported. Recently more work concentrate on gradual transition detection and breakpoint detection in compressed domain[5,6,7,8].

We find that the above methods most rely on intensity/color data in two entire image frames. Due to the volume of video data, the cost of these method is expensive in time and space. Moreover, these method ignore the information in the other frames. Noting the spatio-temporal attribute of video data, we will develop a very fast algorithm to detect scene breakpoint in this paper.

3 Scene Breakpoint Detection Algorithm Using Slice Image of Video Stream

3.1 Define the Slice Image of Video Stream

Video is a sequence of 2-D images. It can be repesented in a 3-D spatio-temporal coordinate system, three axises are x(horizontal),y(vertical) and t(time). Figure 1 shows it.

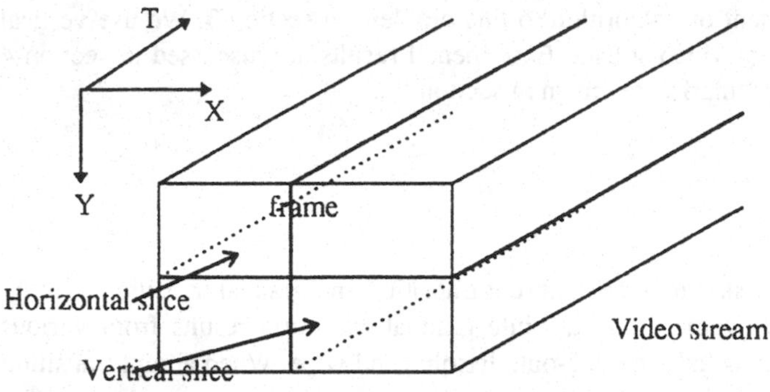

Fig. 1. 3-D video stream

If we keep x at a fixed value, then cut the video stream along t axis, we can get an image in y-t plane, as shown in figure 1. We define this image a vertical slice of the video stream. By using the same method, we can get a horizontal slice. Figure 2 is an example of slice image.

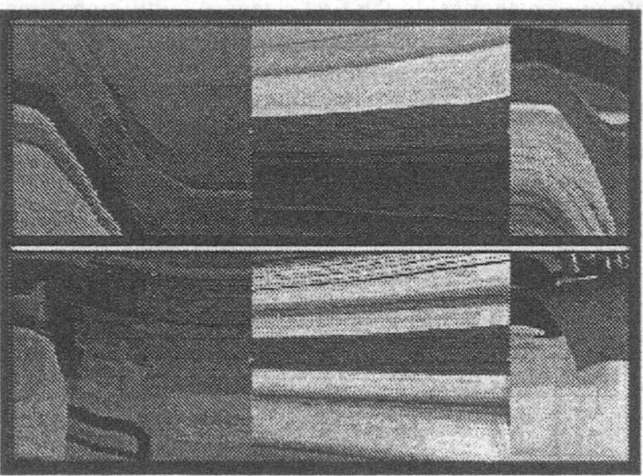

Fig. 2. Example of slice image

From the above figure , we can see that the texture in slice image reflects the motion of scene versus camera in spatio-temporal domain. The color/intensity and texture is continue in the same shot. When the scene changes, the contents of the regions on the two sides of the breakpoint are very different. There is an obvious boundary parallel to Y-axis. By detecting this boundary, we can determine the breakpoint and segment the video into shots.

3.2 Description of Our Algorithm

We first smooth the slice image in Y-direction to reduce the effects of noise and motion. This operation will not weaken the boundrys parallel to Y-axis. The smooth operation is as the following:

$$g(x_i,m)=\sum_{k=-2}^{2} f(x_i,m+k)H(k), H[-2\ \ -1\ \ 0\ \ 1\ \ 2]=\frac{1}{10}[1\ \ 2\ \ 4\ \ 2\ \ 1] \qquad (1)$$

f(x,y) is the original image, g(x,y)is the smoothed image, H is the smooth convolution mask.

After the smooth operation, we can use two methods to detect boundary.

- detect the boundry in real time, compare only two lines in vertical and horizontal slice images. This method needs store only the data of four lines.

$$g(i,j)=\begin{cases}1 & \text{if} \min_{m=-2,-1,0,1,2} |f(i,j)-f(i+1,j+m)| > T_e \\ 0 & \text{otherwise}\end{cases} \qquad (2)$$

$$Boundary(i)=\begin{cases} True & \text{if} \sum_j g(i,j) > T_b \\ False & \text{Otherwise}\end{cases}$$

T_e and T_b are appropriate thresholds, f(i,j) is the smoothed slice image. When Boundary(i)=1, it indicates that the ith frame is a scene break point. Furthermore, only when boundary at the ith frame appears both in vertical and horizontal slice image, scene breakpoint is declared. If the boundary appears in only one image, it is considered to be the result of noise and motion.

- get slice images and use available edge detection algorithm. The slice images are transfered into binary edge map and then we can detect straight lines parallel to Y-axis. These lines correspond to scene breakpoint.

There are many edge detectors now. Considering the computational cost and their performance, we choose the zero-crossing detector: Difference of Exponential(DOE).

$$DOE = K_e\left(e^{-\frac{|x|}{t_1}} - e^{-\frac{|x|}{t_2}}\right) \qquad (3)$$

This operator can be implemented recursively, so its computational cost is very low. In[9],it has been proved that with the scale ratio of 0.3, this edge detector performs best. So in our experiments, we choose t_1=1, t_2=3.

Because this detector can't detect edge direction, we will get edge direction by computing local gradient. Those egdes parallel to Y-axis are saved.

4 Exprimental Results

In our experiments, two image sequences are used to evaluate the performance of the proposed scene breakpoint detection algorithm. We use only abrupt transition. Table 1 summaries the results.

	Real time detection			non-real time detection		
Sequence	Nd	Nm	Nf	Nd	Nm	Nf
No.1	93	3	1	96	0	1
No.2	104	2	2	106	0	1

Table 1. Experimental results

Nd, Nm and Nf are the number of detected, missed and false scene breakpoints respectively. We can see that non-real time method performs better, it can detect all the abrupt scene breakpoints with very few false detection. Real time method is faster and the accuracy rate is satisfactry despite not as well as non-real time method.

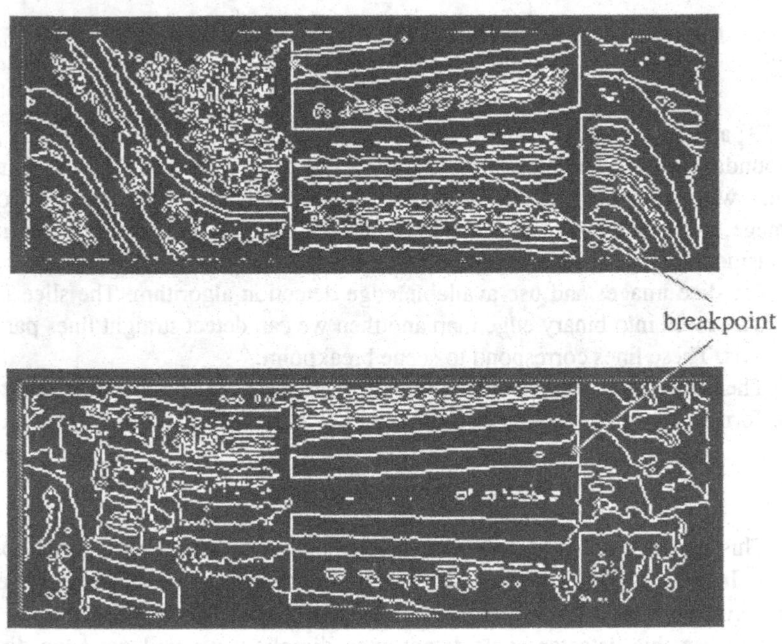

Fig. 3. edge map of slice image

Figure 3 shows an example of the edge map of slice image.

5 Conclusions

In this paper, a new scene breakpoint detection algorithm has been proposed. Compared with the existing detection methods for video, our method is very fast. We use the vertical and horizontal slices of video stream and largely reduce the amount of data that need to be processed. Experimental results show that our method is effective in detecting abrupt scene breakpoints.

References

1. IBM Almaden Research Center, Query by Image and Video Content: The QBIC Systems, IEEE Computer, September 1995.
2. K.Otsuji, Y.Tonomura and Y.Ohba, Video browsing using brightness data, Proc. SPIE Conf. Visual Communications and Image Processing, pp.980-989, November 1991.
3. A.Nagasaka and Y.Tanaka, Automatic video indexing and full-video search for object appearances, Proc. 2nd Visual Database Systems, pp119-133, October 1991
4. Y.Tonomura, Video handling based on structured information for hypermedia systems, ACM Proc. International Conference on Multimedia Information Systems, pp333-344, 1991
5. F.Arman, A.Hsu and M.Y.Chiu, Image Processing on compressed data for large video databases, Proc. First ACM Int'l Conf. Multimedia, August 1993
6. H.Zhang, C.Y.Low, Y.Gong and S.Smoliar, Video parsing using compressed data, Proc. SPIE Conf. Image and Video Processing, vol.2182, pp142-149, 1994
7. I.K.Sethi, N.Patel, A Statistical Approch to Scene Change Detection, Proc. SPIE Conf. Storage and Retrieval for Image and Video Databases, vol.2420, pp329-338, 1995
8. A.Hampapur, R.Jain and T.Weymouth, Production model based digital video segmentation, Journal of Multimedia Tools and Applications, pp1-38, March 1995
9. Bingcheng Li, Songde Ma, Multiscale Filtering Method for Derivative Computation , Proc. SPIE Conf. Visual Communications and Image Processing, pp.1277-1288, 1994.

A Low Latency Hierarchical Storage Organization for Multimedia Data Retrieval

Philip K. C. Tse and Clement H. C. Leung
Victoria University of Technology, School of Communications and Informatics
P.O. Box 14428, MCMC, Melbourne 8001, Victoria, Australia.
Email: {philip, clement}@matilda.vut.edu.au Fax: +61 3 9688 4050

Abstract

Hierarchical storage systems can provide huge storage capacity for multimedia data at very economical cost, but the relative high access latency of tertiary storage devices tends to make them infeasible for many applications. In this paper, we present a storage organization tailored for multimedia data stored in hierarchical storage systems which overcomes the latency limitation. We introduce a novel data segmentation method that reduces the latency of access operations, in which multimedia data objects stored using this storage organization can always be initiated to display at disk access latency. The data segmentation method is analyzed with simulations and it is optimized for minimum disk space and maximum number of segments.

1. INTRODUCTION

Recent advances in computer technologies raise the possibility of new multimedia applications, such as video-on-demand, and electronic shopping. Hierarchical storage systems (HSS) can provide huge storage capacity for multimedia data at very economical cost, but the high access latency of tertiary storage devices makes them very unpopular.

A HSS consists of at least two levels of storage devices. The first level of secondary storage devices may consist of magnetic disks, solid state disks, and other direct access storage devices. They can access data blocks quickly, but they store data at higher media cost. The next level of tertiary storage devices may consist of automatic disk changers, optical jukeboxes and tape libraries. They can store data at a lower media cost, but they perform random accesses at a much slower rate.

In a HSS, the tertiary storage devices provide permanent storage for all multimedia objects. Multimedia objects are copied to the secondary storage devices when they are required using the staging process. When secondary storage space becomes depleted, unpopular or infrequently accessed objects are deleted according to the cache management algorithm. Popular or frequently accessed multimedia objects may reside permanently in the secondary storage devices. Very popular objects may even be replicated or striped into phases.

The staging method, however, has two problems. First, users have to wait for the staging process which takes a long time to complete. Second, the disk space that contains unpopular objects is reserved until the multimedia data are no longer required. The pipeline method has been proposed to rectify these problems [1, 2].

For certain applications such as video-on-demand, users can only view video from beginning to end. Since the time to stage multimedia objects from tertiary store to secondary storage devices can be very long, the system may start to display video to users before the whole object is completely staged to disks. Therefore, the users wait for a much shorter time before they start to view the object.

Some multimedia data applications, such as interactive movie, and electronic shopping, require more interactive capability than video-on-demand. The only existing storage method to provide interactive capabilities is to keep all data resident on magnetic disks. However, this is not economical though magnetic disks are often considered as inexpensive. The traditional staging method provides full random access capabilities at high initial access latency, but the pipeline method limits the access capabilities on the data.

Users of multimedia applications may wish to preview or start to view from the middle of an objects. Although separate preview files may be used, storage space is then wasted. However, the latency of pipeline methods to start display in the middle of an object is also high because the required part may still be in tertiary store. Therefore, a storage method that can provide better access capabilities than the pipeline method is required.

In traditional computer systems, the storage system accesses data and respond to the requests upon completion and the databases are optimized towards minimum response time. In multimedia systems, the storage system needs to guarantee the response time of a number of acceptable streams with each stream consisting of a number of requests that arrives periodically. The databases are then optimized towards the maximum number of acceptable streams at minimum startup latency when resources, including memory, disk space, and disk bandwidth, are limited.

Much research has been done on multimedia data placement and staging data in hierarchical storage systems, and several data striping algorithms on multiple disks have been proposed in the literature. In [3, 4], the feasibility of merging multimedia data streams is studied. Data striping on RAID disks is studied in [5, 6], and movie popularity is used to distribute and replicate data in [7]. In [8, 9], data strips are replicated or placed in a staggered manner in multiple disks to allow simultaneous retrieval. In [10], data strips are placed in a round ribbon manner to neighboring tracks in order to reduce disk seek time for multiple streams. In [11, 12, 13], it is proposed that each data strips in the same disk are stored consecutively to reduce access time for very popular video objects.

In particular, several papers have described the use of hierarchical storage systems. In [14, 15], the storage cost and performance are compared, and in [16], the user latency of staging data is studied using simulations. Video data are allocated to different storage hierarchy according to popularity in [17], and the cache management of the Berkeley VOD system is described in [18-21]. In [22], a data segmentation method is tailored according to the VCR access patterns, where the display time of leaders overlapping with the retrieval time of tails is studied.

In [1], a pipeline method that keeps the leaders in disks and caching the tails is described, and in [2], a circular disk buffer is used in the pipeline method to reduce required disk space. The study that is found most closely related to the present one is the two phase service model (2PSM) described in [23]. The 2PSM segments the video before pipelining to provide preview, fast forward, and fast reverse operations. However, their results are not directly applicable to HSS because their network bandwidth is much lower than the tertiary bandwidth.

We present our storage organization and data segmentation method in Section 2, the performance analysis is carried out in Section 3. The simulation results are summarised in Section 4.

2. A NEW STORAGE ORGANIZATION STRUCTURE

We consider the placement of a multimedia or video object X on secondary storage and tertiary storage, where X may be split into m logical segments (X_1, X_2, ... , X_m). This is done in two steps (Fig. 1). The first step splits the object into c *logical cuts*, such as camera cuts in video, and speaker changes in audio. This provides a list of images to browse the video. Without using the logical cuts, segments may start at undesirable positions and the preview file may miss out important scenes. The second step splits every long logical segments into shorter ones as *length cuts*. This provides a list of logical points to preview the object in fast forward and fast reverse mode. We then merge the two lists to form a list of logical segments. Letting c be the number of logical cuts, s be the number of length cuts, m be the total number of logical segments, then we have

$$m = s + c. \tag{1}$$

In the next stage, each logical segment X_i is split into n slices ($X_{i,1}$, $X_{i,2}$, ... , $X_{i,n}$) such that the display time of $X_{i,1}$ is longer than the retrieval time of $X_{i,2}$, the display time of $X_{i,2}$ is longer than the retrieval time of $X_{i,3}$ and so on. Also, the display time of $X_{i,n}$ is longer than the retrieval time of $X_{i+1,1}$. The size of the last slice in each logical segment, $X_{i,n}$, should occupy at least one media block in order to provide a logical unit for display. Here, we choose one media block to be one Group of Frames (GOF) in MPEG-2 compression.

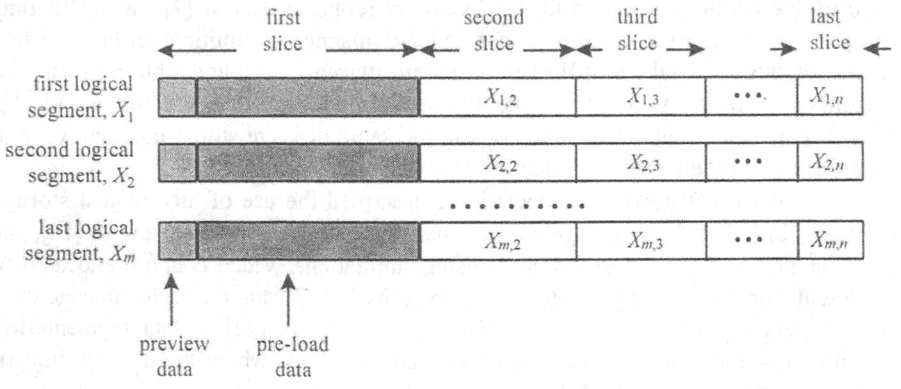

Fig. 1. Logical segments and slices

In the tertiary storage devices, the first media block of all logical segments are stored together as the preview data. The rest of the media blocks of the first slice of all segments are stored together as the pre-loaded data. The second to the last slice of each segment, $X_{i,2}$ to $X_{i,n}$, are then stored together as the tail of the segment. Each segment should be stored together in only one media unit.

In the secondary storage devices, the preview data and the pre-load data should be downloaded beforehand and kept resident in secondary storage to reduce user latency. The tails of the segments are then downloaded on user request and cached according to the cache management algorithm. When multiple disks are used together, the aggregate secondary bandwidth of N disks is

$$\beta = \sum_{i=1}^{N} \beta_i , \tag{2}$$

where β_i is the retrieval bandwidth of the ith disk.

When the user displays an object, the first slice of the first segment is firstly retrieved from disk. While this first slice is being displayed, the tertiary storage device is exchanging media and retrieving the second slice to secondary storage. Before the first slice is displayed completely, the second slice has already been retrieved. While the second slice is being displayed, the third slice is being retrieved. While the last slice of the first segment is being displayed, the first slice of the second segment is being retrieved and so on. This pipeline process continues until the whole object is displayed. Thus, the user only needs to wait for a short time while the secondary storage devices are retrieving the first slice of the first segment before display begins.

We allow two preview methods in this model. The first preview method displays the first media block of all logical segments like a fast forward video. When the user resumes normal display, the pre-load data of the current segment is retrieved from disks. While the first slice is being displayed, the second slice is being retrieved from the tertiary store and so on. The second preview method shows the scene cuts as images. The first I-frame of every scene cut segment is displayed. When the user browses and selects a scene to view, the first slice of the selected scene is retrieved. While the first slice is being displayed, the second slice is being retrieved and so on. In both preview methods, the user only needs to wait while the secondary storage devices are retrieving the first slice of the required segment.

3. PERFORMANCE ANALYSIS

With current technology, the secondary bandwidth of a single disk is around 10MB/s, the tertiary bandwidth of a tertiary storage device is of the order of 100 KB/s, and the display bandwidth requirement for MPEG-2 compressed video is around 1.5 MB/s. Therefore, it is reasonable to make the following assumptions.

1. The secondary bandwidth is high enough to support many display streams.
2. The tertiary bandwidth is lower than the display bandwidth. Otherwise all slices on tertiary store can be one media block in size.
3. The number of streams directed to the tertiary storage is not more than the number of tertiary drives so that there is no queue contention in the tertiary drives.

4. There is no contention at the robot during media exchange. This can be achieved by desynchronizing the media exchange, but the topic will be treated in a separate paper.
5. Each segment is stored in only one media unit to avoid media exchange during the pipeline process. Long segments can be stored in separate media units.

If these conditions can be met, then this analysis is applicable. It can provide closed form solutions to the maximum user latency and the amount of resident disk space.

3.1 Continuous Display Requirements

In order to provide instant response to user requests, the display time of the first slice should be longer than the media exchange time plus the time to retrieve the second slice. Letting ω be the media exchange time of the tertiary storage device, δ be the display bandwidth, and γ be the tertiary bandwidth, we have

$$\frac{X_{i,1}}{\delta} \geq \omega + \frac{X_{i,2}}{\gamma}, \qquad 1 \leq i \leq m\text{-}1. \tag{3}$$

In order to display the whole object continuously, the display time of each slice should be at least the retrieval time of the successive slice. Therefore, we have

$$\frac{X_{i,j}}{\delta} \geq \frac{X_{i,j+1}}{\gamma}, \qquad 2 \leq j \leq n\text{-}1,\ 1 \leq i \leq m, \tag{4}$$

and

$$\frac{X_{i,n}}{\delta} \geq \frac{X_{i+1,1}}{\beta}, \qquad 1 \leq i \leq m\text{-}1. \tag{5}$$

Since the segment sizes and slice sizes can be very different, we find it useful to use the mean segment size and the mean slice size. Letting $\overline{X}$ be the mean average size of all logical segments, we have

$$\overline{X} = \frac{X}{m}. \tag{6}$$

Letting $\overline{X_j}$ be the mean sizes of the jth slice of all logical segments, we have

$$\overline{X_j} = \frac{1}{m}\sum_{i=1}^{m} X_{i,j}. \tag{7}$$

Applying the mean to Eq. (3) to Eq. (5), we get

$$\frac{\overline{X_1}}{\delta} \geq \omega + \frac{\overline{X_2}}{\gamma}, \tag{8}$$

$$\frac{\overline{X_j}}{\delta} \geq \frac{\overline{X_{j+1}}}{\gamma}, \qquad 2 \leq j \leq n\text{-}1, \tag{9}$$

and

$$\frac{\overline{X_n}}{\delta} \geq \frac{\overline{X_1}}{\beta}. \tag{10}$$

3.2 Number of Slices and Size of the Last Slice

The pipeline method is more efficient when more slices are created in each segment. However, the continuous display requirements impose limitations on the maximum number of slices per segment and the size of the slices. Applying Eq. (9) recursively, we get

$$\overline{X_n} \le \frac{\gamma}{\delta}\overline{X_{n-1}} \le \cdots \le (\frac{\gamma}{\delta})^{n-2}\overline{X_2},$$

$$\Rightarrow \overline{X_j} \ge \left(\frac{\delta}{\gamma}\right)^{n-j}\overline{X_n}, \qquad 2 \le j \le n\text{-}1. \tag{11}$$

From Eq. (8), we have

$$\frac{\overline{X_1}}{\delta} \ge \omega + \frac{\overline{X_2}}{\gamma},$$

$$\Rightarrow \overline{X_1} \ge \omega\delta + \left(\frac{\delta}{\gamma}\right)^{n-1}\overline{X_n}. \tag{12}$$

From Eq. (10) and Eq. (12), we get lower and upper bounds on the first slice

$$\omega\delta + \left(\frac{\delta}{\gamma}\right)^{n-1}\overline{X_n} \le \overline{X_1} \le \frac{\beta}{\delta}\overline{X_n},$$

$$\Rightarrow \overline{X_n} \ge \frac{\omega\delta}{\beta/\delta - (\delta/\gamma)^{n-1}}. \tag{13}$$

As both $\omega\delta$ and $\overline{X_n}$ must be greater than zero, we have

$$\frac{\beta}{\delta} > (\frac{\delta}{\gamma})^{n-1}. \tag{14}$$

Hence,

$$n < \frac{\log(\beta/\delta)}{\log(\delta/\gamma)} + 1.$$

In order to achieve the maximum pipeline efficiency, we may use the maximum number of slices. Hence, we have

$$n = \left\lfloor \frac{\log(\beta/\delta)}{\log(\delta/\gamma)} \right\rfloor + 1. \tag{15}$$

We have used the floor function because n is an integer. When the object is cut into short scenes, the actual number of slices in these segments can be less.

However, the size of the slices could become very large when $\frac{\beta}{\delta}-\left(\frac{\delta}{\gamma}\right)^{n-1}$ is small. In such a condition, only a few segments are created. The size of the first slice is large, and the user latency is high. The data segmentation method becomes not useful. Therefore, we trade pipeline efficiency for data segmentation usefulness by reducing the number of slices being used.

Since the size of each slice should be at least one media block, we have $\overline{X_n} \geq M$, where M is the size of one media block. If we have sufficient secondary bandwidth such that $\beta \geq \frac{\omega\delta^2}{M}$, it is desirable to use slightly less slices in each segment so that the last slice is only one media block. Letting η be the number of slices in each segment, we have

$$M \geq \frac{\omega\delta}{\beta/\delta-(\delta/\gamma)^{\eta-1}},$$

$$\Rightarrow \eta = \left\lfloor \frac{\log\ (\beta/\delta-\omega\,\delta/M)}{\log\ (\delta/\gamma)} \right\rfloor + 1. \qquad (16)$$

3.3 Number of Segments and Segment Size

If the length of each segment is too short, a large amount of disk space would be required. If the length of each segment is too long, user can preview only a few media blocks. In order to use minimum disk space and to provide the most preview data, the segment size should be minimized. We substitute $\overline{X_j}$ from Eq. (11), and the smallest mean segment size, $\overline{X}$, should be

$$\overline{X} = \sum_{j=1}^{\eta} \overline{X_j},$$

$$\Rightarrow\ \overline{X} \geq \omega\delta + \sum_{j=1}^{\eta} \left(\frac{\delta}{\gamma}\right)^{j-1} \overline{X_\eta},$$

$$\Rightarrow\ \overline{X} \geq \omega\delta + \frac{(\delta/\gamma)^{\eta}-1}{(\delta/\gamma)-1}\overline{X_\eta}. \qquad (17)$$

This is the lower bound on the segment size when η slices are used. Substituting $\overline{X}$ into Eq. (6), we obtain the maximum number of logical segments

$$m \leq \frac{X}{\omega\delta + \frac{(\delta/\gamma)^{\eta}-1}{(\delta/\gamma)-1}\overline{X_\eta}}. \qquad (18)$$

3.4 User Latency and Resident Disk Space

From Eq. (10), the mean size of the first slice of all segments, $\overline{X_1}$ is $\leq \frac{\beta}{\delta}\overline{X_\eta}$. Hence, if $\beta \geq \frac{\omega\delta^2}{M}$, then the user latency is $\leq \frac{M}{\delta}$.

Since the first slice of all segments should reside in disks permanently, the total size of first slices of all segments, $m\overline{X_1}$, is

$$m\overline{X_1} = m\left[\overline{X} - \sum_{j=2}^{\eta}\overline{X_j}\right].$$

Substituting $\overline{X_j}$ from Eq. (11), the required disk space is then bounded by

$$m\overline{X_1} \leq m\left[\frac{X}{m} - \sum_{j=2}^{\eta}\left(\frac{\delta}{\gamma}\right)^{j-1}\overline{X_\eta}\right],$$

$$\Rightarrow m\overline{X_1} \leq X - m\frac{(\delta/\gamma)^{\eta-1}-1}{(\delta/\gamma)-1}\overline{X_\eta}. \tag{19}$$

4. SIMULATION RESULTS

We have performed simulations on the model using values in the fixed parameter column of Table 1. We varied only one parameter in each simulation using values in the range column.

Parameter	Range	Fixed Parameter
Tertiary bandwidth	0.4 to 1.30 MB/s	0.9 MB/s. This is bandwidth of retrieving one media block consecutively from a 24X CD-ROM drive. The tertiary bandwidth can be very low when repositioning is required in tapes or it may be higher when an array of optical drives is used.
Object size	20 to 200 minutes	two hours of video at 30 frames/second.
Secondary bandwidth	20 to 110 MB/s	80 MB/s. This is the bandwidth of about 8 current disks.
Display bandwidth	1.35 to 2.25 MB/s	1.35 MB/s. This is the data bandwidth of MPEG-2 compressed SVGA video.
Media exchange time	5 to 15 seconds	10 seconds. This is the media exchange time of optical jukeboxes and tape libraries. Automatic CD changers can exchange media faster at around 5 seconds.
Media block size	Not varied	30 images of 24 bit color SVGA (1024 x 768) frames compressed using MPEG at the ratio of 50 to 1.

Table 1. List of simulation parameters

4.1 Disk Space and User Latency

Preview data and pre-load data are kept resident in disks. The amount of disk space required in our method is compared with the size of the first slice in the pipeline method (Fig. 2). In general, the amount of necessary secondary storage space drops gradually when the tertiary bandwidth increases. Our method uses some extra disk space to cater for the media exchange time which is not considered in other papers.

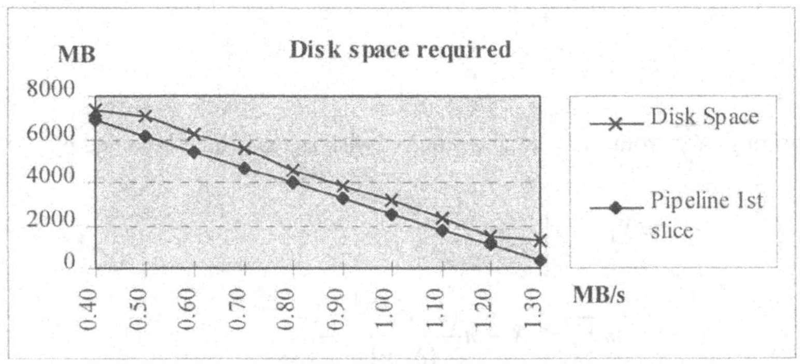

Fig. 2. Disk space required Vs tertiary bandwidth

When user requests for an object, latency is required in the pipeline method to load the first slice from tertiary store. This latency is minimized in our method by keeping all the first slices resident in disks (Fig. 3). This may seem to be an unfair comparison with the pipeline latency. However, the first slices in the two methods are similar in size and they can be loaded in similar time. As latency is a key concern of users, it is a worthwhile tradeoff to reduce user latency with a small amount of resident disk space.

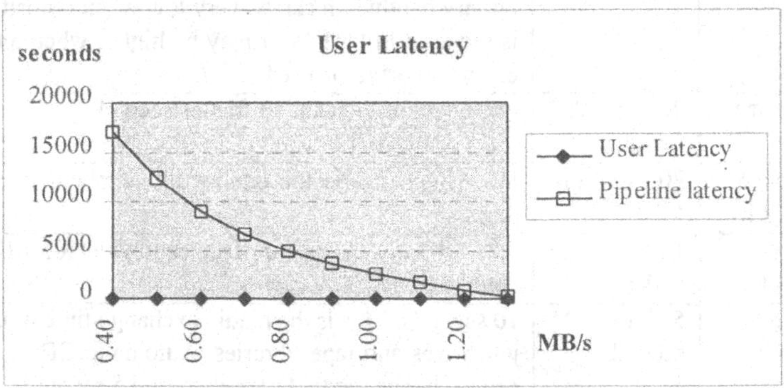

Fig. 3. User latency Vs tertiary bandwidth

When user begins normal display after previewing, the system may need to retrieve some data before it can resume display. The time incurred is called the *reposition latency.* We compare the reposition latency with the fast forward resumption latency of 2PSM in Fig. 4. It can be seen that our method responds much faster. It can usually respond in one second whereas the 2PSM takes much longer to respond. When tertiary bandwidth is nearly the same as the display bandwidth, the 2PSM requires over 30 seconds to resume normal display whereas our method requires less than 4 seconds (not shown). Our method is also more tolerant to changes in tertiary bandwidth.

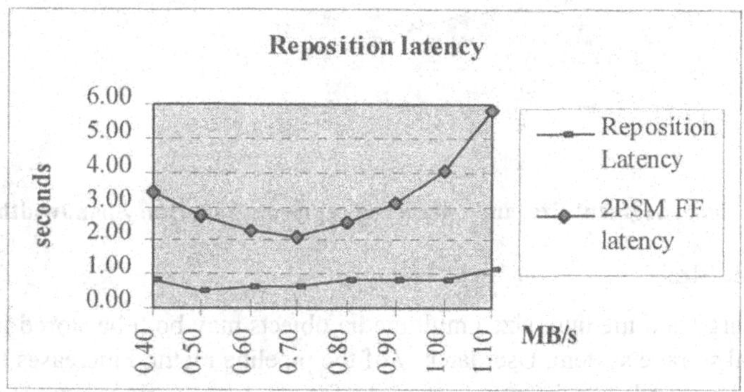

Fig. 4. Reposition latency Vs tertiary bandwidth

4.2 Slices and Segments

When the tertiary bandwidth is low, only a few slices are created in each segment (Fig. 5), and the size of the first slice is small. When the tertiary bandwidth is very close to the display bandwidth, a large number of slices exists in each segment. The first slice becomes larger and the pipeline methods become more efficient at the expense of less preview data and longer reposition latency.

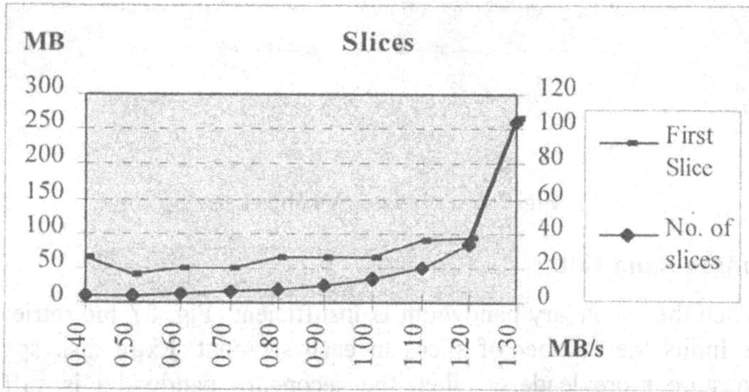

Fig. 5. First slice size and number of slices Vs tertiary bandwidth

When the tertiary bandwidth is low, a large number of small segments are created (Fig. 6). When the tertiary bandwidth is high and close to the display bandwidth, the segment size increases rapidly and the number of logical segments hence decreases. The number of segments also determines the amount of preview data during fast forward display.

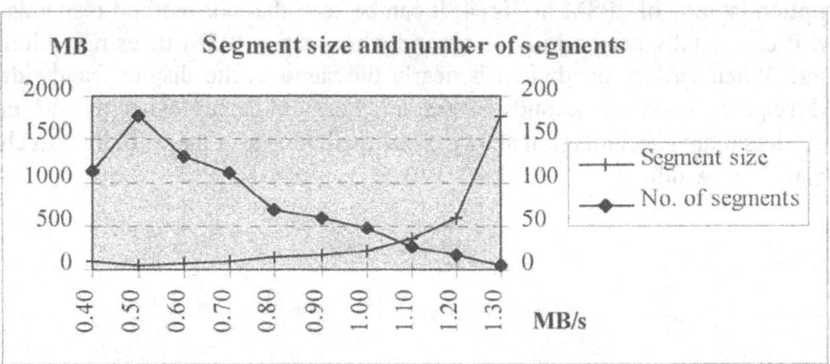

Fig. 6. Segment size and number of segments Vs tertiary bandwidth

4.3 Object Size

Large and medium sized multimedia objects may both be stored in the same hierarchical storage system. User latency of the pipeline method increases linearly but our method responds consistently to requests for objects of various sizes (Fig. 7). This is because we keep sufficient data in secondary storage devices to enable normal display to start before any data are retrieved from tertiary store.

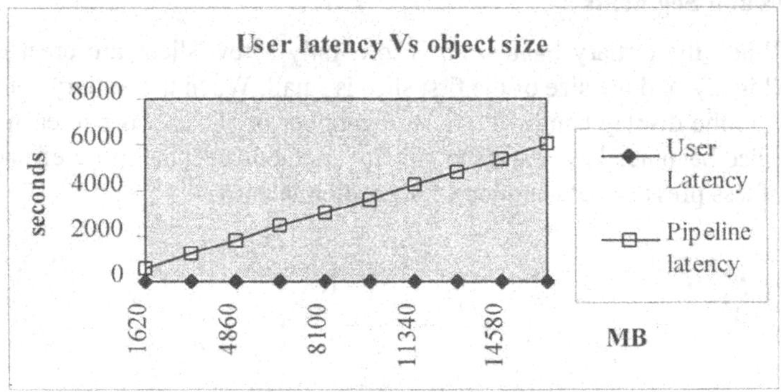

Fig. 7. User latency Vs object size

4.4 Secondary Bandwidth

When the secondary bandwidth is insufficient (Fig. 8), the retrieval time of the leaders limits the number of slices in each segment. Extra disk space is then required to store more leaders. When the secondary bandwidth is sufficient, the leaders are smaller in size. It is therefore desirable to use multiple disks to provide sufficient secondary bandwidth as well as secondary storage space. The secondary bandwidth only affects the reposition latency slightly (Fig. 9) because we create longer first slice when the secondary bandwidth is higher.

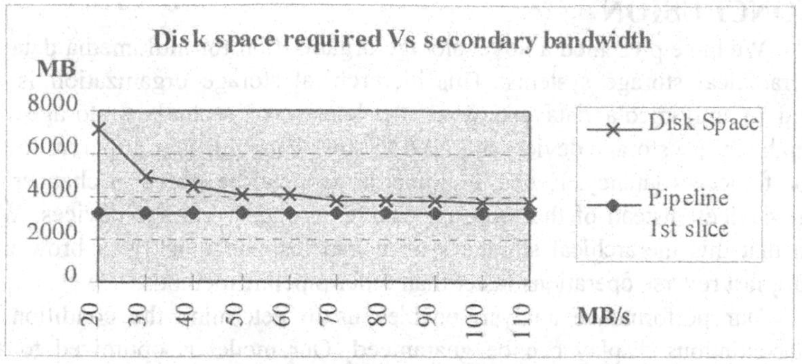

Fig. 8. Disk space required Vs secondary bandwidth

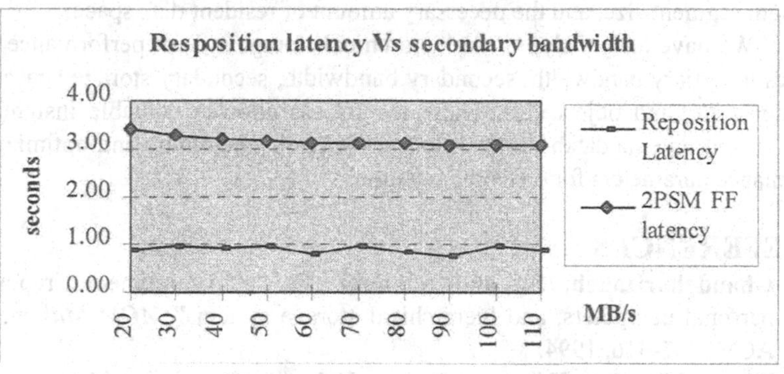

Fig. 9. Reposition latency Vs secondary bandwidth

4.5 Display Bandwidth

When higher quality multimedia data are retrieved, this would increase the display bandwidth. Shorter segments would be created and more disk space is required. Although the pipeline efficiency is reduced, the reposition latency is also reduced (Fig. 10). Our method also responds consistently faster than the 2PSM.

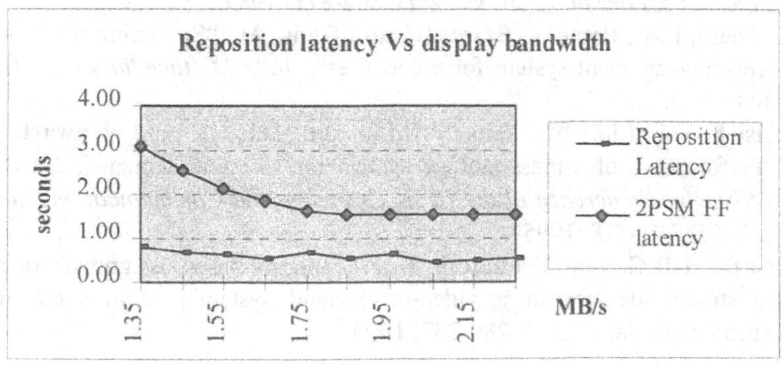

Fig. 10. Reposition latency Vs display bandwidth

5. CONCLUSION

We have presented a novel storage organization for multimedia data storage on hierarchical storage systems. This hierarchical storage organization is able to respond to multimedia data access at the latency of secondary storage devices. Although tertiary storage devices are used to store data, this is transparent to users in relation to access latency. Hence, multimedia data can be stored in cheaper tertiary storage devices instead of the more expensive secondary storage devices. We have shown that this hierarchical storage system can respond quickly to browsing, fast forward, fast reverse operations faster than other pipeline methods.

Our performance analysis enables us to determine the conditions under which continuous display can be guaranteed. Our model is optimized to achieve minimum resident disk space, shortest user latency, and most amount of preview data. Closed form solutions are obtained for the maximum number of slices per segment, minimum segment size, and the necessary amount of resident disk space.

We have also analyzed the hierarchical storage system performance against changes in tertiary bandwidth, secondary bandwidth, secondary storage space, media exchange time, and object size. These results has allowed valuable insights to be gained on multimedia database operations and permits the tuning and optimization of performance parameters for different situations.

6. REFERENCES

[1] **Ghandeharizadeh, S., and Shahabi, C.**, "On multimedia repositories, personal computers, and hierarchical storage systems", *ACM Multimedia'94*, ACM, 407-416, 1994.

[2] **Wang, J.Z., Hua, K.A., and Young, H.C.**, "SEP: a space efficient technique for managing disk buffers in multimedia servers", *Proc. of the International Conference on Multimedia Computing and Systems*, IEEE, 598-607, 1996.

[3] **Rangan, P.V., and Vin, H.M.**, "Efficient storage techniques for digital continuous multimedia", *IEEE Transactions on Knowledge and Data Engineering*, 5(4), 564-573, 1993.

[4] **Yu, C., Sun W., Bitton, D., Yang, Q., Bruno, R., and Tullis, J.**, "Efficient placement of audio data on optical disks for real-time applications", *Communications of the ACM*, 32(7), 862-871, 1989.

[5] **Tobagi, F.A., Pang, J., Baird, R., and Gang, M.**, "Streaming RAID - a disk array management system for video files", *ACM Multimedia'93*, ACM, 393-400, 1993.

[6] **Hsieh, J., Lin, M., Liu, J.C.L., Du, D.H.C., and Ruwart, T.M.**, "Performance of a mass storage system for video-on-demand", *Proc. of the 14th Joint Conference of the IEEE Computer and Communications Societies*, 2, IEEE, 771-778, 1995.

[7] **Little, T.D.C., and Venkatesh, D.**, "Popularity-based assignment of movies to storage devices in a video-on-demand system", *Multimedia Systems*, Springer-Verlag, Vol. 2, 280-287, 1995.

[8] **Ghandeharizadeh, S., and Ramos, L.**, "Continuous retrieval of multimedia data using parallelism", *IEEE Transactions on Knowledge and Data Engineering*, 5(4), 658-669, 1993.

[9] **Berson, S., Ghandeharizadeh, S., Muntz, R., and Ju, X.**, "Staggered striping in multimedia information systems", *Proc. of SIGMOD*, ACM, 79-90, 1994.

[10] **Ghandeharizadeh, S., Kim, S. H., and Shahabi, C.**, "On configuring a single disk continuous media server", *ACM Multimedia'95*, ACM, 37-46, 1995.

[11] **Özden, B., Biliris, A., Rastogi, R., and Silberschatz, A.**, "A low-cost storage server for movie on demand databases", *VLDB Conference*, 594-605, 1994.

[12] **Özden, B., Rastogi, R., and Silberschatz, A.**, "On the design of a low-cost video-on-demand storage system", *Multimedia Systems*, Springer-Verlag, Vol.4, 40-54, 1996.

[13] **Chua, T.S., Li, J., Ooi, B.C., and Tan, K.L.**, "Disk striping strategies for large video-on-demand servers", *ACM Multimedia'96*, ACM, 297-306, 1996.

[14] **Chervenak, A.L., Patterson, D.A., and Katz, R.H.**, "Storage systems for movies-on-demand video servers", *Proc. of the 14th IEEE Symposium on Mass Storage Systems*, IEEE, 246-256, 1995.

[15] **Doganata, Y.N., and Tantawi, A.N.**, "A cost/performance study of video servers with hierarchical storage", *Proc. of the International Conference on Multimedia Computing and Systems*, 393-402, 1994.

[16] **Chan, S.H.G., and Togabi, F.A.**, "Hierarchical storage systems for on-demand video servers", *Proc. of Conference on High Density Data Recording and Retrieval Technologies*, SPIE, Vol.2604, 103-120, 1996.

[17] **Suzuki, H., Nishimura, K., Uemori, A., and Sakamoto, H.**, "Storage hierarchy for video-on-demand systems", *Proc. of Storage and Retrieval for Image and Video Databases II*, SPIE, Vol.2185, 198-207, 1994.

[18] **Rowe, L.A., Boreczky, J.S., and Berger, D.A.**, "A distributed hierarchical video-on-demand system", *Proc. of IEEE International Conference on Image Processing*, IEEE, 334-337, 1995.

[19] **Brubeck, D.W. and Rowe, L.A.**, "Hierarchical storage management in a distributed VOD system", *IEEE Multimedia*, IEEE, 37-47, 1996.

[20] **Dan, A., and Sitaram, D.**, "An online video placement policy based on bandwidth to space ratio (BSR)", *Proc. of SIGMOD*, ACM, 376-385, 1995.

[21] **Federighi, C., and Rowe, L.A.**, "A distributed hierarchical storage manager for a video-on-demand system", *Proc. of Conference on Storage and Retrieval for Image and Video Databases II*, SPIE, Vol. 2185, 185-195, 1994.

[22] **Tse P.K.C.**, "Tertiary storage for multimedia system", *Proc. of Computer Science Postgraduate Students Conference*, Royal Melbourne Institute of Technology, Melbourne, 1996.

[23] **Tavanapong, W., Hua, K.A., and Wang, J.Z.**, "A framework for supporting previewing and VCR operations in a low bandwidth environment", *ACM Multimedia'97*, ACM, 303-312, 1997.

Exploiting Image Indexing Techniques in DCT Domain*

C. W. Ngo**, T. C. Pong & R. T. Chin

The Hong Kong University of Science & Technology,
Clear Water Bay, Kowloon, Hong Kong
Email: {cwngo, tcpong, roland}@cs.ust.hk

Abstract. This paper is concerned with the indexing and retrieval of images based on features extracted directly from JPEG discrete cosine transform (DCT) domain. We examine possible ways of manipulating DCT coefficients by standard image analysis approaches to describe image shape, texture, and color. Through Mandala transformation, our approach groups a subset of DCT coefficients to form ten blocks. Each block represents a particular frequency of the original image. We use two blocks to model rough object shape; nine blocks to describe subband properties; and one block to compute color distribution. As a result, the amount of data used for processing and analysis is reduced significantly. This can lead to simple yet efficient ways, of indexing and retrieval in a large scale image database. Experimental results show that it only takes approximately 6ms to index shape features, 5ms to index texture features, and 8ms to index color features from an image the size of 128×128 on a Sun Sparc Ultra-1 machine.

1 Introduction

With the wide spread use of WWW, future digital libraries are expected to manipulate huge amounts of image and video data. Due to the limitations of space and time, most of the data are represented in compressed forms. As a result, techniques in editing, segmenting, and indexing images directly in the compressed domain have become one of the important topics in digital libraries. In this paper, we investigate the use of DCT coefficients, which are the major components of JPEG and MPEG, in content-based image retrieval (CBIR). In general, CBIR emphasizes rough image matching rather than exact matching. Our approach can capture the rough and global content of an image with very few DCT coefficients. Since other techniques such as relevancy feedback [1] and query expansion [2] can be used to fine tune the retrieval results, CBIR should not suffer greatly as long as a majority of relevant images are retrieved.

In our proposed approach, we extract ten DCT coefficients from each 8×8 JPEG image block. By applying Mandala transformation [3], we group these

* This work is supported in part by RGC Grants HKUST661/95E and HKUST6072/97E

** Please direct all enquires to C. W. Ngo, Email:cwngo@cs.ust.hk

coefficients and form ten blocks. Each block represents a particular frequency content of the original image. We apply the appropriate techniques to each block and generate features to describe the shape, texture, and color properties of the original image. Since the first block conveys color information, we use it to compute color histograms. To model object shape, we combine two blocks to compute its image gradient. With this, we track the contour of the underlying object and compute moments to estimate its global shape. Finally, we calculate the intensity variances of these blocks to describe their texture properties.

2 Related Work

The recent works on the processing of compressed data include algebraic operations [4], geometric transformation [6], image segmentation [7], feature extraction [8], indexing [9], and camera break detection [11]. Smith & Rowe [4] show how pixel addition, pixel multiplication, scalar addition and scalar multiplication can be implemented in DCT domain. With these algorithms, one can dissolve a sequence of compressed images and overlay a subtitle on a compressed image. Chang & Messerschmitt [5] further propose algorithms to manipulate compressed videos using the DCT coefficients with or without motion compensation. Shen & Sethi [6] describe the methods on performing geometric transformations such as rotation and diagonal flip by manipulating DCT coefficients. In addition, Soltane *et al.* [7] suggest an adaptive edge operator selection scheme for image segmentation based on mean, variance, and entropy of DCT coefficients. Shen & Sethi [8] further present an edge detector with twenty times speed up compared to conventional methods. Their proposed approach determines edge strength and orientation through the analysis of the patterns of DCT coefficients, however, this method assumes that the edge in each 8×8 block is a straight line. As a result, disconnected and broken edges, which are not suitable for contour extraction or segmentation, happen at the boundaries of blocks. In contrast to Shen & Sethi [8], our proposed method generates a reduced yet smooth edge map by manipulating two Mandala blocks. Since CBIR requires only rough matching, our method provides adequate visual cues and is suitable for further image analysis.

Chang [9] reports several possible ways of extracting low level features from the compressed domain. For instance, the texture feature is formed by computing the statistical measures of the DCT coefficients. To reduce the dimensionality of the feature space, one can employ the Fisher Discriminant techniques to maximize the separability among the known texture classes. Similarly, Seales *et al.* [10] employ principle component analysis to obtain eigenvectors from DCT coefficients. Since DCT is linear and orthogonal, the distance in eigenspace is preserved. The authors project the DCT space to the first few principle axes and perform object recognition directly in the compressed domain. These approaches may not be suitable for databases of large volume due to the training and updating costs, moreover, the retrieval accuracy may be degraded when new data arrives. Since future digital libraries are targeted to tackle dynamic environments in the Internet, re-training of the huge amount of data will also be infeasible.

The properties of DCT coefficients have also been investigated in the pattern recognition field. One of the earlier research, where DCT coefficients are applied to classified ship and cloud images, is reported by Hsu *et al.* [3]. The authors have claimed that DCT domain has some unique scale invariance and zooming characteristics that provide certain insight into object and texture process identification. In addition, Hou *et al.* [13] demonstrate that DCT behaves approximately like subband filters and their impulse responses closely relate to wavelets. In fact, it is also worthwhile to investigate the possibility of using DCT coefficients especially in describing object shape, which is not well understood in the current literature.

DCT coefficients also play a major role in video segmentation. Patel & Sethi [11] propose approaches to detect camera cuts by analyzing the DCT coefficients extracted from I-frames of MPEG. The DCT coefficients of P and B-frames can be estimated using algorithm developed by Yeo & Liu [14]. In addition, Ariki & Saito [15] report that DCT coefficients are insensitive to abrupt intensity change due to camera flushing. Motivated by these research efforts, it is obvious that image indexing techniques in the DCT domain can also be employed directly in video segmentation.

3 JPEG DCT Coefficients

3.1 Compression Scheme

In JPEG the original image is divided into 8 × 8 blocks, then, each block is transformed independently by DCT. The transformed coefficients are quantized and Huffman coded. The only information loss is due to the quantization step. The quantization factors will perturb but not destroy the essential characteristics of the DCT coefficients. The DCT is defined as

$$F_{u,v} = \frac{1}{4}C(u)C(v)\sum_{i=0}^{7}\sum_{j=0}^{7}\cos\frac{(2i+1)u\pi}{16}\cos\frac{(2j+1)v\pi}{16}f_{i,j} \tag{1}$$

where $C(u), C(v) = \frac{1}{\sqrt{2}}$ if $u, v = 0$, otherwise $C(u), C(v) = 1$. $F_{u,v}$ is the 2D DCT coefficient, and $f_{i,j}$ is the image spatial value. $F_{0,0}$ is normally called DC, while the rest are referring to as AC coefficients. The basis vectors of DCT are linear and orthogonal.

In this paper, we only extract $F_{0,0}$, $F_{0,1}$, $F_{1,0}$, $F_{2,0}$, $F_{1,1}$, $F_{0,2}$, $F_{0,3}$, $F_{1,2}$, $F_{2,1}$ and $F_{3,0}$ for indexing and retrieval. All coefficients are quantized but not Huffman coded. Experiments in the following sections are conducted on a Sun Sparc Ultra 1 machine.

3.2 Basic Properties

Denote $\mathbf{F}_j = \cos\frac{(2j+1)\pi}{16}\sum_{i=0}^{7} f_{i,j}$ for $0 \le j \le 3$ and $\mathbf{F}_j = \cos\frac{15-2j\pi}{16}\sum_{i=0}^{7} f_{i,j}$ for $4 \le j \le 7$, which is the sum of column j multiplied by a constant. Then,

$F_{0,0} = \frac{1}{8}\sum_{i=0}^{7}\sum_{j=0}^{7} f_{i,j}$ and $F_{0,1} = \frac{1}{4}\{(\mathbf{F}_0+\mathbf{F}_1+\mathbf{F}_2+\mathbf{F}_3)-(\mathbf{F}_4+\mathbf{F}_5+\mathbf{F}_6+\mathbf{F}_7)\}$. Notice that $F_{0,0}$ is actually 8 times of the block intensity mean and $F_{0,1}$ is the horizontal block intensity difference. Similarly, $F_{1,0}$ is the vertical block intensity difference.

One can project these coefficients from the DCT domain to Mandala domain. Mandala transformation simply groups the coefficients with the same u, v as a block, where each block represents a particular frequency content of the original image. Denote I as the Mandala block with $u, v = 0$, $I_{x^u} = \frac{\partial^u I}{\partial x^u}$ as the Mandala block with $v = 0$ and $I_{x^v} = \frac{\partial^v I}{\partial x^v}$ as the Mandala block with $u = 0$. Then, $I_{x^u y^v} = \gamma \frac{\partial^u I}{\partial x^u}\frac{\partial^v I}{\partial y^v}$ represents the Mandala block with $u, v \neq 0$, where γ is a variable depending upon I. We can regard the 64 blocks as $I, I_x, I_y, I_{y^2}, I_{xy}, I_{x^2}, \ldots, I_{x^7 y^7}$ in zig-zag order.

I is normally called DC image. We can express the gradient ∇f, edge direction θ, and zero crossing $\nabla^2 f$ of I as,

$$\nabla f = \sqrt{F_{0,1}^2 + F_{1,0}^2} \tag{2}$$

$$\theta = \tan^{-1}\frac{F_{1,0}}{F_{0,1}} \tag{3}$$

$$\nabla^2 f = F_{0,2} + F_{2,0} \tag{4}$$

Figure 1 shows the resulting 64 × 64 images in Mandala domain, computed from the 512 × 512 lena and airplane images.

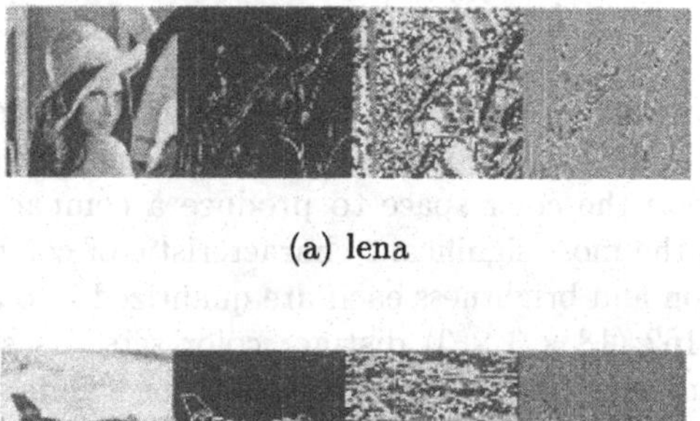

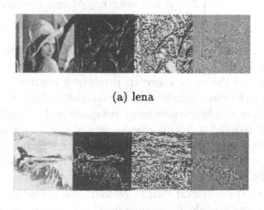

(a) lena

(b) airplane

Fig. 1. Images constructed directly from DCT coefficients. (*from left to right*) intensity or DC, image gradient, edge direction and zero crossing images.

In fact, DCT is an *approximated* Karhunen-Loeve Transform (KLT) for any image. It is asymptotically equivalent to KLT for stationary Markov-1 signals [12]. Its auto-covariance matrix, Toeplitz matrix, is predetermined. In this case, selecting the first few DCT coefficients is approximately equivalent to selecting the KLT transformed values that exhibit significant variance.

4 Color Histogram

We group the zero frequency, $F_{0,0}$, coefficient of JPEG image to form a reduced and smooth DC image. We then transform the color space from RGB (red, green, blue) to HSV (hue, saturation, brightness). HSV is widely used in color histograms because of its uniformity, compactness, completeness and naturalness [16]. [2]

Let $v_c = (r, g, b)$ be the color triple in RGB space, and $w_c = (h, s, v)$ be the corresponding transformed triple in HSV space, where $r, g, b, s, v \in [0 \ldots 1]$ and $h \in [0 \ldots 6]$. Denote $\mathbf{T}$ as the transformation, then $w_c = \mathbf{T}(v_c)$. $\mathbf{T}$ is,

$$v = \max(r, g, b) \quad s = \frac{v - \min(r, g, b)}{v} \tag{5}$$

$$x = \frac{v - r}{v - min(r, g, b)} \quad y = \frac{v - g}{v - \min(r, g, b)} \quad z = \frac{v - b}{v - \min(r, g, b)}$$

$$h = \begin{cases} 5 + z; \ r = max(r, g, b) \text{ and } g = \min(r, g, b) \\ 1 - y; \ r = max(r, g, b) \text{ and } g \neq \min(r, g, b) \\ 1 + x; \ g = max(r, g, b) \text{ and } b = \min(r, g, b) \\ 3 - z; \ g = max(r, g, b) \text{ and } b \neq \min(r, g, b) \\ 3 + y; \ b = max(r, g, b) \text{ and } r = \min(r, g, b) \\ 5 - x; \text{ otherwise} \end{cases}$$

The color histograms are obtained by summing the number of pixels with similar values in the HSV components. To reduce the length of the histogram features, we quantize the color space to produce a compact set of colors. Because hue conveys the most significant characteristic of color, we quantize it to 18 levels. Saturation and brightness each are quantized into 3 levels. The quantization provides 162 ($18 \times 3 \times 3$) distinct color sets. As stated in [16], such representations can yield greater perceptual tolerance while separating the hues such that red, green, blue, yellow, magenta, and cyan are each represented by three subdivisions.

Figure 2 shows the retrieval results of four image queries from the VisTex database [18]. We use Euclidean distance for the similarity measure. The results indicate that these color indices are also useful for retrieving images with similar texture; in addition, it takes less than 2 seconds to index 228 JPEG images.

[2] Note that JPEG uses YCrCb color space. Through the software package in [17], one can obtain the RGB components directly from JPEG images.

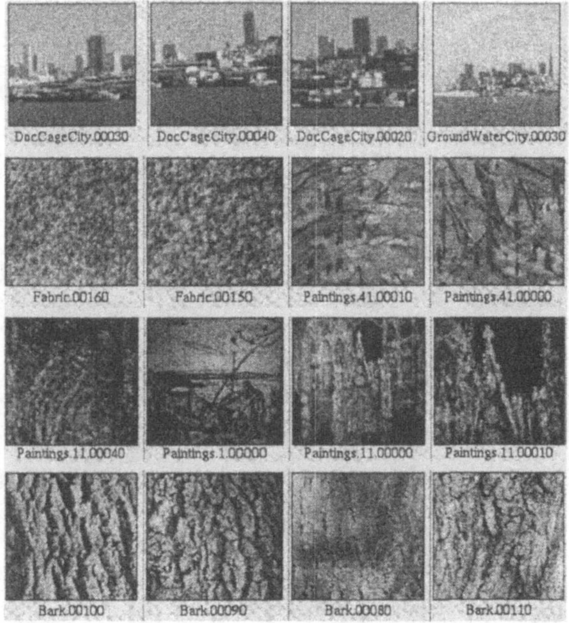

Fig. 2. Retrieval results of four image queries by using color features. The left column consists of query images. The results are ranked from left to right, starting with the most similar image.

5 Rough Shape Modeling

The basic idea in our shape modeling scheme is to generate an image gradient $\bigtriangledown f$ from $F_{0,1}$ and $F_{1,0}$, track the contour of the underlying object, and then compute the invariant contour moment features for indexing. We use the shape database suggested in [20] for our experiments. The database consists of 83 JPEG animal images. Figure 3 shows some sample images and their corresponding extracted contours. The size of a sample image is 128×128, while the contour image is only 16×16!

Given a contour $\mathbf{V} = [\mathbf{v}_1, \mathbf{v}_2, \ldots, \mathbf{v}_n]$, where $\mathbf{v}_i$ is defined on the finite grid: $\mathbf{v} \in \mathbb{E}^2 = \{(x, y) : x, y = 1, 2, \ldots, M\}$. Denote $\mathbf{g} = (\bar{x}, \bar{y})$ as the contour centroid. The central moment of the $(p+q)$th order is

$$\mu_{p,q} = \sum_{(x,y)} \sum_{\in \mathbf{v}_i} (x - \bar{x})^p (y - \bar{y})^q \tag{6}$$

The centre moments are invariant to translation. They can be normalized to scale invariance by [19]

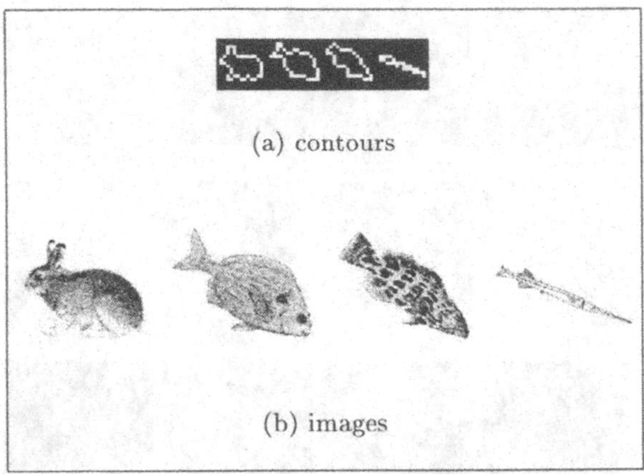

Fig. 3. The contours (a) extracted from the corresponding images in (b).

$$\eta_{pq} = \frac{\mu_{pq}}{\mu_{00}^{\gamma}} \text{ where } \gamma = p + q + 1, \text{ for } p + q = 2, 3, \ldots \quad (7)$$

Seven moment invariants based on the 2nd and 3rd-order moments are computed as,

$$\begin{aligned}
\phi_1 &= \eta_{20} + \eta_{02} \quad (8)\\
\phi_2 &= (\eta_{20} - \eta_{02})^2 + 4\eta_{11}^2\\
\phi_3 &= (\eta_{30} - 3\eta_{12})^2 + (\eta_{03} - 3\eta_{21})^2\\
\phi_4 &= (\eta_{30} + \eta_{12})^2 + (\eta_{03} + \eta_{21})^2\\
\phi_5 &= (\eta_{30} - 3\eta_{12})(\eta_{30} + \eta_{12})[(\eta_{30} + \eta_{12})^2 - 3(\eta_{21} + \eta_{03})^2] +\\
&\quad (3\eta_{21} - \eta_{03})(\eta_{21} + \eta_{03})[3(\eta_{30} + \eta_{12})^2 - (\eta_{21} + \eta_{03})^2]\\
\phi_6 &= (\eta_{20} - \eta_{02}[(\eta_{30} + \eta_{12})^2 - (\eta_{21} + \eta_{03})^2] + 4\eta_{11}(\eta_{30} + \eta_{12})(\eta_{21} + \eta_{03})\\
\phi_7 &= (3\eta_{21} - \eta_{03})(\eta_{30} + \eta_{12})[(\eta_{30} + \eta_{12})^2 - 3(\eta_{21} + \eta_{03})^2] +\\
&\quad (3\eta_{12} - \eta_{30})(\eta_{21} + \eta_{03})[3(\eta_{30} + \eta_{12})^2 - (\eta_{21} + \eta_{03})^2]
\end{aligned}$$

Generating moment, by its nature, is computationally expensive. Nevertheless, due to the significant amount of data reduction, it takes less than 0.5 second to index 83 JPEG images. Figure 4 shows the retrieval results of four image queries. We employ the cosine similarity measure to rank the results. More than half of the top twelve retrieved images are similar to human perception. In most cases, the top four images are ranked correctly. Since [20] adopts the deformable model for exact shape modeling in the spatial domain, their results are better than ours. Nevertheless, our approach is computationally efficient and is capable of including similar images in the top rankings. As a compromise, one can decompress these top ranked images and apply the deformable model to fine tune the retrieval results.

Fig. 4. Retrieval results of four image queries by using shape features. The left column consists of query images. The results are ranked from left to right, starting with the most similar image.

6 Texture Description

Because DCT compresses the image energy into lower order coefficients, we only consider the first nine AC coefficients. The texture feature is,

$$S_{u,v} = \mathbf{E}[F_{u,v}^2] - \mathbf{E}[F_{u,v}]^2 \tag{9}$$

where $S_{u,v}$ and $\mathbf{E}[F_{u,v}]$ are the variance and expectation of $F_{u,v}$ respectively, for $0 < u + v \leq 3$. Notice that the resulting texture feature vector is actually the variance of $< I_x, I_y, I_{x^2}, I_{xy}, I_{y^2}, I_{x^3}, I_{x^2y}, I_{xy^2}, I_{y^3} >$ in the Mandala space.

We set up two experiments to investigate how the performance changes as we increase the database size. The databases composed of images from the Brodatz Album [21]. In the first experiment, we cut 22 images of size 512 × 512 into 352 non-overlapping images of size 128 × 128. In the second experiment, we cut 36 images of size either 1024 × 1024 or 512 × 512 into 2064 images of size 128 × 128. Euclidean distance is employed for the similarity measure. In addition, we evaluate the performance in terms of recall and precision where

$$recall = \frac{\text{number of relevant images retrieved}}{\text{total number of relevant images in database}} \tag{10}$$

$$precision = \frac{\text{number of relevant images retrieved}}{\text{total number of images retrieved}} \tag{11}$$

This standard evaluation mechanism is widely used in a series of TREC evaluations for document retrieval [22]. *Recall* measures a system's ability to present all relevant items, while *precision* measures the ability of a system's ability to present only relevant items. Sub-images originated from the same image are classified as similar and relevant images. We use every image in the database as the query image and measure its recall and precision. Then, we calculate the 11-point average precision values for recall levels $0.0, 0.1, \ldots, 1.0$ over all images. The recall-precision curves for the databases of 2064 images and 352 images are shown in Figure 5. Their mean precision values, over the 11-point precision values, are 0.593 and 0.775 respectively.

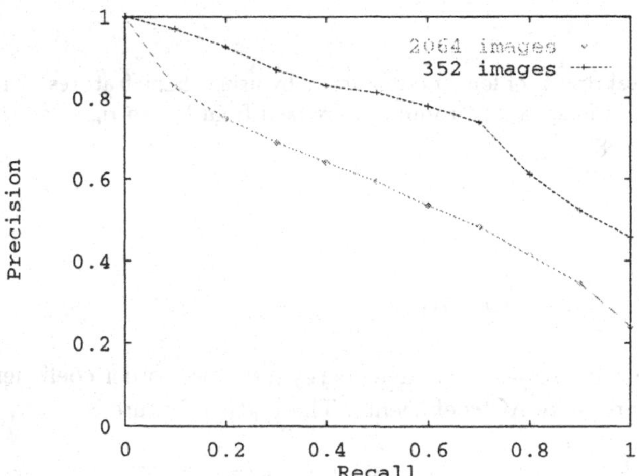

Fig. 5. Average recall and precision curves for the databases of 2064 images and 352 images. Their mean precision values are 0.593 and 0.775 respectively.

Figure 6 shows the retrieval results from the Brodatz database [21] of 2064 images. Figure 7 shows the retrieval results from the VisTex database [18] of 228 images. The results indicate that our approach works reasonably well with both homogeneous and inhomogeneous texture patterns. In addition, it only takes approximately 10 seconds to index 2064 JPEG images.

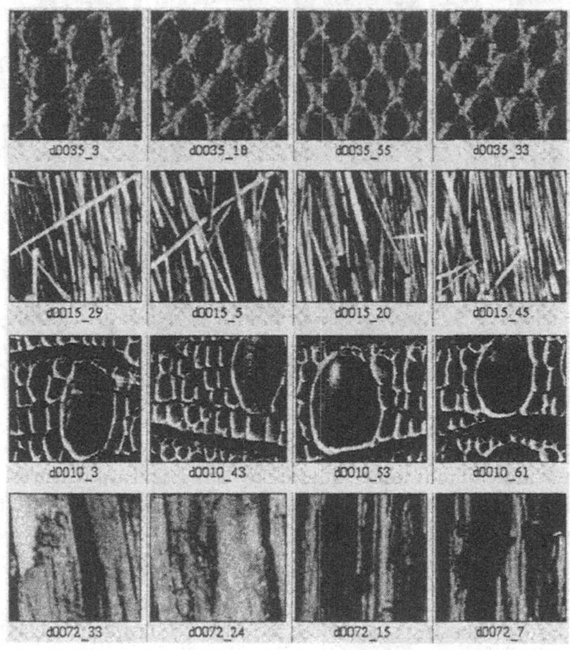

Fig. 6. Retrieval results of four image queries by using texture features. The left column consists of query images. The results are ranked from left to right, starting with the most similar image. (Database: Brodatz)

7 Conclusion and Future Work

We have presented approaches for indexing shape, texture, and color features directly in the DCT domain. In order to speed up shape indexing in large scale image databases, we propose a rough shape matching technique capable of achieving acceptable retrieval accuracy. One can further refine the retrieval results by applying the deformable model to discard spurious matches of the currently retrieved images, or including other appropriate features into the original query through relevance feedback. Besides shape modeling, we calculate the variance of nine Mandala blocks to describe image texture properties, such as

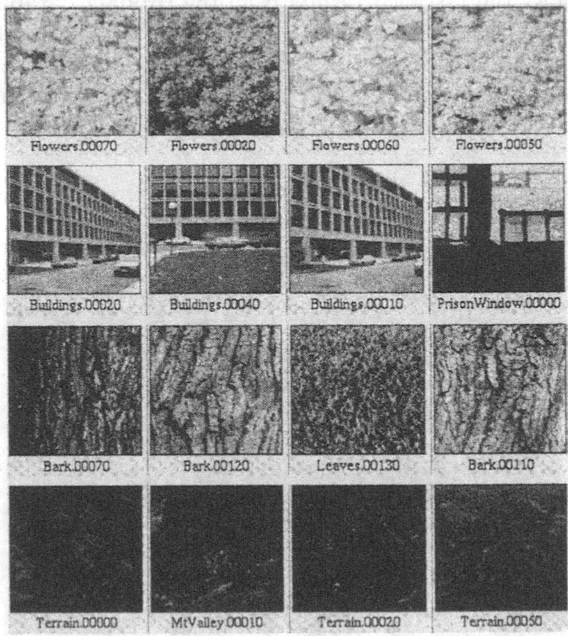

Fig. 7. Retrieval results of four image queries by using texture features. The left column consists of query images. The results are ranked from left to right, starting with the most similar image. (Database: VisTex)

rough edge and zero crossing statistics. To index color features that are perceptually uniform, we transform the color space of DC values from YCbCr to HSV. Experimental results demonstrate that our methods are able to compromise two conflicting factors: *indexing speed* and *retrieval accuracy*.

Our future goals include combining shape, texture, and color features for retrieval. We will incorporate relevancy feedback techniques to model the subjectivity of human perception, and hence, investigate the learning capability of retrieval systems that can automatically select the appropriate features for retrieval.

References

1. Y. Rui, T. S. Huang & S. Mehrotra.: Content-based Image Retrieval with Relevancy Feedback In Mars. Proc. of IEEE Int. Conf. on Image Processing, pp. 815-818, 1997.
2. Kok F. Lai, H. Zhou, & S. Chan.: Query Expansion by Raw Image Features and Text Annotations in Image Retrieval. Third Asian Conf. on Computer Vision, vol. 1, pp. 402-409, 1998.

3. Y. S. Hsu, S. Prum, J. H. Kagel, and H. C. Andrews.: Pattern Recognition experiments in the Mandala/cosine domain. IEEE Trans. Pattern Anal. Machine Intell., vol. PAMI-5, no. 5, pp. 512-520, Sept, 1983.
4. B. C. Smith, & L. A. Rowe.: Algorithms for manipulating compressed images. IEEE Computer Graphics and Applications, vol. 13, no. 5, pp. 34-42, Sept 1993.
5. Shih-Fu Chang & D. G. Messerschmitt.: Manipulation and Compositing of MC-DCT Compressed Video. EEE Journal on Selected Areas in Communications, vol.13, no.1, pp. 1-11, Jan 1995.
6. B. Shen & I. K. Sethi.: Inner-Block Operations on Compressed Images. Proc. ACM Intl. Conf. Multimedia'95, pp. 490-499, Nov, 1995.
7. S. Soltane, N. Kerkeni & J. C. Angue.: The use of Two Dimensional Discrete Cosine Transform for an Adaptive Approach to Image Segmentation. Proc. SPIE Image and Video Processing IV, pp. 242-251, 1996.
8. B. Shen & I. K. Sethi.: Direct Feature Extraction from Compressed Images. Proc. SPIE Storage and Retrieval for Image and Video Database IV, vol. 2670, pp. 404-14, 1996.
9. Shih-Fu Chang.: Compressed Domain Techniques for Image/Video Indexing and Manipulation. IEEE Intern. Conf. on Image Processing, ICIP 95, pp. 314-317, 1995.
10. W. Brent Seales, C. J. Yuan & W. Hu.: Content Analysis of Compressed Video. Technical Report 265-96, University of Kentucky, 1996.
11. N. V. Patel, I. K. Sethi.: Compressed Video Processing for Cut Detection. IEE Proc. Visual Image Signal Process, vol. 143, no. 5, pp. 315-23, Oct 1996.
12. K. R. Rao & P. Yip.: Discrete Cosine Transform : Algorithm, Advantage, Applications. The Univeristy of Texas, Academic Press, 1990.
13. H.S.Hou, D.R.Tretter, M.J. Vogel.: Interesting Properties of the Discrete Cosine Transform. Journal of Visual Communication and Image Representation, vol. 3, no. 1, pp. 73-83, March 1992.
14. B.L. Yeo and B. Liu.: On the Extraction of DC Sequence from MPEG Compressed Video. IEEE Int. Conf. on Image Processing, vol. 2, pp. 260-3O, Oct 1995.
15. Y. Ariki & Y.Saito.: Extraction of TV News Articles Based on Scene Cut Detection Using DCT clustering. Proc. Int. Conf. on Image Processing, vol. 3, pp. 847-50, 1996.
16. J. R. Smith.: Integrated Spatial and Feature Image Systems: Retrieval, Analysis and Compression. Ph.D Thesis, Chapter 2, Columbia University, 1997.
17. ftp.uu.net/graphics/jpeg/jpegsrc.v6a.tar.gz, The Independent JPEG Group's JPEG software.
18. www-white.media.mit.edu/vismod/imagery/VisionTexture/vistex.html, Vision Texture.
19. Chaur-Chin Chen.: Improved Moment Invariants for Shape Discrimination. Pattern Recognition, vol. 26, no. 5, pp. 683-86, 1993.
20. Stan Sclaroff.: Deformable Prototypes for Encoding Shape Categories im Image Databases. Pattern Recognition, vol. 30, nol. 4, pp. 627-41, April 1997.
21. P. Brodatz.: Textures: A Photographic Album for Artists and Designers. New York:Dover, 1966.
22. D. K. Harman.: The First Text REtrieval Conference (TREC-1). Information Processing and Management, vol. 29, no 4, pp. 411-414, 1993.

Content-Based Image Indexing and Retrieval in an Image Database for Technical Domains

Petra Perner

Institute of Computer Vision and Applied Computer Sciences
Arno-Nitzsche-Str. 45, 04277 Leipzig, Germany
Email: ibaiperner@aol.com

Abstract. The availability of a variety of sophisticated data acquisition instruments has resulted in large repositories of imagery data in different applications like non-destructive testing, technical drawing, medicine, museums and so one. Effective extraction of visual features and contents is needed to provide meaningful index of and access to visual data. In the paper, we proposed an image database architecture, which can be used for most industrial problems. The image database is able to handle structural representations of images. Indexing is possible object based, spatial relation based, and by a combination of both. The query can be a textual query or an image content-based query. We describe how the image query is processed, how similarity based retrieval is performed over images, and how the image database is organized. Results are presented based on an application of ultra sonic images from non-destructive testing.

Keywords: Image Database, Query-by-Image-Content, Structural Similarity Measure, Indexing, Learning

1 Introduction

In the paper, we describe an approach, which combines both concepts, the concept of textual queries with query-by-image content, for image indexing and retrieval. The approach allows a user to use both types of queries: to index by textual description if no proper images for indexing are available and to index by image content if the content based query can be processed from an available image. For the textual keywords is the user provided by the system with standard vocabulary for shape, size, gray level and spatial location, which he can input to the system via a keyword mask. The same semantic content can be processed from the image by image processing and signal-symbol transformation unit. This ensures high flexibility, completeness and consistency in the textual description and the usage of the same algorithm for similarity determination and image retrieval.

The signal-symbol transformation unit and the feature extraction unit are generic. The image-processing unit in the system uses domain dependent algorithm. This requires from the user to specify the domain he is considering before indexing and retrieval of images, but allows automatic processing of image queries.

The system considers two main data types: objects and scenes. Objects are described by their shape, size, location and gray level. A scene comprises of objects and their spatial relation to each other, which makes up a high-level description of an image. The high-level description is a structural representation of the image realized as attributed graph.

Such a description is sufficient for most technical domains like mechanical objects, defect images of welding seams, ultra sonic images of vessels, fingerprint images [12] and technical drawings [13].

In Section 2, we describe the overall architecture of our image database. The content based query features are described in Section 3. The similarity measure as well as the algorithm used for similarity determination is presented in Section 4 and Section 5. Section 6 and Section 7 deal with the index structure. Retrieval is presented in Section 8. Finally, we show our results in Section 9 and give conclusions in Section 10.

2 Overview of Image Database

Our image database mainly consists of five functional units (see Figure 1):

- the image database itself with the index structure,
- the image processing and interpretation unit with the image query construction unit,
- the textual user interface,
- the automatic image query construction unit,
- the similarity determination and retrieval unit, and
- the learning unit for index structure updating.

Two query types are possible: textual keywords and image content based queries. For inputting the textual queries the user is provided by the system with a vocabulary for image description. The same terms are used by the signal-to-symbol transformation unit, which automatically transforms the numeric features extracted from the image to a symbol.

The content-based query is automatically processed from the image by the image processing algorithm and feature extraction unit. This results in another problem with image databases caused by pattern recognition. An accurate automatic detection and recognition algorithm never works for all kind of images. Such an algorithm is rather domain dependent than generic. Therefore, Lee et al [15] suggest to provide an image database with standard procedures for image enhancement, image analysis and feature extraction. The application of these procedures to the image is left to the user. He can process the query in an interactive fashion with the help of the procedures provided by the system. However, it requires by the user knowledge about the image processing technology. He needs to be trained on image processing and pattern recognition technology in order to understand what effect image processing and pattern recognition procedures produces on the image.

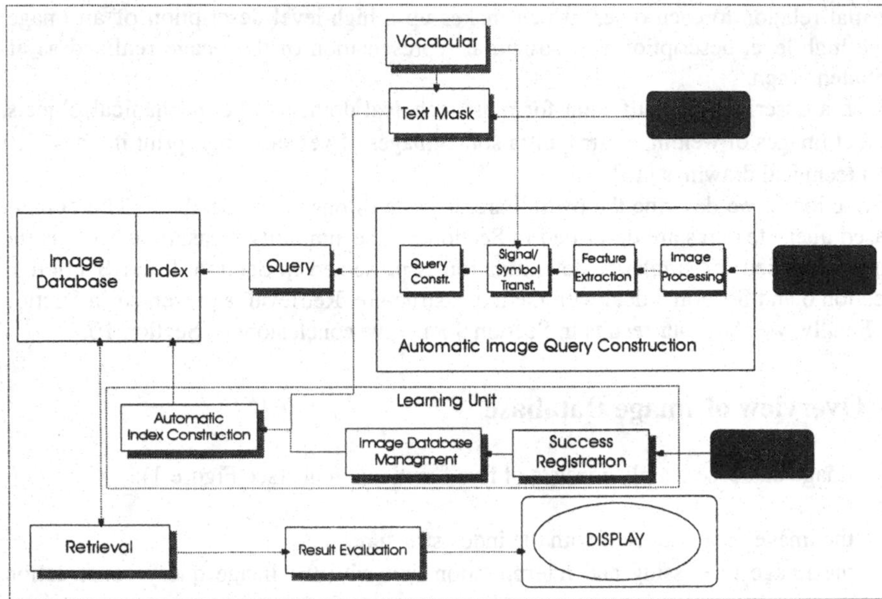

Fig. 1 Architecture of the Image Database

Such an approach is not appropriate in most technical and medical domains where the image databases are used in day to day practice. It is sufficient to provide the system with domain dependent image processing and pattern recognition algorithm for time efficiency purposes.

Therefore is our system equipped with domain dependent image processing facilities, described in Section 9 on the special domain. That requires from the user to specify to the system the domain he is working on but only in case that images from more than one application is contained in database. Based on that information the system selects the proper image-processing algorithm for automatic image query processing which had been developed and installed beforehand. Such an approach ensures no interaction with the user, which makes it easier for him to employ the database.

After the numeric features are transformed to symbolic features the high-level representation of the image is formed and used for query. The high-level representation is an attributed graph. The similarity measure should work for that kind of representation and should allow exact and partial match retrieval.

The index structure is automatic constructed from the high-level representation of the image. It should allow indexing the whole image as well as part images and single objects.

New images not contained in database should easily be incorporated to the image database as well as to the index structure. Therefore, a learning unit observes the success or failure of the database and activates the automatic index construction unit for incremental learning of index structure.

The result of the retrieval process is shown on display to the user.

An entry of the image database consists of: non-image information like date of image acquisition, sensor parameter and so one, the high-level representation of the image and the image itself, see Fig. 2.

3 Content Based Query Features

The system considers two main data types: objects and scenes. Objects have features like gray level, location, size and shape. The scenes comprises of objects and their spatial relation to each other and make up the high-level description of the image. Generally, an automatic image processing algorithm will consist of the following steps: image preprocessing, image segmentation (meaning labeling of object pixel and background pixel), morphological operations for noise reduction, labeling of objects, contour following method, numeric feature extraction and signal-symbol conversion. Once the object has been labeled (see Fig. 3) all the next operations are generic enough to get used for processing of other kind of images.

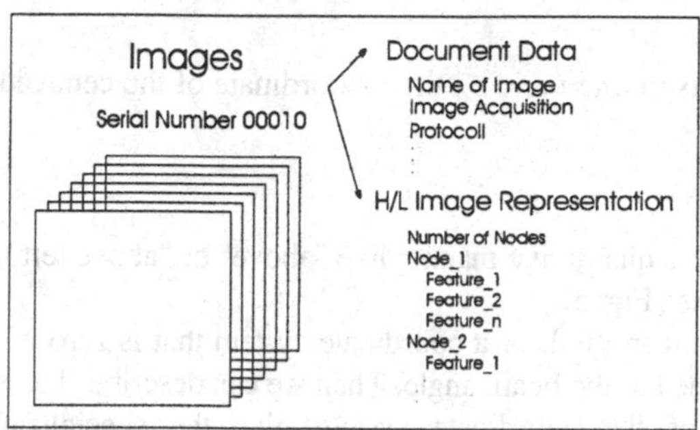

Fig. 2 Data Structure

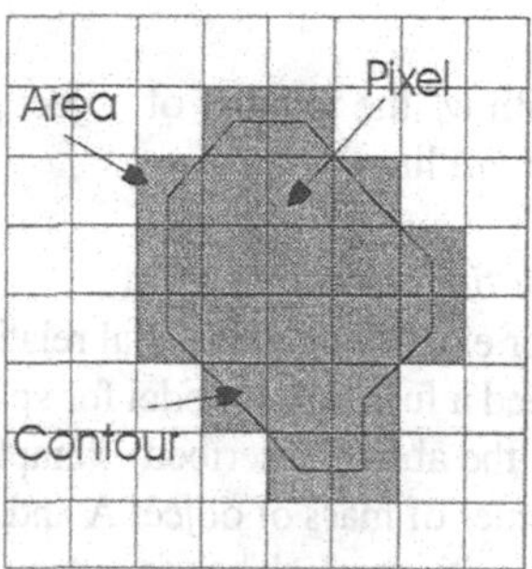

Fig. 3 Labeled Object and Object Contour

3.1 Features

The computed features are the following:

Gray level:
From the area inside the extracted object boundary the mean gray level is computed and taken as the gray level feature for the object. We quantize the gray level space into k levels and associates to each level a symbolic term like white, very light gray, light gray, gray, dark gray, very dark gray, black.

Size:
The size of an object is computed from the area A inside the extracted object boundary.
Shape:
The shape feature is computed based on the following formulae:

$$F = 10 \bullet \frac{A}{u^2}$$

with u for contour length.
The following symbols are associated to the values of *F:* round, longelongated, non-round.
This is a simple shape measure, which cannot describe complex objects but is accurate enough for most technical applications. If there is a need for more complex shape measure, we can change our measure to other measures in our system; e.g. moment based shape measure [16] or fractal dimension[17].

Location:
The centroid for an object is calculated and the coordinates s_x, s_y associated to that pixel are taken for the location.

$$s_y = \frac{1}{A} \sum_{i=y\min}^{y\max} w_i \cdot i \quad s_x = \frac{1}{A} \sum_{y\min}^{y\max} w_i \cdot s_{xi} \, ,$$

with w_i the number of object points in line i and s_{xi} the x-coordinate of the centroid of the *ith* line.

Spatial Location:
For expressing the spatial relation in a qualitative manner like "above" or "above left" we need a functional model for space, see Fig. 5.
In the above-described example, we can think of a coordinate system that is zero in the center of mass of object A and aligned to the beam angle. Then we can describe "behind" and "above". The four square of the coordinate system give the specialization "left_behind", "right_behind" and "left_infront", "right_infront". We can shift the coordinate system from one object to another object. Then, we look from that focal point to the spatial relations to all other objects in the image. There are various levels of granularity [18]:

Projection		
disjointness	no_contact	
tangency		
overlap	contact	no_projection_info
inclusion		

Orientation
0. no orientation
1. a. left right
   b. back in_front
2. back, left, right, in_front
3. back, left, right, in_front, ..., right_back, left_back , right_in_front,....

Note that the meaning of back e.g. varies depending on the level of granularity.
The line, for example S_O and S_U, gives the interval for the spatial attribute right_in_front that will play a role in the later described similarity. An abstraction for two spatial attributes, one is "right" and the other is "right_in_front", would be "right, in_front".
For the representation of "more_left_behind", we need to quantize our model or a representation of a fuzzy area [19].

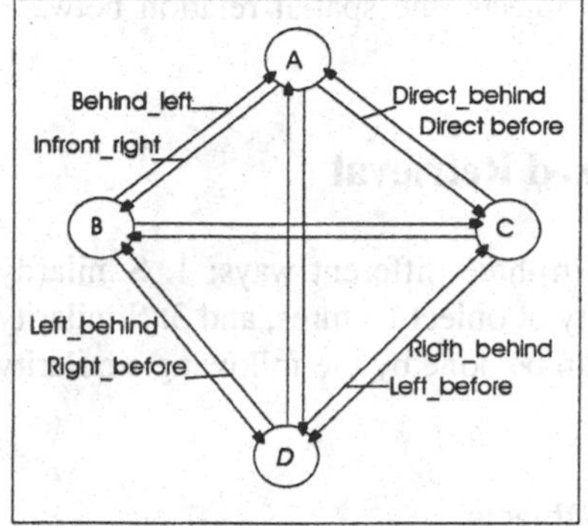

Fig. 4 Image Graph

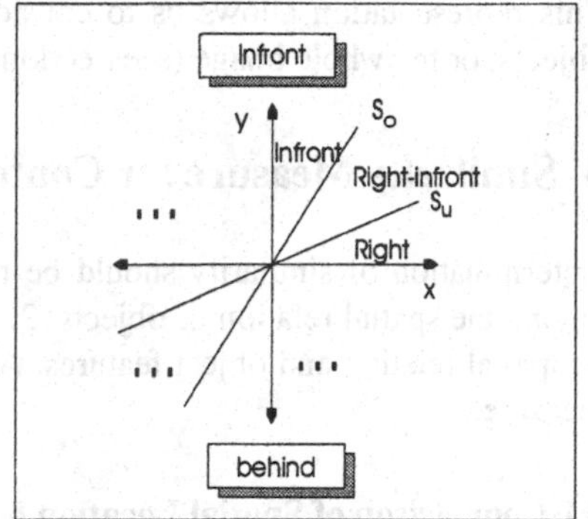

Fig. 5 Model for Spatial Relations

3.2 High-level Representation of the Image

The high-level description of an image comprises of objects, their object features, and the spatial relation between the objects. The intern representation is an attributed graph like shown in Figure 4. The graph is defined as:

Definition 1:

W ... set of attribute values
e.g.: W = {"dark_grey", "left_behind", "directly_behind", ...}
A ... set of all attributes
e.g.: A = {shape, object area, spatial_relationship, ...}
b: $A \rightarrow W$ partial mapping, called attribute assignments
B ... set of all attribute assignments over A and W.

A graph G = (N, p, q) consists of
N ... finite set of nodes
$p : N \rightarrow B$ mapping of attributes to nodes
$q : E \rightarrow B$ mapping of attributes to edges, where $E = (N x N) \backslash I_N$ and I_N is the Identity relation in N.

The nodes are the objects and the edges are the spatial relation between the objects. Each object has attributes, which gets associated to the corresponding node within the graph.

The explicit specification of the set of edges can be abandoned. The spatial relation between two objects is determinable between two objects any time. That means, the edges between the nodes of a graph do exist. The assumed symmetry of the set of edges is redundant according to spatial relations, but is advantageous for the part isomorphism algorithm. The attribute assignment of the opposite direction can be done without any problem by negation of edge labels, e.g. "/behind = in_front". Note, the set of edges is N*(N -1).

This representation allows us to consider only objects, the spatial relation between objects, or the whole image (see Section 4).

4 Similarity Measure for Content-Based Retrieval

Determination of similarity should be possible in three different ways: 1. Similarity among the spatial relation of objects, 2. Similarity of object features, and 3. Similarity of spatial relation and object features. All this can be done by the following similarity measure.

4.1 Comparison of Spatial Location between Objects

We may define our problem of similarity as to find structural identity or similarity between two structures. If we ask for structural identity, we need to determine isomorphism. That is a very strong requirement. A relaxation of the requirements is to ask for part isomorphism.

Based on part isomorphism, we can introduce a partial order over the set of graphs:

Definition 2:

Two graphs $G_1 = (N_1, p_1, q_1)$ and $G_2 = (N_2, p_2, q_2)$ are in the relation $G_1 \leq G_2$ iff there exists a one-to-one mapping f: $N_1 \rightarrow N_2$ with

(1) $p_1(x) = p_2(f(x))$ for all $x \in N_1$

(2) $q_1(x) = q_2(f(x), f(y))$ for all $x, y \in N_1$, $x \neq y$.

Is a graph G_1 included in another graph G_2 then the number of nodes of graph G_1 is not higher than the number of nodes of G_2.

4.2 Similarity Measure for Spatial Location and Object Features

Similarity between attributed graphs can be handled in many ways. We propose the following way for the measure of closeness.

In the definition of part isomorphism, we may relax the required correspondence of attribute assignment of nodes and edges in that way that we introduce ranges of tolerance:

If $a \in A$ is a attribute and $W_a \subseteq W$ is the set of all attribute values, which can be assigned to a, then we can determine for each attribute ***a*** a mapping:

$distance_a : W_a \rightarrow [0,1]$.

The normalization to a real interval is not absolute necessary but advantageous for the comparison of attribute assignments.
For example, let a be an attribute a = spatial_relationship and

W_a = {behind_right, behind_left, infront_right, ...}.

Then we could define:

$distance_a$ (behind_right, behind_right) = 0
$distance_a$ (behind_right, infront_right) = 0.25
$distance_a$ (behind_right, behind_left) = 0.75 .

Based on such distance measure for attributes, we can define different variants of distance measure as mapping:

distance: $B^2 \rightarrow R^+$
(R^* ... set of positive real numbers) in the following way:

$$distance(x,y) = 1/D \sum_{a \in D} distance_a (x(a), y(a))$$

with D = domain (x) $\cap$ domain(y).

Usually, by the comparison of graphs not all attributes have the same priority. Thus, it is good to determine a weight factor v_a and then, define the distance as following:

$$distance(x,y) = \sum_{a \in D} v_a * distance_a (x(a), y(a))$$

For definition of part isomorphism, we get the following variant:

Definition 3

Two graphs $G_1 = (N_1, p_1, q_1)$ and $G_2 = (N_2, p_2, q_2)$ are in the relation $G_1 \leq G_2$ iff there exists a one-to-one mapping f: $N_1 \rightarrow N_2$ and threshold's C_1, C_2 with
(1) distance(p_1(x), p_2(f(x)) $\leq C_1$ for all $x \in N_1$
(2) distance(q_1(x,y), q_2(f(x), f(y)) $\leq C_2$ for all x,y $\in N_1$, $x \neq y$.

Another way to handle similarity is the way how the L-set's are defined and particularly the inclusion of K-lists:

Given C a real constant, $n \in N_1$ and $m \in N_2$. K(n) $\subseteq_C$ K(m) is true iff for each attribute assignment b_1 of the list K(n) attribute assignment b_2 of K(m) exists, such that distance(b_1,b_2) $\leq$ C.
Each element of K(m) is to assign to different element in list K(n).

Obvious, it is possible to introduce a separate constant for threshold for every attribute. Depending on the application, the inclusion of the K-lists may be sharpen up by a global threshold:

If it is possible to establish a correspondence g according to the requirements mentioned above, then an additional condition is to fulfill:

$$\sum_{(x,y) \in g.} \text{distance}(x,y) \le C_3 \quad (C_3 \text{ - threshold constant}) .$$

Then, for the L-set we get the following definition (see also Sect. 5.1):

Definition 4

$$L(n) = \{m \_ m \in N_2, \text{distance } (p_1(n), p_2(m)) \le C_1, \; K(n) \subseteq_C K(m) \}.$$

In step 3 of the algorithm for the determination of one-to-one mapping is also to consider the defined distance function for the comparison of the attribute assignments of the edges. The total effort is increasing by this new calculation but the complexity of the algorithm is not changed.
For basic introduction to graph theory and graph grammars see [20] and [21]. For other similarity concepts see [13] and [14].

5 Algorithm for Determining Similarity between Structural Representations

Now, consider an algorithm for determining the part isomorphism of two graphs. This task can be solved with an algorithm based on [20]. The main approach is to find an overset of all possible correspondences f and then exclude non-promising cases. In the following, we assume that the number of nodes of G_1 is not higher than the number of nodes of G_2.
A technical help is to assign to each node n a temporary attribute list K(n) of all attribute assignments of all the connected edges:

$K(n) = (a \mid q(n,m) = a, m N \setminus \{n\}) \quad (n \in N)$.
The order of list elements has no meaning. Because all edges do exist in a graph the length of K(n) is equal to 2*(|N|-1).
For demonstration, consider the example in Fig. 7. The result would be:

K(X) = (bl, bl, br)
K(Y) = (br, br, bl).

The complexity of the algorithm is in the worst case $O(\mid N \mid^3)$.

In the next step, we assign to each node of G_1 all nodes of G_2 that could be assigned by a mapping f that means we calculate the following sets:

$$L(n) = \{ m \mid m \in N_2, p_1(n) = p_2(m), K(n) \subseteq K(m)\} .$$

The inclusion $K(n) \subseteq K(m)$ shows that in the list K(m) the list K(n) is included without considering the order of the elements. Does the list K(n) multiple contains an attribute assignment then the list K(m) also has to multiple contain this attribute assignment.

For the example in Fig. 6 and Fig. 7, we get the following L-sets:

$$L(X) = \{A\}$$
$$L(Y) = \{B_1\}$$
$$L(Z) = \{C\}$$
$$L(U) = \{ D, B_2\} .$$

We did not consider in this example the attribute assignments of the nodes.
Now, the construction of the mapping f is prepared and if there exists any mapping then must hold the following condition:

$$f(n) \in L(n)(n \in N_1) .$$

The first condition for a mapping f regarding the attribute assignments of nodes holds because of the construction procedure of the L-sets. In case of one, set L(n) is empty then there is no part isomorphism.
In addition, in the case of nonempty sets in a third step it is to check if the attribute assignments of the edges do match.
If there is no match then the corresponding L-set is to reduce:

for all nodes n_1 of G_1
  for all nodes n_2 of $L(n_1)$
    for all edges (n_1, m_1) of G_1
    if for all nodes m_2 of $L(m_1)$
    $p_1(n_1,m_1) \neq p_2(n_2,m_2)$
    then $L(n_1) := L(n_1) \setminus \{n_2\}$

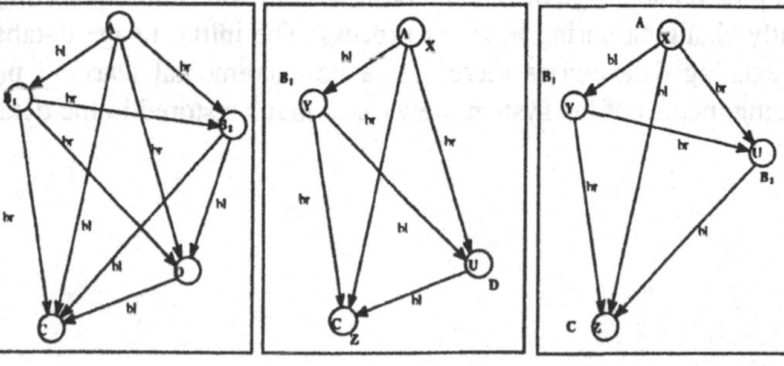

Fig. 6 Graph_1   Fig. 7 Graph_2   Fig. 8 Result

If during this procedure, the L-set of node has been changed then the examinations already carried out are to repeat. That means, this procedure is to repeat until none of the L-sets was changed.
Is the result of this step 3 an empty L-set then there is also no part isomorphism. If all L-sets are nonempty then some mapping's f form N_1 to N_2 are determined. If each L-set contains exactly only one element then there is only one mapping. In a final step all mappings are to exclude, which are not of the type one-to-one.

For example, let's compare the representation of pore_1 and pore_2 in Fig. 6 and Fig. 7. In step 3, the L-set of pore_1 will not be reduced and we get two solutions, shown in Fig. 7 and Fig. 8:

N_1	f_1	f_2
X	A	A
Y	B_1	B_1
Z	C	C
U	D	B_2

If we compare the representation of pore_1 and pore_3 a L-set of pore_1 contains also two elements:

$$L(U) = \{T, P\}$$

However, in step 3, the element T will be excluded because during examination of the node U the attribute assignments of the edges (U,Y) and (T,R) do not match.
If during step 3, the L-set of a node has been changed then the examinations already carried out are to repeat. That means, step 3 is to repeat until there is no change of any L-set.

This algorithm has a total complexity of the order $O(|N_2|^3, |N_1|^3 * |M|^3)$. $|M|$ represents the maximal number of elements in any L-set ($|M| \leq |N_2|$).

6 Index Structure

By the development and usage of the database, it is to consider that the database may be permanently changing during learning process. The initial image database may be built up by existing data entries therefore; a nonincremental learning procedure is required. During the use of the system, new cases may be stored in the database. They

should be integrated in the already existing image database. Therefore, we need an incremental learning procedure [22].
Elements in the index structure are representations between graphs. As an important relation between these graphs, we have considered similarity based on part isomorphism. Because of this characteristic, it is possible to organize the index structure as directed graph.
In the following, we define the index structure of the image database as graph that contains the before described image graphs in the nodes:
Definition 5

Given is H, the set of all image graphs.
A index graph is a Tupel IB = (N,E,p), with
(1) $N \subseteq H$ set of nodes and
(2) $E \subseteq N^2$ set of edges.
This set should show the part isomorphism in the set of nodes, meaning it should be valid $x \leq y \Rightarrow (x,y) \in E$ for all $x,y \in N$.
(3) $p: N \rightarrow B$ mapping of image names to the index graph.

Because of the transitivity of part isomorphism, certain edges can be directly derived from other edges and do not need to be separately stored. A relaxation of top (2) in definition 5 can be reduced storage capacity.

7 Learning of Index Structure

Now, the task is to build up the graphs of IB in a supergraph by a learning environment. Formally, this task is to solve permanently:

Input is:
Supergraph IB = (N, E, p) and
image graph $x \in H$.
Output is:
modified Supergraph IB' = (N', E', p')
with $N' \subseteq N \cup \{x\}, E \subseteq E', p \subseteq p'$
At the beginning of the learning process or the process of construction of index graph N can be an empty set.
The attribute assignment function p' gives as output the value (p'(x), (dd)). This is an answer to the question: What is the image name that is mirrored in the image graph x?
The inclusion $N' \subseteq N \cup \{x\}$ says that the image graph x can be isomorphic to one in the image database contained image graph y, so $x \leq y$ and also $y \leq x$ hold. Then, no new node is created that means the image database is not increased.
The algorithm for the construction of the modified index structure IB' can also use the circumstance that no image graph is part isomorphic to another image graph if it has more nodes then the second one.

For technical help for the algorithm there are introduced a set N_i. N_i contains all image graphs of the image database IB with exactly i nodes. The maximal number of nodes of the image graph contained in the image database is k then it is valid:

$$N = \bigcup_{i=1}^{k} N_i$$

The image graph, which has to be included in the image database, has l nodes (l > 0). By the comparison of the current image graph with all in the image database contained graphs, we can make use of transitivity of part isomorphism for the reduction of the nodes that has to be compared.

Algorithm

E' := E;
Z := N;
for all $y \in N_l$
if $x \leq y$ **then** [IB' := IB; return];
N' := $N \cup \{x\}$;
for all i with $0 < i < l$;
  for all $y \in N_i \setminus Z$;
  for all $y \leq x$ **then** [$Z := Z \setminus \{u \_ u \leq y, u \in Z\}$;
   $E' := E' \cup \{ (y,x)\}$];
for all i with $l < i \leq k$
  for all $y \in N_i \setminus Z$
  if $x \leq y$ **then** [$Z := Z \setminus \{u \_ y \leq u, u \in Z \}$;
   $E' := E' \cup \{ (x,y)\}$];
$p' := p \cup \{ (x, (dd : \text{unknown}))\}$;

If we use the concept of Sect. 4 for similarity handling, then we can use the algorithm of Sect. 4 without any changes. However, we should notice that for each group of image graphs that is approximately isomorphic, the first occurred image graph is stored in the case base. Therefore, it is better to calculate of every instance and each new instance of a group a prototype and store this one in the index structure of the image database.

8 Retrieval

When an image is given as query to the system, first the image graph is constructed by the feature extraction unit. The query can be the high-level representation from the whole image or only one node representing one object. This representation is given as query to the image database. The question is: Is there any similar case in the image database?

This question is answered by matching the current case through the index hierarchy. First the image representation with the same number of nodes like the query image representations is determined (see algorithm in Sect. 6), then between the remaining

$$(x \leq y \vee y \leq x) \wedge \left\| N_x \right| - \left| N_y \right\| \leq d$$

set of images having similar representation and the query the part isomorphism relation is determined. The output of the system is all images y in the image database, which are in relation to the query x as follows:
where d is a constant which can be chosen by the user of the system and N_x and N_y are the sets of nodes of the graph x resp. Y.
Figure 9 shows an example of an index structure and the relation of current a query to images in image database.
The user will see these images on display.

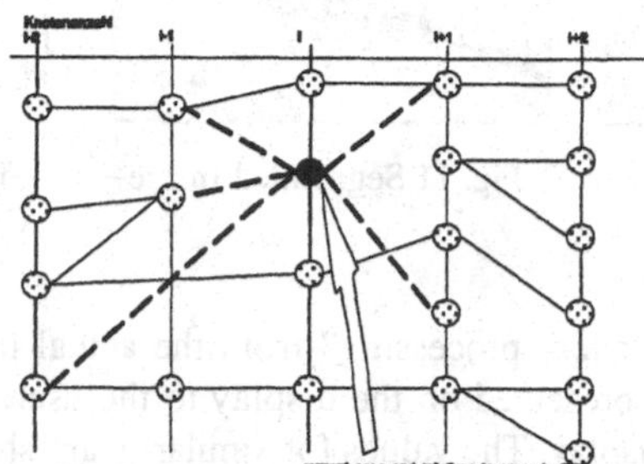

Fig. 9 Schematic Description of the Retrieval Process

9 Results

We used our system for the storage of ultra sonic images from non-destructive testing. The images represent an industrial metallic component having a defect (crack or hole) inside of the component. The images were taken by a SAFT ultra sonic imaging system.
Each entry represents an inspection of an industrial part, consisting of:
an image acquisition protocol: sensor parameters, the parameters of the amplifier,
a protocol about the type or the characteristic of the component,
the information about the defect type, and the image.
It is necessary to keep these images for recourse purposes as certificate and also, for hard inspection cases. Than an image from the difficult inspection problem is compared to the other inspection problems contained in the image database to find the right interpretation of the defect type.
The certification of the condition of technical components, buildings and other industrial parts is one of the main purposes why images need to be stored in image databases.
Figure 10 shows an ultra sonic image of crack inside a flat metallic component as gray level image. The image are posterized for viewing purposes and is then displayed as color image on the display of the database, see Fig. 15. From the original image, a binary image is obtained by thresholding technique. Preprocessing is done by morphological operators like dilation and erosion and afterwards the objects are labeled by contour following procedure [23]. The results after the image processing steps are shown for the segmentation in Figure11 and for the object labeling in Figure 12. The

resulting high-level description are displayed graphically in the image query, see Figure 16. Queries can be: the textual description of the image (see Fig. 13) or the high-level representation of the image (see Fig. 14).

Fig. 10 Original Ultra Sonic Image

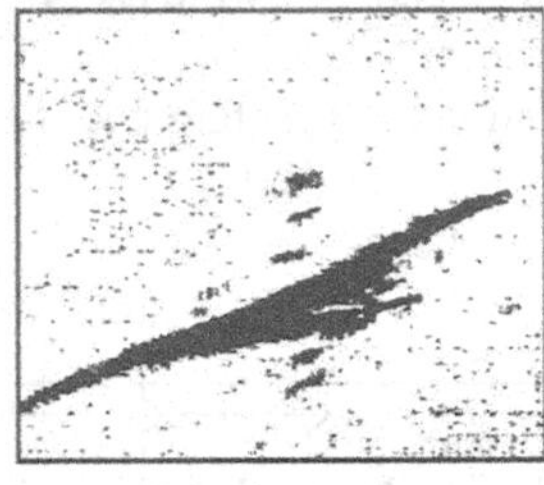

Fig. 11 Segmented Image

Fig. 12 Processed Image Query

By pressing the button "query processing" from the actual image the query is automatically processed and presented on the display to the user. The similar images are shown in a preview on display. The values for similarity are shown in another frame to the user ranked accordingly. If the user selects one of the objects by mouse click in the query image, only images having similar objects are retrieved. If he wants to consider only the spatial relation, he needs to specify this in the menu "interactive query". The system was implemented on PC with C++ Images.

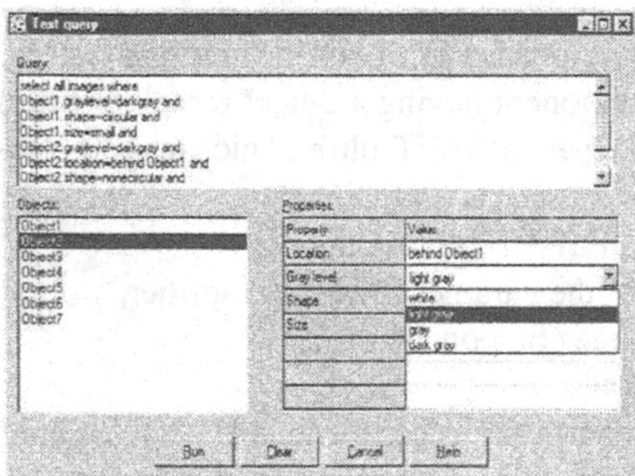

Fig. 13 Text Query

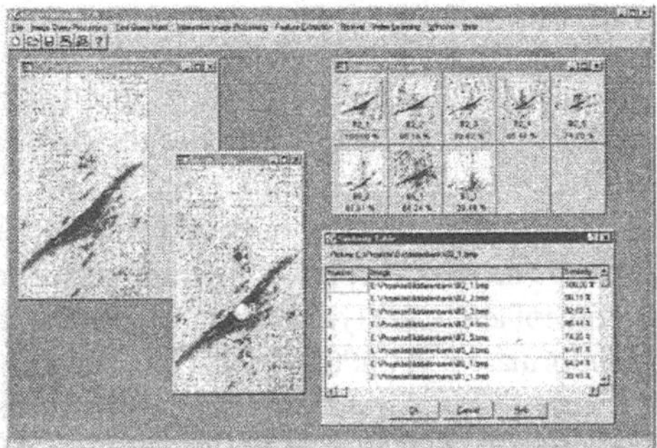

Fig. 14 Hardcopy of the User Interface

10 Conclusion

In our paper, we proposed an image database architecture, which can be used for most industrial problems. The image database is able to handle structural representation of images. Indexing is possible object based, spatial relation based, and by a combination of both. The query can be a textual query or an image content based query. For the later, the system is provided with image processing and pattern recognition facilities. By signal-to-symbol transformation the processed image content is automatically transformed into symbols, which are the same, the user can use for textual query. This

enables the user to understand the image content and ensures that the same algorithm can be user for indexing and similarity retrieval.
The proposed a similarity measure for structural representations is fundamental and flexible enough to get used for a class of different problems. We described the similarity measure in detail and showed the different degree of freedom the similarity measure allows. The algorithm for similarity determination is of polynomial order and fast enough for the considered class of applications.
The index structure is organized in a hierarchical fashion based on the relation of sub graph isomorphism. The index structure can be automatically incrementally learned which allows to incorporate new images in an easy way.
Finally, the performance of the system is presented based on ultra sonic images from non-destructive testing.

References

1. S.-F. Chang,"Content-Based Indexing and Retrieval of Visual Informations," IEEE Signal Processing Magazine (July 1997), vol. 14, no.4, p. 45-48.
2. C., Meghini, "An image retrieval model based on classical logic", SIGIR Forum (1995) spec. Issue, p. 300-308.
3. Th. Whalen, E.S. Lee, and F. Safayeni, "The retrieval of images for image databases: trademarks," Behaviour and information technology, 1995, vol. 14, No. 1, 3-13.
4. A.R. Rao, A Taxonomy for Texture Description and Identification, Springer Verlag 1990.
5. R. Mehrotra, "Integrated Image Information Managment: Research Issues," Proc. of the SPIE 1995, vol. 2488, 168-177.
6. J. Hildebrandt, C. Irving, and K. Tang,"An Implementation of Image Database Systems using differing indexing methods," Proc. of the SPIE (1995), vol. 2606, p. 246-257.
7. R. Samadani, C. Han, and L.K. Katragadda, "Content based event selection from satellite images of the aurora," Proc. of SPIE (1993), vol. 1908, p. 50-59.
8. Q.-L. Zhang, S.-K. Chang, and S. S.-T. Yau," A unified approach to iconic indexing, retrieval, and maintenance of spatial relationships in image databases," Journal of Visual Communiction and Image Representation (Dec. 1996), vol. 7, no.4, p. 307-24.
9. D. Lee, R. Barber, and Wayne Niblack, "Indexing for Complex Queries on a Query-By-Content Image Database," In Proc. of the ICPR`94, Jerusalem Oct. 9-13 1994, Israel, vol. 1, p. 142-146.
10. D. Papadias and T. Sellis,"A Pictorial Query-By-Example Language," Journal of Visual Languages and Computing (1995) 6, p. 53-72.
11. M.S. Lew, D.P. Huijsmans, and D. Denteneer, "Optimal keys for image database," In Proc. of the 9th ICIAP`97, A. Del Bimbo (Eds.) , Springer Verlag 1997, p. 148-55.
12. F.R. Johannesen, S. Raaschou, O.V. Larsen, and P. Jürgensen,"Using Weighted Minutiae for Fingerprint Identification," In Proc: P. Perner, P. Wang, and A. Rosenfeld (Eds.), Advances in Structural and Syntactical Pattern Recognition, Springer Verlag 1996, pp. 289-299
13. H. Bunke, K. Wakimoto, S. Tanaka, and A. Maeda, "A similarity retrieval method of drawings based graph representation," Systems and Computer in Japan, Oct. 1995, vol. 26, no. 11, p. 100-109.
14. G. Vosselman, Relational Matching, Lecture Notes in Computer Sciences, Springer Verlag 1992.
15. D. Lee, R. Barber, W. Niblack, M. Flickner, J. Hafner, and D. Petkovic, "Query by Image Content using multiple objects and multiple features: User Interface Issues. In ICIP, Austin, TX, 1994

16. P. Zamperoni,"Feature Extraction," In: H. Maitre and J. Zinn-Justin, Progress in Picture Processing, Elsevier Science, 1996, pp.123-184.
17. B. B. Mandelbrot, The Fractal Geometry of Nature, W.H. Freeman and Company 1977
18. D. Hernandez,"Relative Representation of Spatial Knowledge: The 2-D Case," Report FKI-135-90, Aug. 1990, TU Muenchen.
19. J.R. Schirra, "A Contribution to Reference Semantics of Spatial Prepositions: the Visualization and its Solution in VITRA," SFB 314, KI-Wissensbasierte Systeme, Dec. 1990.
20. M.I. Schlesinger, "Mathematical Tools of Picture Processing", (in Russian), Naukowa Dumka, Kiew 1989.
21. F. Harary, "Graph Theory," Addison-Wesley 1972.
22. P. Perner and W. Pätzold,"A Incremental Lerning System for Interpretation of Images," In Proc. of SSPR: D. Dori and A. Bruckstein (Eds.) Shape, Structure and Pattern Recognition, World Scientific Publishing Co., pp. 311-323
23. P. Perner,"Ultra Sonic Image Interpretation for Non-Destructive Testing," In Proc. MVA`96, pp. 552-554.

The PRIME Information Retrieval System Applied on a Medical Corpus

C. Berrut[1], P. Mulhem[1], F. Fourel[1] and M. Mechkour[2]

[1] CLIPS IMAG, BP 53, F-38041 Grenoble Cedex - France,
Tel: (33) (0)4 76 51 46 29, Fax: (33) (0)4 76 44 66 75,
email:(Catherine.Berrut, Philippe.Mulhem, Franck.Fourel)@imag.fr

[2] The Robert Gordon University, St Andrew Street, Aberdeen, AB25 1HG, UK,
Tel : (44) (0)1224 262785, Fax: (44) (0)1224 262727,
email:mrm@scms.rgu.ac.uk

Abstract. PRIME is a precision oriented information retrieval system, managing multimedia structured documents. An information retrieval system provides access to documents based on their semantic content. The precision of such a system is measured as its capacity to give not necessary all, but only good answers. An information retrieval system is precision oriented when the information need of the users requires precise answers from the system; this leads the users to give also precise queries. Such systems are based on complex and controled representation languages, allowing a deep semantic understanding of the documents and the queries. The precision of an information retrieval system is quantified when instanciating it onto an application and a category of users. This paper describes the PRIME system, as we applied it on a medical corpus of documents for specialist physicians. The structured corpus contains medical texts, radiologic images and medical records. The software architecture of PRIME is built on the O2 system; the data structures dealing with the retrieval are conceptual graphs; the interfaces are developed in X-Motif and HTML using forms.

1 Introduction

1.1 Goals of an information retrieval system

The goal of an information retrieval system is to allow the access of documents using their semantic content: the user expresess its need by indicating the content he wants find in the retrieved documents. To achieve this goal, information retrieval systems are based on a formal model of matching. The efficiency of the query processing and the ergonomy of the query formulation are priorities when implementing the theoretical model into an operational system.
An information retrieval is usually composed of three parts [1], [2] :

- a query processing, that allows a user to query the corpus. This processing provides a matching model between a query and the corpus, respectively represented in a model of query and a model of document (the indexing language),

- an indexing process that is dedicated to the extraction of an homogeneous representation of the semantic content of the documents. This representation is composed of indexing terms that are elements of an indexing language,
- a modeling of knowledge. This modeling usually takes the form of a thesaurus, and is used by both of the query processing and the indexing process.

At the operational level, an information retrieval system can only be evaluated by observing answers given for a set of queries. Tests underline the duality between the system relevance (i.e. "what the system gives") and the user relevance (i.e. "what the user wishes to find"). The most common measures to evaluate this difference are the recall and precision measures having their values in [0,1]: the recall describes the ability of the system to give all the relevant documents, and the precision describes the ability of the system to give only relevant documents. With an ideal system, recall and precision are both equal to 1. In contexts managing highly specialized concepts, the precision has to be priveleged; it implies to manage specific indexing terms.

1.2 The medical context of PRIME

An hospital generates annually several terabytes of data. Such amount of information requires systems able to store and to retrieve the data in an efficient way. The current medical systems (based on DBMSs) do not allow content-based retrieval. If a specialist wants to retrieve a medical report containing a specific medical information, he has to know the identifier (or an external attribute) of the document. The content-based access functionality is i) a complement to "pure" DBMS functionalities, and ii) a necessity when we manage huge amount of data that can be retrieved and created by several people.
On one hand, the administrative data (like the patient name, the examination dates, id numbers, ...) are external attributes (i.e. we only have to know them to retrieve documents). On the other hand, the textual medical reports, the medical images has to be content-based retrieved. These textual and image data are created by experts (physicians) for specialists (physicians). The information need in this context is precise, and the information retrieval must cope with this expertise. A system in this context has then to be precision oriented.
Based on the above points, we define the PRIME system (standing for "Prototype de Recherche d'Informations MEdicales", i.e. medical information retrieval prototype) dedicated to provide storage, navigation, indexing and content-based retrieval of medical data to physicians.

1.3 The PRIME prototype

On one hand, PRIME must offer access to medical data using external criteria (name, id number, ...), as well as navigation on such data. On the other hand, PRIME has to provide content-based access to medical data.
In this highly specialized context, PRIME must fulfill the following requirements : i) the use of an exhaustive and specific indexing language. The exhaustivity

of a language is its ability to describe all the concepts of a topic that appear in a document. The specificity of a language is described by the accuracy of an indexing language concept to represent one document concept, ii) the use indexing tools (interfaces, control processes, ...), iii) a query interface that allows the user to express precisely what he needs, iv) a matching process able to manage the complexity of the language.

As we show later, the indexing language and the matching process are respectively described by conceptual graphs [3] and the project operator on such graphs. We chose the conceptual graphs formalism at the operational level in respect to the theoretical model of IR used [4]. Because of the equivalence between the projection (of the conceptual graphs formalism) and the implication of the first order logic, we keep the theoretical properties of the system, and therefore we validate them during experiments.
This close relationship between a formal and a theoretical model is rare in the context of information retrieval. The existing approaches (except the boolean model) are usually only theoretical, or only ad-hoc (thus, not provable) implementations.
The objective of this paper is to present each part of PRIME. We present in a first part the related works on medical systems and prototypes. Then, the data model and software architecture of PRIME are described in Section 3. Sections 4, 5 and 6 are respectively dedicated to the navigation in the database, the indexing tools, and the retrieval process. We conclude in Section 7, by explaining the current limits of PRIME and the future works on this prototype that is the result of a 10 years work [5], [6], [7], [8].

2 Related Works

The PRIME system is not dedicated to tackle the problem of decision making as in [9], due to the complexity of the analysis task of brain MR images. Our approach, more humble, is to help the physician when making a diagnosis, by providing access to relevant clinical data. He can then refine its diagnosis by viewing similar cases. This work takes place then in the management of clinical data and a help in the context of clinical decision support as described in [10].
Even if PRIME provides a Web based interface, we did not yet focus on network considerations as it done in the LUX-PACS project [11] or in [12] with virtual patient records or to cooperative medicine as in [13], but more on the way the retrieval of relevant information is feasible in a medical context.

Our work is more closer to what is done in the Kmed Project [14], according to the fact that the retrieval of data in KMed is based on precise descriptions. We also integrate navigation and retrieval features in PRIME, as it is done by Ip, Law and Chan in [15]; we focus on the representation of the semantic content of multimedia structured documents that is automatically generated from their components. Our work focuses on the different aspects of integration of the semantic content of multimedia structured medical documents: we provide precise representation of the semantic content of documents parts in a way to retrieve them, and on the other hand we integrate presentations of the index of documents in a way to support more semantic-based navigation in structured documents.

3 The data model

We model in the following the data managed by PRIME using a BNF grammar. In respect to the information system of the University Hospital of Grenoble, we separate the administrative data and the medical data of the patients. The administrative data management is done by the administrative department, and these data are not accessible by medical departments.

```
1. < Patient Record >          ::= < Adm. Rec. of the Pat. >
                                   < Med. Rec. of the Pat. >
2. < Adm. Rec. of the Pat. >   ::= < Id >< L_Name >< F_Name >
                                   < Address >< Sex >
                                   < Birth Place >< Citizenship >
3. < Med. Rec. of the Pat. >   ::= < Common Med. Rec. of the Pat. >
                                   1* < Specialty Rec. of the Pat. >
```

The common medical data (i.e. of a potential interest to each medical department of the hospital) may be accessed by all department of the hospital, using a CMRP (Common Medical Record of the Patient).

```
4. < Common Med. Rec. of the Pat. > ::= * < Chornical Illness >
                                        * < Allergy >
                                        * < Med. Past Hist. >
                                        * < Other Remarks >
```

Each medical department manages its own SMRP (Specialty Medical Record of the Patient).

```
5. < Specialty Med. Rec. of the Pat. > ::= < Date of Examination >
                                           < Examination >
```

The current corpus of PRIME contains only data related to the department of nuclear magnetic imaging. We will see in the following that the corpus is specific to encephalon and neck pathologies. This limited corpus has no fundamental implication on PRIME. In the context of magnetic resonance, the examinations contain NMR (nuclear magnetic resonance) images and are modeled by :

```
6. < Examination >       ::= 1* < Asking Physician > 1* < Specialist >
                             < Report > 1* < Series of Images >
                             0 * 1 < Examination_Index >
7. < Report >            ::= < Text > 0 * 1 < Report_Index >
8. < Serie of Images >   ::= 1* < NMR >
9. < NMR >               ::= < Image > 0 * 1 < Image_Index >
```

A we said already, physicians have access to examinations (e.g., "examinations that describe tumors on the brain"), to textual reports (e.g., "reports describing an axial cut showing a tumor"), and to MNR images ("images showing a tumor that touches the right orbit"). To address these points, the query processing on the semantic content on examinations, reports and images uses respectively the <Examination_Index>, <Report_Index> and <Image_Index>, that appear in the rules 7, 8 and 9. We describe these elements after the indexing language.

3.1 The general architecture of PRIME

PRIME allows the storage, the management, the indexing and the retrieval of the multimedia medical data described above. More specifically, PRIME is based on the O2 system, on top of which we put additional functionalities (especially a toolbox for the images manipulations), and this extended system is a base for several applications.

PRIME uses the O2 DBMS for the storage and the consistency of data, the concurrent accesses to the data, and the query processing on external attributes. For the management of the semantic content of data, PRIME is splitted up into two parts:

- the Indexing part extracts the semantic content of the documents. This extraction is automatic or manual, depending on the documents,
- the Retrieval part deals with queries based on the semantics of the documents (records, reports or images). It also integrates the navigation (an adapted interface to medical data according to [16]) along well defined links based on the structure and on the semantic content of documents.

From a software architecture point of view, the elements that compose PRIME are given in figure 1[3]. In this Figure, we see that the design principle is directed through three levels of abstraction on top of the DBMS: the first one is to provide extensions to the DBMS in a way to manage the multimedia document parts (images) and the indexing language data in a context free approach; the second one (i.e., the management level) defines in wich context the extensions above are used; the third one handles the elements dedicated to the presentation (Queying, Indexing and Navigation). The main consideration for such separations is the fact that the core of PRIME (the DBMS extended with PGGC and Multimedia features) is independant from any context. From another point of view, the "vertical" split that separates documents from their index allows modifications and

[3] In the figure, the modules (1) are coded in O2C and make calls to C functions. The module (2) is coded entirely in C and Motif. The modules (3) are coded in O2C and communicate with a Web client using the O2Web software. The other modules are coded in O2C.

extentions of the indexes without a lot of modification of the document management. Such separation allows furthermore to have several index associated to documents.

The conceptual graphs extension (PGGC) provides operands and operators used for the creation, management and consistency of conceptual graphs (creation operator for graphs, definition and verification of canonicity of graphs, ...). This framework provides the software level for the representation and retrieval of indexed documents [17]. The document and index managers deal with the representation of the corpus data according to the model defined above. <Image_Index>, <Report_Index> and <Examination_Index> depend on the index manager, and the remaining of the data model depends on the document manager. Other data are managed by the DBMS, and are accessed using navigation or external attribute querying.

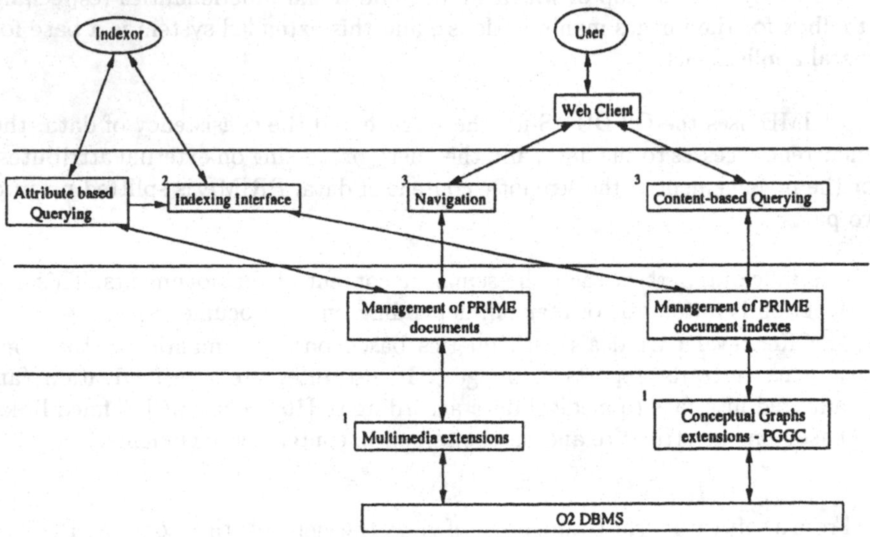

Fig. 1. PRIME Architecture

The navigation uses the document manager to present documents to a user, as the query processing and the indexing uses the index manager. In fact, this separation is not so clear, because we will show in the following that the navigation display documents using their semantic content (i.e. using a relationship between the document manager and the index manager) and the component dedicated to the indexing also needs to access to the documents themselves. These points explain the link between the two managers in figure 1.
Based on the architecture described above, we will focus on each of these parts in the following. The principles of the navigation will be described. Then, we explain in detail the indexing process of the documents, and finally we address the retrieval of documents.

4 The navigation

4.1 The levels of freedom during navigation

The navigation on PRIME explicitly uses the data model defined previously. The user then can navigate using the composition links between document parts. The navigation can be "in width" between the documents of the same type of one composed document, and also allows the access to indexing elements.
To avoid disorientation during the navigation, PRIME always displays a contextual information with a medical data. The user has moreover always the possibility to reach two entry points, that are: the first screen of the application, and the starting point of the current patient data of which we present the medical data. The first screen of the application offers the choice to begin a new navigation, or to define a content-based query. When going back to the starting point of a patient data, the user "reorients" himself, but he can also consult more easily the data related to the current patient. These two entry points appear in each interface of PRIME under the form of yellow arrows.

The navigation at the beginning of the application At the beginning of the application, the user chooses between three modes of access to the data :

begin_of_application → choice_patient_by_name
begin_of_application → choice_patient_by_id
begin_of_application → choice_patient_by_content

(we will talk of the choice_patient_by_content in part 5).
These rules can be described as: the interface "begin_of_application" contains three anchors allowing to choose between one navigation (choice_patient_by_name and choice_patient_by_id) or the definition of a query (choice_patient_by_content). From the choice of a patient P using his name or his identification number, the user accesses the medical or administrative data of P:

choice_patient_by_name → <Adm. Rec. of the Pat.>
choice_patient_by_name → <Med. Rec. of the Pat.>
choice_patient_by_id → <Adm. Rec. of the Pat.>
choice_patient_by_id → <Med. Rec. of a Pat.>

The navigation using structural links The navigation uses the composition links of the data model in five cases:

<Med. Rec. of the Pat.> ↔ <Examination>
<Examination> ↔ <Serie of Images>
<Examination> ↔ <Report>
<Serie of Images> ↔ <Image>
<Serie of Images> ↔ <Image_Index>

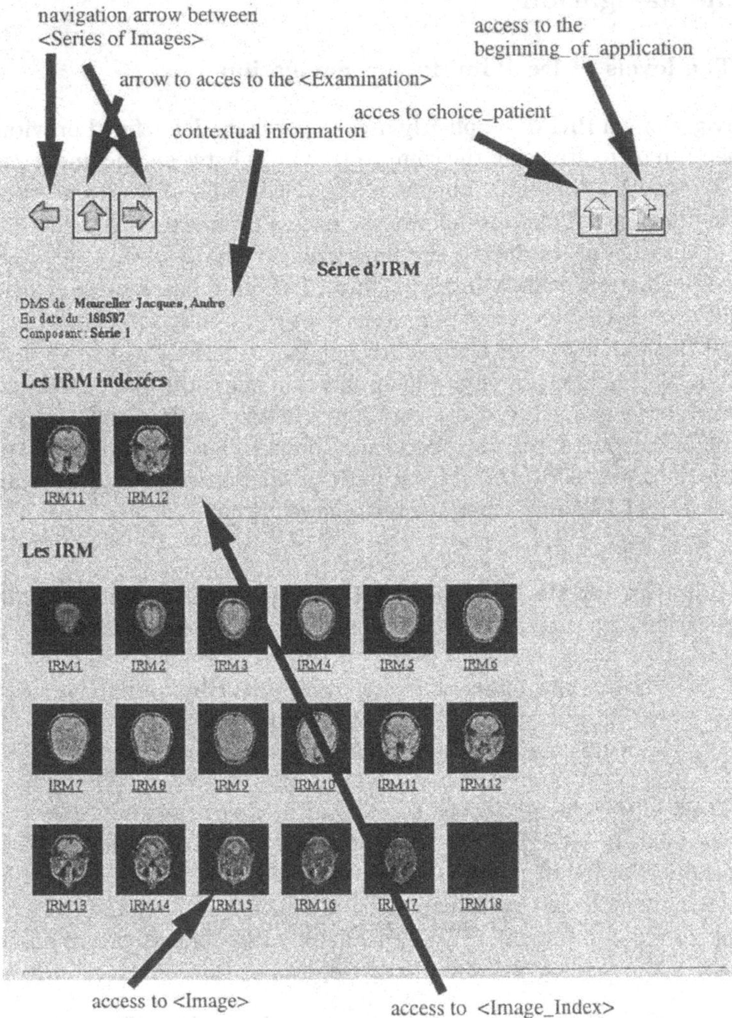

Fig. 2. Description of the interface for a Serie of medical images

From an interface point of view, the descending navigation (i.e., from a composed document to a composing one) is visualized by anchors, when the ascending navigation (i.e., from the composing document to the composed one) is reached by an arrow in the upper left of the interface. This arrow is only active in these five contexts, and remains inactive otherwise.

The navigation allows also to visualize "in width" the data, using given n-ary associations between data. This point arises in the <Medical Record of a Patient> of the same patient, the <Serie of Images> of an <Examination>, the <Image> of a <Serie of Images> or the <Image_Index> of a <Serie of Images>. This navigation is reached by two arrow anchors ← (predecessor) and → (successor) in the upper left part of the interface. These arrows are active only in these 4 contexts.

Navigation using indexes Finally, five additional types of navigation exist. They address the presentation of the indexes (cf. the following part) of a document, and the navigation between the documents and their indexes:

<Image> → <Image_Index>
<Image_Index> → <Image> (navigation between one image and its index)
<Image_Index> → <Image_Index> (display of the index of an image)
<Examination> → <Element_Index_Examination>
<Element_Index_Examination> → list of the <Image_Index> containing this <Element_Index_Examination>

4.2 An interface example : <Serie of images>

Figure 2 shows an example of what is presented to a user.

5 The Indexing

The term "Indexing" involves, in an information retrieval system, two distinct problems :

- the definition of an indexing language, allowing the representation of the concepts of the corpus documents;
- the definition of an indexing process that extracts the indexing terms (a representation that conforms to the indexing language) from the documents.

As we said in the introduction, the users of PRIME formulate their queries in a precise way, and wait for precise answers of the system (i.e. only good answers). A complex and controlled indexing language provides a way to find a solution to this problem.
An indexing language L can be simple if L = {term ∈ Vocabulary } or complex if the indexing terms can be related. For instance, L = {term, term ∈ Vocabulary ∪ (L × Operator × L)} defines a complex indexing language. The Vocabulary set is the basic vocabulary for the expressions of indexing terms, Operator is the set of the semantic connectors between indexing terms. An indexing language is closed or controlled if Vocabulary and, possibly Operator, cannot be modified. On the other case the indexing language is called open.
The complexity is a direct result of the precision need of PRIME : "a tumor on the left lung" must not be confused with "a tumor on the right lung", or as "a tumor that is near the right lung". So, the terms "tumor", "lung" , "left" must be represented, as well as the operators stands_on, is_near, in a way to distinguish all this different notions.
The indexing language of PRIME is controlled. In fact, there is no need for

a closed vocabulary, but in our case we ascertain that the medical specialty language is well bounded. This property has been underlined during a work on the indexing process on medical reports [7].

5.1 Index of <Report> : <Report_Index>

The indexing language of medical reports is defined in [18] as a language L, complex and controlled, defined as term, term ∈ Vocabulary ∪ (L × Operator × L). The definition of the sets Vocabulary and Operator has been done in cooperation with physicians of the University Hospital of Grenoble, and a grammar describes which subset of L × Operator × L is really used.
Concretely, we have defined a set Vocabulary of basic terms, a set Semantic_Type = {report, lesion, organ, examination, ...}, a set Operator = {bears_on, touches, ...}, a grammar G that validates the associations between terms and operators: every rule of G is like:
Left_Part := Right_Part1 Operator Right_Part2
where Left_Part, Right_Part1 and Right_Part2 ∈ Semantic_Type.
Then, a relation Semantic_Relationship(V,T) associates to each term of V or of G one or several semantic types. For instance, we have Semantic_Relationship(lung, organ), Semantic_Relationship(cancer, lesion) and Semantic_Relationship(cancer bears_on lung, lesion) because of the rule lesion ::= lesion bears_on organ ∈ G.
An indexing term t is valid if and only if :

t ∈ Vocabulary
or t = t1 op t2, so that t1 and t2 are valid, and ∃r∈G, r = p1 op p2
so that Semantic_Relationship(t1,p1)
and Semantic_Relationship(t2,p2) exists.

The indexing process allows an automatic generation of the indexing terms of a textual report. This generation uses four linguistic processes (morphology, syntax, semantic, pragmatic). On the other hand, a rewriting system deals with a canonical representation of the indexing terms that partially solves the paraphrasing problems. The indexing process is written in Prolog, for corpus of lung tomodensimetry documents.
In the current state of PRIME, we did not yet integrate the automatic processing of the reports. The current corpus of PRIME deals with pathologies of the encephalon and of the neck, and we do not have now a thesaurus and especially no validated grammar needed for an automatic indexing. Then, the current indexing language for the reports is limited to a set Vocabulary of terms of the domain of encephalon and neck diseases:

10. <Report_Index> ::= *<Term>
11. <Term> ::= orbit | metastasis | cancer | ...

During the navigation in the medical data, the terms that index a <Report> are showed in bold in the <Text> of the <Report>.

5.2 Indexing of <Image> : <Image_Index>

The images can offer several interpretations, their indexing in then not an easy task. The problem is then to define a language or a model that formalizes the information, and a environment that helps the indexing process.
Then, the first objective in PRIME was to define a model, called EMIR2, that formalizes the images indexing language, and to define an interface that handles the indexing. The interface, dedicated to assist manual indexing, is described in [19], and is integrated in the PRIME prototype.

EMIR2 is formally described in [17], where the translation into conceptual graphs, and its use in others contexts than PRIME are also depicted. We are currently using a subset of EMIR2 defined syntactically as :

12. < Image_Index > ::= < Phys_View >< Log_View >
13. < Log_View > ::= < Struct_View >< Symb_View >< Spat_View >
14. < Symb_View > ::= ∗ < Term >
(<Term> defined for <Report_Index> in rule 11)
15. < Spat_View > ::= < Spat_Object >
| ∗ (< Spat_Objet >< Spat_Operator >
< Spat_Object >)

The physiscal view is the pixel matrix, the symbolic view deals with the symbols related to elements of the spatial view, these spatial elements having spatial relationship between each others. During the navigation, an <Image_Index> is showed at the same time as an image; geometrical shapes (rectangles and points) symbolize the spatial objects of an image, each one associated to one term. The user can then see some of the term by making a selection (see above the rule <Image_Index> → <Image_Index> in the navigation).

5.3 Indexing of <Examination> : <Examination_Index>

The physicians can also query <Examinations>. The queries do not look like a query Q1 so that "Give me an examination that contains an image showing a tumor". This kind of queries is already solved: it is a part of a query result on medical images, when we use an ascendant navigation. The physicians can look for "examinations that identify a tumor" (query Q2). The query Q2 is more precise that Q1: an examination identifies a tumor if the tumor is an important part of the examination, but an image can show a tumor that is not a fundamental element of the examination.
The <Examination> are complex data that do not contain direct indexed, i.e. semantic data. The index of the examination must then be obtained using the related indexed data: <Report_Index> and <Image_Index>. We must then define a dynamic indexing for these data. The syntax of the <Examination_Index> is the following one :

16. <Examination_Index> ::= 1*<Term>

In the current version of PRIME, the dynamic indexing is simple, because we only keep, for the index of an examination, the terms in common between two <Image_Index> or between one <Image_Index> and a <Report_Index> for each <Examination>.

During the navigation, the index list of an <Examination> is displayed while the <Examination> is presented. Each list element is an anchor, allowing the access to the image(s) that contains this indexing term (navigation rule :
<Examination> → <Index_Examination_Element>
and
<Index_Examination_Element> → list of <Image_Index> containinig the <Index_Examination_Element>).

5.4 The PGGC framework

The <Examination_Index>, <Report_Index> and <Image_Index> are represented by conceptual graphs, using the index manager that depends on the framework PGGC (for conceptual graphs manager). PGGC includes two main parts: the core that handles the internal representation and operators on conceptual graphs, and the interface that deals with the manipulation of the graphs. Because of the lack of space, we do not explain more this point here. The reader should refer to [20] for more details about PGGC.

6 The retrieval process

In an information retrieval system, the retrieval process is dedicated to find all the documents that answer a given query, eventually associated to a weight (the relevance value) in a way to rank the documents. According to this description, a retrieval process needs a representation of the query language and a matching function that finds the relevant documents to a query.
In PRIME, the retrieval process is an implementation of the logic model of IR, as proposed in [21] and [4], by using the projection operator on conceptual graphs. The logical model of IR consists in proving that a document D is an answer for a query Q iff D→Q, where → is the logical implication. The projection predicate of a graph A into a graph B, namely Π_A(B), is a predicate that indicates if A projects in B, i.e. if B is a specific graph of A. Sowa showed the equivalence between the projection operation and the first order logic implication, when the concepts are simple. Then, D→Q ⇒ $\Pi_{Pgc(Q)}$gc(D), where gc(Q) and gc(D) are the conceptual graph representation of Q and D, respectively.

6.1 What query language?

In PRIME, because of the three types of index, the implication D⇒Q is translated into three rules:
<Report_Index>→<Q_{report}>
<Image_Index>→<Q_{Image}>

<Examination_Index>→<$Q_{Examination}$>
These three types of queries are defined as :
<Q_{Report}> ::= *<Term>
<Q_{Image}> ::= *<Term>
<$Q_{Examination}$> ::= * (<Term> | <Term> <Spatial_Operator> <Term>)
The users can query PRIME through a query interface (see the screen shot in figure 3). Because the interface is adapted to each query type, it has three facets. The <Term> and <Spatial_Operator> are dynamically generated according to the current query type, using the index of the same type in the base. This explains why, when a user asks a query <$Q_{Examination}$>, he does not use the same vocabulary than when he makes a <$Q_{Examination}$>. This also explains why we distinguish queries <Q_{Report}> and <$Q_{Examination}$>.

6.2 What matching function?

For <Q_{Report}> and <$Q_{Examination}$> The present answer to a query Qr ∈ <Q_{Report}> is splitted into two categories defined as :

- i) the first set, named Total_Answer(Qr), contains each Report R described by an index in which the query can be projected
- ii) the second set, named Answer_1(Qr), contains each Report R' not in Total_Answer(Qr) described by an index in which at least one query term can be projected.

The result is then composed of two categories: Total_Answer(Qr) firstly and Answer_1(Qr) secondly.

The matching above will soon rewritten to integrate an thinner ordering according to the number of query terms that projects in the Report index. This means that, in the contrary to the present matching results, all document that answer partially to a query is in the answer.
The same approach is also used for a query <$Q_{Examination}$>.

For <Q_{Image}> The answer for a query Qim ∈ <Q_{image}> contains :

- Ans_im_1(Qim) : the set of documents in which the whole query Qim is projected,
- Ans_im_2(Qim) : the set of documents not previously retrieved in which the symbolic and spatial views of the query are projected,
- Ans_im_3(Qim) : the set of documents not previously retrieved in which the symbolic view of the query is projected,
- Ans_im_4(Qim) : the set of documents not previously retrieved in which the spatial view of the query is projected,
- Ans_im_5(Qim) : the set of documents not previously retrieved in which a term of the symbolic view of the query is projected.

The ranking of the result list (Ans_im_1(Qim) , Ans_im_2(Qim), Ans_im_3(Qim), Ans_im_4(Qim) , Ans_im_5(Qim)) reflects the relevance relationship between the documents that belong to the query result.

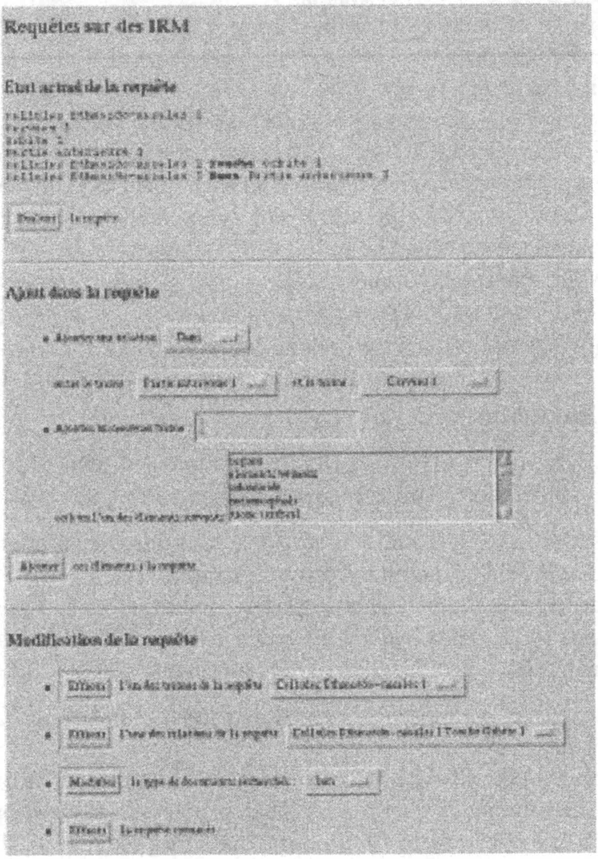

Fig. 3. The query interface of PRIME.

6.3 What answer to present to the user?

Using the matching functions <Report_Index> → <Q_{Report}>, <Image_Index> → <Q_{Image}> and <Examination_Index> → <$Q_{Examination}$>, the results given to a user are respectively <Report>, <Image>, <Examination>. We provide a part of the structural context of the retrieved information, in a way to avoid additional navigation.

On the screen, the result data appears in an array having in the first column anchors to respectively <Report>, <Image> and <Examination>, according to the relevance ranking when it exists. The other columns of the result provide the structural context of the answers. For a <Q_{Report}> query, we can access by clicking to (except the <Report>) the <Specialty Record of the Patient> and the <Medical record of the Patient>. For a <Q_{Image}> query, we provide access to the <Serie of Images>, to the <Specialty Record of the Patient> and the <Medical record of the Patient> for each image of the result. For a

<$Q_{Examination}$> the direct access to the <Medical record of the Patient> is possible.

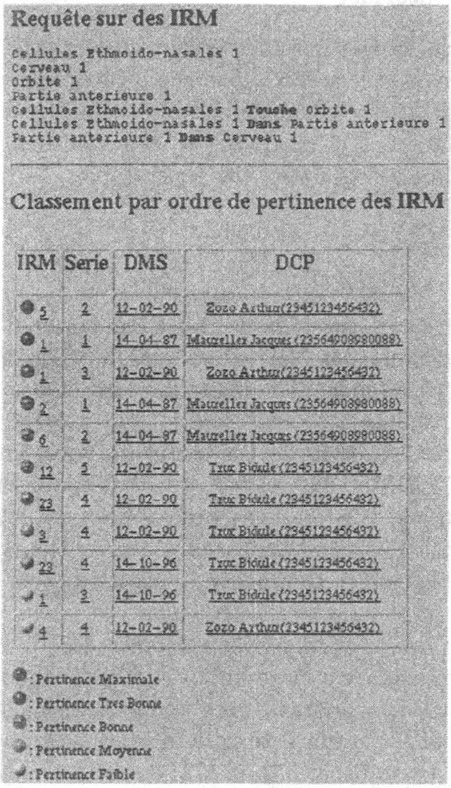

Requête sur des IRM

Cellules Ethmoido-nasales 1
Cerveau 1
Orbite 1
Partie anterieure 1
Cellules Ethmoido-nasales 1 Touche Orbite 1
Cellules Ethmoido-nasales 1 Dans Partie anterieure 1
Partie anterieure 1 Dans Cerveau 1

Classement par ordre de pertinence des IRM

IRM	Serie	DMS	DCP
5	2	12-02-90	Zozo Arthur(2345123456432)
1	1	14-04-87	Maurellez Jacques (23564908980088)
1	3	12-02-90	Zozo Arthur(2345123456432)
2	1	14-04-87	Maurellez Jacques (23564908980088)
6	2	14-04-87	Maurellez Jacques (23564908980088)
12	5	12-02-90	Truc Bidule (2345123456432)
23	4	12-02-90	Truc Bidule (2345123456432)
3	4	12-02-90	Truc Bidule (2345123456432)
23	4	14-10-96	Truc Bidule (2345123456432)
1	3	14-10-96	Truc Bidule (2345123456432)
4	4	12-02-90	Zozo Arthur(2345123456432)

: Pertinence Maximale
: Pertinence Tres Bonne
: Pertinence Bonne
: Pertinence Moyenne
: Pertinence Faible

Fig. 4. The answer interface.

6.4 The query interface

The figure 3 shows the query interface of PRIME. The HTML interface allows to choose terms when querying Reports, Examinations and Images and to define relationships between terms when querying Images. The example of figure 3 is dedicated to retrieve images, and the user wants to retrieve images that contain "ethmoido-nasal cells, in the anterior part of the brain, that touch an orbit". The terms are "Cellules Ethmoido-nasales" (Ethmoido-nasal cells'), "Cerveau" (brain), "Partie anterieure" (anterior part of the Head). The Relationships are "Touche" (touch) and "Dans" (in). The interface put a number on each term to allow their disambiguisation: we are then able to distinguish "a tumor that touches one orbit and another tumor that touches the brain" and "a tumor that touches one orbit and also the brain". Such precision in the expression of the query is needed in the medical context.

6.5 The answer interface

The figure 4 shows the answer interface to the query of the part 5.4 above. The result is ranked by relevance according to the 5 categories described above; the relevance levels are related to colors going from red to white. The anchors to the images or to the contextual elements of the retrieved images (cf. part 5.3) are presented in the different columns of the result table.

7 Conclusion

The database currently on use contains about one hundred medical reports and 1300 images. The size of the database is about 470Mb. The major part of PRIME is written in O2C (10000 lines). The indexing interface is done in C++ (3500 lines) and in C (24000 lines).
The database is small according to the features above, but it is a testbed for the current state of PRIME. We verified thus that the interfaces and dialogs of PRIME satisfy the users. The recall and precision ratios obtained seem moreover to validate our precision oriented approach. However the number of queries and documents are not big enough to provide a definitive proof of the validity of the approach.
In order to provide a real sized prototype (and also to validate qualitatively PRIME), we have to manage a more important base (about a thousand of documents). We must then collaborate more closely with physicians to create and verify the indexing of the medical texts and images. We also have to define a test collection with a set of solved queries (i.e. queries and their results according to the physicians). In fact, such long-term work is hard, and the validation mechanisms are also difficult to define.

The current limitations of PRIME are related to the manual indexing of MR images, and to the subset of the conceptual graphs formalism used. A known aspect in the information retrieval field is that usually indexing usually differs from one person to another. However, in a very specific context like MR images of the brain, such inter-indexor inconsistency does not occur. The conceptual graphs formalism allows to identify uniquely concepts. For now we do not use this feature; so we are not yet able to indicate that the tumor on one image is the same than the tumor on another image.

Except this qualitative verification of PRIME, we will enhance the system in a way to integrate: i) more interactive interfaces, ii) a more expressive power in the query language, and more refined matching functions, iii) more interleaving between navigation and query retrieval. The users need a more reactive interaction to visualize documents and to express queries. Java programs will be used to add interactivity into the existing parts of PRIME.

The query languages must be enhanced. The richness of the indexing languages are underused, and we should formulate queries as document indexes. We have then to develop new matching functions. The current matching functions are too rigid to provide what the physicians expect in any case.
Finally, we have to develop strong associations between navigation and query-based retrieval, and this point becomes more obvious on huge databases. This tight integration should be used as the two sides of a same coin: the query-based information retrieval being a focusing tool, and the navigation being a discovering tool. To associate these two tools means to define dynamically consistent groups of data based on static links (a priori links like usually in hyperbases) or on dynamic links (defined by a query processing), between the data.

References

1. C. J. VAN RIJSBERGEN, Information Retrieval, Butterworth, London, 1979.
2. G. SALTON, Introduction to Modern Information Retrieval, McGraw-Hill, New-York, 1983.
3. J. F. SOWA, Conceptual Structures : Information Processing in Mind and Machine, Addison-Wesley publishing company, 1984.
4. Y. CHIARAMELLA and J. NIE, A Retrieval Model Based on Extended Modal Logic and Its Application to the RIME Experimental Approach, ACM SIGIR Conference on Research and Development in Information Retrieval, Brussel, Belgium, 1990, pp. 25-43.
5. G. MUNOZ BACA, Stockage et exploitation de dossiers médicaux multimédia au moyen d'une base de données généralisée, Thèse de l'Universite Joseph Fourier, Grenoble, 1987.
6. C. BERRUT and Y. CHIARAMELLA, Indexing Medical Reports in a Multimedia environment : the RIME Experimental Approach, ACM SIGIR Conference on Research and Development of Information Retrieval, Boston, Massachussetts, USA, 1989, pp. 197-197.
7. C. BERRUT, Indexing Medical Reports : The RIME Approach, Information Processing & Management, Vol 1, No. 26, 1990.
8. J. NIE, An Outline of a General Model for Information Retrieval Systems, 11th ACM SIGIR International Conference on Research and Development in Information Retrieval, Presses Universitaires de Grenoble, Grenoble, France, 1988, pp. 495-506.
9. G. FORGIONNE and R. KOHLI, Integrated MSS Effects: an Empirical Health Investigation, Information Processing & Management, Vol. 31, No. 6, 1995, pp. 879-896.
10. J. G. Anderson, Learing the Way for Physicians' Use of Clinical Information Systems, Comm. of the ACM, Vol. 40, No. 8, August 1997, pp. 83-90.
11. D; Bandon, R. Kanz, V. Boissart, C. Debas, G. Evers, J. Duchêne and C. Wehenkel, Medical Image Storage and Retrieval Strategies in the LUX-IMACS Project, Proc. of RIAO'94, New York, USA, October 1994, pp. 439-448.
12. D. G. Kilman and D. W. Forslund, An Interbational Collaboratory based on Virtual Patient Records, Comm. of the ACM, Vol. 40, No. 8, August 1997, pp. 111-117.
13. L. Kleiholz and M. Ohly, Supporting Cooperative Medicine: The Bermed Project, IEEE Multimedia, Vol. 1, No. 4, winter 1994, pp. 44-53.

14. W. CHU, A. GARDENAS and R. TAIRA, KMed: a Knowledge-Based Multimedia Medical Distributed Database System, Information Systems, Vol. 20, No. 2, Elservier Science, 1995, pp. 75-96.
15. H. H. Ip, K. C. Law and S. L. Chan, An open framework for a multimedia medical document system (a multimedia document system framework), Journal of Microcomputer Applications, Vol. 18, 1995, pp. 215-232.
16. J. RAMIREZ, L. SMITH and L. PETERSON, Medical Information Systems : Characterization and Challenges, SIGMOD Record, Vol. 23, No. 3, 1994.
17. M. MECKOUR, EMIR2 : Un modèle de représentation et de correspondance d'images pour la recherche d'informations. Application á un corpus d'images historiques, Thèse de l'Université Joseph Fourier, Grenoble, 1995.
18. C. BERRUT, Une méthode d'indexation fondée sur l'analyse sémantique de documents spécialisés, Thèse de l'Université Joseph Fourier, Grenoble, 1988.
19. C. BERRUT, P. MULHEM and P. BOUCHON, Modelling and Indexing Medical Images : the RIME Approach, Proceedings of HIM'95 (Hypertext - Information Retrieval - Multimedia), Schriften zur Informationswissenschaft 20, Universitat Verlag Konstanz, Konstanz, Germany, 1995, pp. 105-115.
20. M. MECHKOUR, P. MULHEM, F. FOUREL and C. BERRUT, A Medical Information Retrieval Prototype on the WEB, Seventh International Workshop on Research Issues in Data Engineering: High performance Database Management for Large Scale Applications, Birmingham, England, April 7-8, pp2-9, 1997.
21. C. J. VAN RIJSBERGEN, A new Theoretical Framework for Information Retrieval, ACM SIGIR Conference on Research and Development in Information Retrieval, Italy, Pisa, 1986.

Envelope Parameter Calculation of Similarity Indexing Structure

Bai Xuesheng , Xu Guangyou , Shi Yuanchun

Department of Computer Science and Technology,
Nation Laboratory of Intelligent Technology and System,
Tsinghua University, Beijing 100084, P.R.C
xsbai@vision.cs.tsinghua.edu.cn
shyc@tsinghua.edu.cn
xugy@tsinghua.edu.cn

Abstract. Similarity Indexing is very important for content-based retrieval on large multimedia databases, and the "tightness" of data set envelope is a factor that influences the performance of index. For equidistant envelope (bounding sphere), calculation of envelope is somewhat difficult because of the complexity of direct computation in high dimension space. In this paper we summarize the envelope used in similarity indexing structures, and discuss the envelope parameter calculation problem. We improved the γ-spatial search algorithm proposed by R. Kuniawati and J.S.Jin for equidistant envelope parameter calculation, and apply it to various distance spaces. Basic thoughts and theorem are provided in the paper, also with algorithm implementation.

1 Introduction

With multimedia databases being widely used in various areas, management and retrieval of media objects (images, videos) is becoming a problem. In most cases, media objects are required to be retrieved by their content, which is represented as a feature vector using specific media understanding technology (image processing, computer vision, etc). Thus similarity indexing technologies which retrieve object based on distance in feature space are required.

Similarity indexing is a k-nearest-neighbor search (K-N-N) problem in nature, thus index structures have been used in K-NN research can be adopted. The structures can be classified into two categories: R*-tree based (such as R*-tree [1], TV-tree [2] (Telescopic Vector Tree), X-tree [3] and SS-tree [4] and k-d-tree based (for example, k-d-tree [5], VAMSplit k-d-tree[6], etc.). But no matter what similarity index is being used, reduction of overlapped areas is a key factor to promote index performance. This means that the envelope should provide a "tight" bounding for the data set.

Range envelope (or bounding box, bounding rectangle) has been used in many index structures, because their envelope parameters can be calculated easily. On the other hands, equidistant envelope provides a more "tight" envelope since the sub data set generated by clustering method tend to be of sphere shape, but the difficulty of equidistant envelope parameter calculation (the time complexity of direct calculation is exponential in the number of dimensions) inhibits it from being widely used. To solve this problem, R. Kuniawati and J.S.Jin [7] proposed the γ -spatial search algorithm which can calculate equidistant envelope parameters efficiently. But their algorithm needs preserve all pre-computed plane parameters and is only designed to work in Euclidean distance space. In this paper, we summarize the envelope calculation problem, improve the γ -spatial search algorithm and apply it to quadric distance space and block distance space as well. Basic thoughts and theorem are provided in the paper, also with algorithm implementation.

The sections below are arranged as followed. In section 2, we briefly introduce the basic concepts of similarity indexing. Section 3 focuses on the envelope parameter calculation problem, where parameter calculations of both range envelope and equidistant envelope in various distance space (Euclidean distance space, Quadric distance space, Maximum Distance and Block distance space) are discussed. In section 4, experiment results and analysis are given.

2 Basic Concepts of Similarity Indexing

2.1 Definition of similarity indexing

There are many different definitions for similarity indexing. In this paper, we adopt the similarity selection given by D. White and R. Jain, which considered approximate search:

Definition 1: (similarity selection)

The similarity selection takes a form of **Query(f, n, th, e)**, parameter of which is defined as:

- **f** is the query vector, where d is the dimension of corresponding feature space,
- **n** is a positive integer that specifies the maximum number of (approximate) nearest neighbors returned in the query result,
- **th** is a positive real number that specifies the maximum distance from **f** to any vector in query result,
- **e** is a non-negative real number that specifies a bound on the approximation error:

$$\mathbf{D} \le (1+\mathbf{e})\mathbf{D}^* \tag{1}$$

where **D** is either equal to **th** if **n** result vectors have not been found, or is the maximum of distances from **f** to nearest neighbor vectors in the approximate query result, and where $\mathbf{D}^*$ is the distance from the closest missed vector to **f** . A missed vector is a vector that is not present in the result set of an approximate query but is present in the result set of the corresponding exact query. The query is exact when **e** =0 since **D**= **D***.

2.2 Composition of similarity indexing structure

Essentially, similarity indexing is a k-nearest-neighbor (k-NN) search problem. Current k-NN methods are all based on the thoughts followed: divide the data set into several sub data sets and establish description for each sub data set, search procedure can avoid some sub sets through comparison of descriptions, thus reduces the search cost. In general, similarity indexing structures are all composed of the two parts below:

1. description of data set, describe general properties of the data set, usually contains
   - object number, the total number of objects in the data set,
   - envelope , a camber enclosing all data set points inside, used to give a constraint of the position and distribution of the data set,
   - representative object ID, the object id of a object which can represent the data set.
2. pointers to sub data sets, contains pointers to sub data set node (corresponding to the internal node of index tree) or pointers to arrays of object id and feature (corresponding to the leaf node of index tree).

2.3 Similarity index searching

Different distance space, envelope type and data set splitting method lead to different index structures, but the basic thought of index searching is similar. The general scheme is shown as below:

Algorithm 1: (Similarity Index Searching Scheme)
Input: query vector **f**, rootnode of similarity index, maximum object number M, distance threshold T, approximation error E
Output: object Id and correspondent distance.
Algorithm:

1. Establishing two lists——a node list with fields (nodeID, maxdist, mindist) and a object list with fields (objectID, dist). Empty the two list and put the root node in node list, calculate its maximum and minimum distance from **f** according to its envelope, fill in its maxdist and mindist filed.
2. Select the node with minimum mindist and remove it from node list. If the node is a leaf node, go to 3; else all its child nodes in the node list and calculate their maxdist and mindist, remove node whose mindist is larger than T/(1+E).
3. Sort the nodelist by maxdist in ascending order and calculate n which is the smallest number satisfying first n nodes have at least M objects. Denote the

maxdist of n-th node as T', remove all nodes whose mindist is larger than T'/(1+E). Go to 5.
4. Put all objects in the leaf node in object list, calculate their distance from **f** and fill in dist field. Sort the object list by dist in ascending order, remove object whose index is larger than M or whose dist is larger than T. If there are M objects in the object list, denote the M-th object's dist as T', remove all nodes in node list whose mindist is larger than T'/(1+E).
5. If the node list is not empty, go to 2.
6. Sort the object list by dist in ascending order. If there are more than M objects, return the first M objects; else return all of them.

3. Parameter Calculation of Envelope

Envelope is a closed camber enclosing all correspondent data set points. From algorithm 1 we know that the maximum and minimum distance calculated from query feature **f** to envelope are used to discard irrelevant sub data sets, thus they should be a good approximation to the maximum and minimum values of distances between points in data set to **f**. Besides, the calculation of maximum distance and minimal distance from envelope parameters should be simple, since in most cases the query need a real-time response.

According to the bounding characteristics, envelopes currently used can be classified into two types: range envelope and equidistant envelope. Range envelope exploits data set's distribution range on each coordinate axes, and is often called bounding rectangle or bounding box. Equidistant envelope uses a equidistant camber to a point in the feature space, and is often called bounding sphere since the equidistant camber is a sphere in the mostly used Euclidean distance space. Other types of envelope are seldom used because of their computation complexity.

3.1 Range envelope

3.1.1 the form of range envelope

As mentioned above, range envelope uses distribution range on coordinate axes. In a N dimension space, the envelope can be represented by a N*2 array env[N][2], where env[i][0] and env[i][1] are the minimum and maximum of the point coordinate in the data set on the i-th axe. Then the i-th coordinate of any point in the data set belongs to the interval [env[i][0], env[1][1]], and the envelope becomes a "box" in the space with all points inside or on the surface.

3.1.2 parameter calculation of range envelope
For range envelope, parameter calculation reduces to the calculation of minimum and maximum value on each dimension. Reading values of each points and simple comparison can do this work.

3.1.3 maximum and minimum distance calculation of range envelope
The key point to the calculation of maximum and minimum distance is to find the nearest and farthest points to the querying vector in the envelope. Denoting querying vector as f=f[N], nearest point as near[N] and farthest point as far[N], the calculation is as followed:

$$\text{near}[i] = \begin{cases} \text{env}[i][0]\ , & \textit{if}\quad f[i] < \text{env}[i][0] \le \text{env}[i][1] \\ f[i], & \textit{if}\quad \text{env}[i][0] \le f[i] \le \text{env}[i][1] \\ \text{env}[i][1]\ , & \textit{if}\quad \text{env}[i][0] \le \text{env}[i][1] < f[i] \end{cases} \tag{2}$$

$$\text{far}[i] = \begin{cases} \text{env}[i][0]\ , & \textit{if}\quad \left|\text{env}[i][0] - f[i]\right| > \left|\text{env}[i][1] - f[i]\right| \\ \text{env}[i][1]\ , & \textit{if}\quad \left|\text{env}[i][1] - f[i]\right| > \left|\text{env}[i][0] - f[i]\right| \end{cases} \tag{3}$$

3.2 Equidistant envelope

3.2.1 the form of equidistant envelope
Equidistant envelope can be characterized by its center and the distance from center to the equidistant camber (radius). In a N dimension space, that will be a N dimension array c[N] and a radius r. All points in the data set are enclosed by this camber.

3.2.2 maximum and minimum distance calculation of range envelope
Maximum and minimum distance calculation is quite simple for equidistant envelope. The distance from querying vector to envelope center plus and minus radius give the maximum and minimum distance. Denoting envelope centered at c[N] with radius r as env(c[N], r), the calculation of maximum and minimum distance is

$$\text{maxdist}(f[N], \text{env}(c[N], r)) = \text{dist}(f[N], c[N]) + r \tag{4}$$

$$\text{mindist}(f[N], \text{env}(c[N], r)) = \begin{cases} \text{dist}(f[N], c[N]) - r, & \textit{if}\quad \text{dist}(f[N], c[N]) \ge r \\ 0, & \textit{if}\quad \text{dist}(f[N], c[N]) < r \end{cases} \tag{5}$$

3.2.3 parameter calculation of equidistant envelope
For equidistant envelope, parameter calculation is relative complex and depends on the distance space used. Direct calculation in high dimension space is impossible in real applications because of its high computation complexity. R. Kuniawati and J.S.Jin proposed a spatial search algorithm utilizing the golden ratio

($\gamma = (\sqrt{5} - 1)/2 = 0.6180339887\ ...$), but their algorithm is only designed for Euclidean distance space and need preserving all past plane parameters. We improved the algorithm and applied it to various distance space (including quadric distance space and block distance space).

Euclidean distance space

In Euclidean space, distance between vector **x** and **y** is defined as

$$\mathbf{d(x, y)} = \sqrt{\sum_{i=1}^{n} (\mathbf{x}_i - \mathbf{y}_i)^2} \tag{6}$$

The equidistant camber is a sphere camber, as shown in Fig. 1 (a).

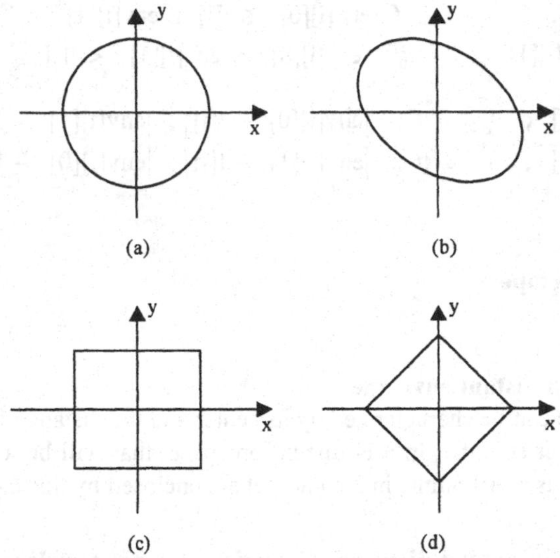

Fig. 1. Equidistant Camber in different distance space (2 dimension case)
(a) Euclidean distance space (b) Quadric distance space
(c) Maximum distance space (d) Block distance space

In Euclidean space, the envelope of a data set is just its smallest enclosing sphere (SES), thus calculation of envelope parameters changes to the calculation of SES. The γ spatial search algorithm is based on the lemma below:

Lemma 1: Given a data set S and a point A in Euclidean space, F is the farthest point to A in S, P is the plane through A and perpendicular to AF, then the center of the smallest enclosing sphere (SES) C and F must be on the same side of P (including P).
Proof: Since F is the farthest point to A in data set S, then the radius of the SES of S must be no greater than |AF|. So the center of SES C must be in the sphere (including the surface) centered at F with radius |AF|, thus C and F must be on the same side of P (including P).

Proven.

Based on this lemma, the γ spatial search algorithm proposed by R. Kuniawati and J.S.Jin is shown in Fig. 2.

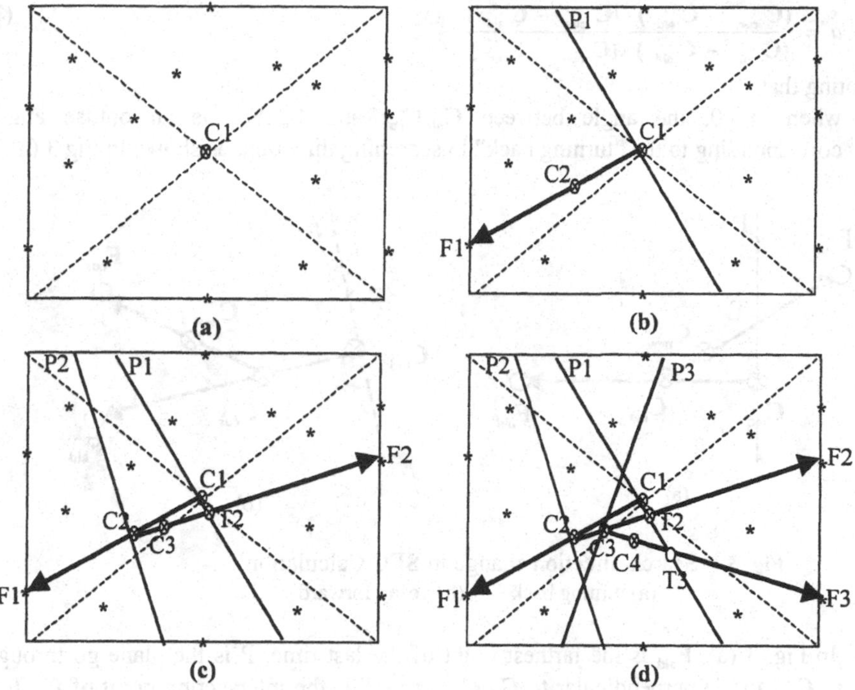

Fig. 2. γ spatial search method for SEC Calculation

First, select the center of range envelope —— $\mathbf{C}_1$ as the initial guess of the SEC center (Fig.2 a). Find the farthest point in the data set form $\mathbf{C}_1$, denoted as $\mathbf{F}_1$, then the second guess of SEC center $\mathbf{C}_2 = \gamma\, \mathbf{C}_1 + (1-\gamma)\mathbf{F}_1$ (Fig.2 b). Denoting the plane through $\mathbf{C}_1$ and perpendicular to $\mathbf{C}_1\mathbf{F}_1$ as P_1, select the farthest point in the data set form $\mathbf{C}_2$, denoted as $\mathbf{F}_2$, and the line segment $\mathbf{C}_2\mathbf{F}_2$ intersect plane P_1 at point $\mathbf{T}_2$, then the new center $\mathbf{C}_3 = \gamma\, \mathbf{C}_2 + (1-\gamma)\mathbf{T}_2$ (Fig.2 c). Denoting the plane through $\mathbf{C}_2$ and perpendicular to $\mathbf{C}_2\mathbf{F}_2$ as P_2, select the farthest point in the data set form $\mathbf{C}_3$ and denote it as $\mathbf{F}_3$. Consider if the line segment intersect $\mathbf{C}_3\mathbf{F}_3$ intersect current planes (P_1 and P_2), we see that it does not intersect with plane P_2 but intersect with plane P_1 at point $\mathbf{T}_3$. Select the nearest intersect point (in this case is $\mathbf{T}_3$) then the new center $\mathbf{C}_4 = \gamma\, \mathbf{C}_3 + (1-\gamma)\mathbf{T}_3$ (Fig.2 d). Repeat this process until the distance between old center and new center is less than a given threshold, the new center is the center of the envelope, and the distance of the farthest point in the data set to the center is the radius.

It should be noted that the above algorithm need preserving past planes (such as and P_1 and P_2) in calculation, which may cause difficulties if the procedure runs many steps before it converges. To solve this problem, we proposed a improved algorithm which depends on the change of searching directions:

Denoting the center of last time as $\mathbf{C}_{old}$, current center as $\mathbf{C}_{cur}$, and the farthest point from $\mathbf{C}_{cur}$ as $\mathbf{F}_{cur}$, calculating

$$a = \frac{(\mathbf{C}_{cur} - \mathbf{C}_{old}) \cdot (\mathbf{C}_{cur} - \mathbf{C}_{old})}{(\mathbf{C}_{cur} - \mathbf{C}_{old}) \cdot (\mathbf{C}_{cur} - \mathbf{F}_{cur})} \tag{7}$$

Noting that

a) when $\alpha > 0$, the angle between $\mathbf{C}_{old}\mathbf{C}_{cur}$ and $\mathbf{C}_{cur}\mathbf{F}_{cur}$ is an obtuse angle, corresponding to the "turning back" in searching direction, as shown in Fig 3 (a).

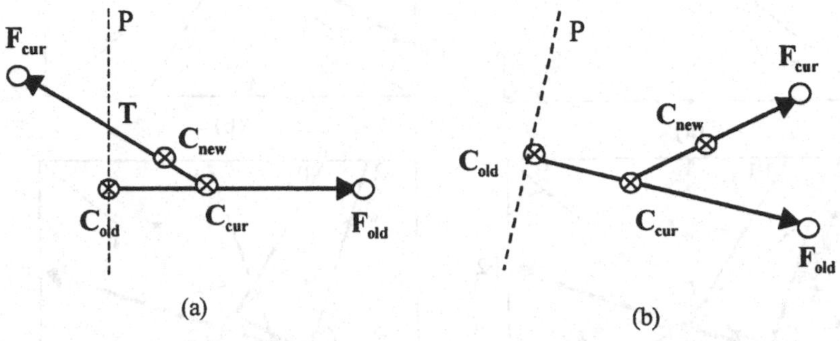

Fig. 3. Search direction change in SEC Calculation
(a) turning back (b) going forward

In Fig. 3 (a), $\mathbf{F}_{old}$ is the farthest point of the last time, P is the plane go through point $\mathbf{C}_{old}$ and is perpendicular to $\mathbf{C}_{old}\mathbf{C}_{cur}$, and T is the intersecting point of $\mathbf{C}_{cur}\mathbf{F}_{cur}$ and plane P. Obviously,

$$\mathbf{T} = \mathbf{C}_{cur} + a(\mathbf{F}_{cur} - \mathbf{C}_{cur}) \tag{8}$$

Similar to the γ spatial search algorithm, in this case

$$\mathbf{C}_{new} = \gamma\mathbf{C}_{cur} + (1-\gamma)\mathbf{T} \tag{9}$$

b) when $\alpha < 0$, the angle between $\mathbf{C}_{old}\mathbf{C}_{cur}$ and $\mathbf{C}_{cur}\mathbf{F}_{cur}$ is an acute angle, corresponding to the " going forward" in searching direction, as shown in Fig 3 (b).

In this case, line segment $\mathbf{C}_{cur}\mathbf{F}_{cur}$ has no intersecting point with plane P, as shown in Figure 3 (b). We adopt the method of advancing the center to the direction of current farthest point with a length same as the distance between $\mathbf{C}_{cur}$ and $\mathbf{C}_{old}$. That is, moving $\mathbf{C}_{cur}$ to $\mathbf{F}_{cur}$ along $\mathbf{C}_{cur}\mathbf{F}_{cur}$ with length $|\mathbf{C}_{old}\mathbf{C}_{cur}|$, which gives

$$\mathbf{C}_{new} = \mathbf{C}_{cur} + \frac{|\mathbf{C}_{cur} - \mathbf{C}_{old}|}{|\mathbf{F}_{cur} - \mathbf{C}_{cur}|}(\mathbf{F}_{cur} - \mathbf{C}_{cur}) \tag{10}$$

Combining the two cases above, our improved algorithm for envelope parameter calculation in Euclidean space is as followed.

Algorithm 2: (Improved algorithm for equidistant envelope parameter calculation in Euclidean space)

Input: the data set, distance threshold.

Output: the parameter (center and radius) of the data set's equidistant envelope.

Algorithm:

1. select the range envelope center as the initial guess of envelope center, calculate the farthest point to $\mathbf{C}_{old}$ in the data set and denote it as $\mathbf{F}_{old}$, calculate the current guess of center

$$\mathbf{C}_{cur} = \gamma \mathbf{C}_{old} + (1-\gamma)\mathbf{F}_{old}$$

2. calculate the farthest point to $\mathbf{C}_{cur}$ in the data set and denote it as $\mathbf{F}_{cur}$, then calculate α according to equation (10).
3. calculate new guess of center using α :

$$\mathbf{C}_{new} = \begin{cases} \gamma \mathbf{C}_{cur} + (1-\gamma)[\mathbf{C}_{cur} + a(\mathbf{F}_{cur} - \mathbf{C}_{cur})], & if\, a > 0 \\ \mathbf{C}_{cur} + \dfrac{|\mathbf{C}_{cur} - \mathbf{C}_{old}|}{|\mathbf{F}_{cur} - \mathbf{C}_{cur}|}(\mathbf{F}_{cur} - \mathbf{C}_{cur}), & if\, a < 0 \end{cases}$$

4. if the distance between $\mathbf{C}_{new}$ and $\mathbf{C}_{cur}$ is less than the threshold given, go to 6.
5. Replace $\mathbf{C}_{old}$ with $\mathbf{C}_{cur}$, $\mathbf{C}_{cur}$ with $\mathbf{C}_{new}$, go to 2.
6. $\mathbf{C}_{new}$ is the center of the envelope, and distance from the farthest point in the data set to $\mathbf{C}_{new}$ is the radius.

Quadric distance space

In quadric distance space, distance between **x** and **y** is defined as

$$\mathbf{d}(\mathbf{x}, \mathbf{y}) = \sqrt{(\mathbf{x}-\mathbf{y})^T \mathbf{A}(\mathbf{x}-\mathbf{y})} \tag{11}$$

The quadric distance is often used when there is similarity between dimensions (for example, the color histogram). A is a symmetric matrix, its entry $a_{i,j}$ represents the similarity between dimension i and dimension j. Obviously, quadric distance reduces to Euclidean distance when A is identity matrix I.

The equidistant camber in quadric distance space is an ellipsoid depending on A, as shown in Figure 1 (b).

To apply γ spatial search algorithm to quadric distance space, we need introduce the definition of inner product:

Definition 2: (Inner product in quadric distance space)

In quadric distance space, the inner product of vector **x** and **y** is defined as

$$(\mathbf{x}, \mathbf{y})_{quadric} = \mathbf{x}^T \mathbf{A}\mathbf{y} \tag{12}$$

Based on this definition, perpendicular concept can be derived and α can be calculated as followed:

$$a = \frac{(\mathbf{C}_{cur} - \mathbf{C}_{old})^T \mathbf{A}(\mathbf{C}_{cur} - \mathbf{C}_{old})}{(\mathbf{C}_{cur} - \mathbf{C}_{old})^T \mathbf{A}(\mathbf{C}_{cur} - \mathbf{F}_{cur})} \tag{13}$$

The algorithm of envelope parameter calculation in quadric distance space is the same as in Euclidean space except the calculation of α, thus replace equation (7) with equation (13) in algorithm 1 will work.

Maximum distance space

In maximum distance space, distance between **x** and **y** is defined as

$$\mathbf{d}(\mathbf{x},\mathbf{y}) = \max_{\mathbf{i}} |\mathbf{x}_i - \mathbf{y}_i| \tag{14}$$

The equidistant camber in Maximum distance space is a square in 2-dimenson case, as shown in Figure 1 (c).

Envelope parameter calculation in quadric distance space is quite simple and does need γ spatial search —— the parameters can be adopted directly from range envelope.

Block distance space

In block distance space, distance between **x** and **y** is defined as

$$\mathbf{d}(\mathbf{x},\mathbf{y}) = \sum_{i=1}^{n} |\mathbf{x}_i - \mathbf{y}_i| \tag{15}$$

The equidistant camber in block distance space is a square rotated 45 degree in 2-dimenson case (as shown in Fig.1 d) and is a regular octahedron in 3-dimenson case.

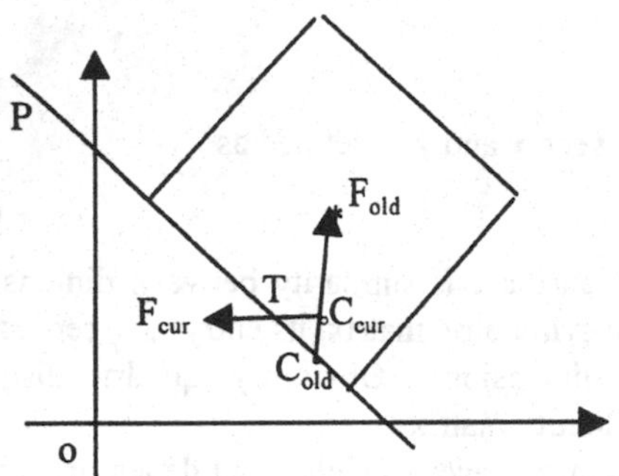

Fig. 4. Constraints on center location in block distance space

In block space, the calculation of envelope parameters is a bit more complex. Denoting the guess of center last time as $\mathbf{C}_{old}$, farthest point as $\mathbf{F}_{old}$, then similar to Lemma 1, the center C must be in the equidistant camber centered at F with radius |FC| (including surface), as shown in Fig. 4. Denoting the current guess of center as $\mathbf{C}_{cur}$, farthest point as $\mathbf{F}_{cur}$, then we need calculate the intersecting point (denoted as **T**) of $\mathbf{C}_{cur}\mathbf{F}_{cur}$ and the equidistant camber side plane going through $\mathbf{C}_{old}$ (denoted as P), then $\mathbf{T}=\mathbf{C}_{cur}+\alpha(\mathbf{F}_{cur}-\mathbf{C}_{cur})$, and vector $\mathbf{T}-\mathbf{C}_{old}$ is perpendicular to the normal vector of plane P (denoted as **D**).

We have

$$(\mathbf{C}_{cur} + \alpha(\mathbf{F}_{cur} - \mathbf{C}_{cur}) - \mathbf{C}_{old}) \cdot \mathbf{D} = 0$$

which gives

$$\alpha = \frac{(\mathbf{C}_{cur} - \mathbf{C}_{old}) \cdot \mathbf{D}}{(\mathbf{C}_{cur} - \mathbf{F}_{cur}) \cdot \mathbf{D}} \tag{16}$$

The normal vector $\mathbf{D}=(d_i)_M$ can be calculated from $\mathbf{C}_{old} = (c_i^{old})_M$ and $\mathbf{F}_{old} = (f_i^{old})_M$ using:

$$d_i = \begin{cases} 1, & \text{if } f_i^{old} \geq c_i^{old} \\ -1, & \text{if } f_i^{old} < c_i^{old} \end{cases} \qquad (i = 0,1,2,\ldots, M-1) \qquad \textbf{(17)}$$

Thus algorithm for envelope parameter calculation in block distance space can be derived:

Algorithm3: (Algorithm for equidistant envelope parameter calculation in Block distance space)
Input: the data set, distance threshold.
Output: the parameter (center and radius) of the data set's equidistant envelope.
Algorithm:

1. select the range envelope center as the initial guess of envelope center, calculate the farthest point to $\mathbf{C}_{old}$ in the data set and denote it as $\mathbf{F}_{old}$, then calculate the current guess of center

$$\mathbf{C}_{cur} = \gamma\mathbf{C}_{old} + (1-\gamma)\mathbf{F}_{old}$$

2. select the farthest point to $\mathbf{C}_{cur}$ in the data set, denote it as $\mathbf{F}_{cur}$.
3. calculate normal vector D with $\mathbf{C}_{old}$ and $\mathbf{F}_{old}$, according to equation (17).
4. calculate α from D according to equation (16).
5. calculate new guess of center using α:

$$\mathbf{C}_{new} = \begin{cases} \gamma\mathbf{C}_{cur} + (1-\gamma)[\mathbf{C}_{cur} + a(\mathbf{F}_{cur} - \mathbf{C}_{cur})], & \text{if } a > 0 \\ \mathbf{C}_{cur} + \dfrac{|\mathbf{C}_{cur} - \mathbf{C}_{old}|}{|\mathbf{F}_{cur} - \mathbf{C}_{cur}|}(\mathbf{F}_{cur} - \mathbf{C}_{cur}), & \text{if } a < 0 \end{cases}$$

6. if the distance between $\mathbf{C}_{new}$ and $\mathbf{C}_{cur}$ is less than the threshold given, go to 8.
7. Replace $\mathbf{C}_{old}$ with $\mathbf{C}_{cur}$, $\mathbf{C}_{cur}$ with $\mathbf{C}_{new}$, go to 2.
8. $\mathbf{C}_{new}$ is the center of the envelope, and distance from the farthest point in the data set to $\mathbf{C}_{new}$ is the radius.

3.3 Comparison of envelope

From the analysis above, it is shown that the range envelope is simple in form and can be calculated easily. But in Euclidean space (which is mostly often used) and quadric distance space, the envelope can not provide a "tight" constraint on the data set since the sub data set generated by clustering method tends to be of sphere shape in distribution. Similarly, range envelope can not provide a good constraint in block distance space either.

Compared with range envelope, the parameter calculation of equidistant envelope need iterative calculation thus need relatively longer time. But the parameter calculation is in the data base production stage, thus does need to be real-time. In most cases, using equidistant envelope can improve performance in speed efficiently through its tight constraint which reduces overlap areas between different sub data set. Only the maximum distance space is an exception. In this case, equidistant envelope take a form similar to range envelope, but it can not provide a constraint as efficient as range envelope because of the lack of flexibility on each dimension.

4. Implementation and Demonstrations

Based on the analysis above, we implement a similarity indexing structure called GSS-tree (General Similarity Search tree). The GSS-tree is designed to handle similarity indexing in four distance spaces: Euclidean space, Quadric distance space, Max distance space and Block distance space, using equidistant envelope, rectangle envelope or the combination of two . The definition of GSS-tree is given in Fig. 5. Envelope parameter calculation algorithms presented in this paper have been implemented in GSS-tree's index-generating module.

```
typedef struct envelop {
  int type;                  /*available envelop type, range, equidistant, or both */
  SPHERE      *sph;          /* the equidistant envelop */
  RECTANGLE *rect;           /* the rectangle envelop */
} ENVELOP;

typedef struct treenode {
  long objectnum;            /* total number of object in the node and subnode */
  ENVELOP *env;              /* envelop of thsi node */
  int treenum;               /* number of subtree */
  struct treenode **child;   /* subtree pointer       */
  OB_DESP *obdesp;
} TREENODE;

typedef struct distance {
  int     type;         /* distance type, can be max,cityblock, euclidian, or quadric */
  float *coef;          /* matrix coef of type quadric */
} DISTANCE;

typedef struct rootnode {
  int dim;                   /*dimension of the feature space */
  DISTANCE *distdef;         /* definition of distance function */
  TREENODE *tree;            /*the real tree */
} ROOTNODE;
```

Fig. 5. Definition of GSS-tree

We used the GSS-tree as similarity indexing structure for our content-based image retrieval system ImgRetr, which is now available for Chinese domestic Internet users during daytime (Beijing TimeZone) at http://gnome.sist.tsinghua.edu.cn/~cbir/ or http://166.111.78.172/~cbir/ (Users outside may access the system through proxies on Chinese domestic Internet network). English and Chinese edition are in the directory imgretr-e and imgretr-c respectively. Among the features used, most adopt Euclidean distance space (eg. , dominating color, color distribution and texture), some use Quadric distance space (such as the sketch feature). The database currently contains 11804 images, and in most cases the system give a real-time response (the response time depends on the dimension and distance space used, for example, all other features can response immediately except the sketch feature, which is of 16*16 dimension and uses quadric distance space).

References

1. Beckmann, N., Kriegel, H.P., Schneider, R., Seeger, B.: The R*-tree: An Efficient and Robust Access Method for Points and Rectangles. Proceedings of the ACM SIGMOD International Conference on the Management of Data, 322-331.

2 Lin, K.-I., Jagadish, H.V., Faloutsos, C.: The TV-tree: An index structure for high-dimensional data. VLDB Journal,Vol 3, No. 4, (1994)517-549

3 Berchtold, S., Keim, D.A., Kriegel, H.P.: The X-tree: An index structure for high-dimensional data. Proc. 22th Int. Conf. On Very Large Data Bases, Bombay, India(1996).

4 White D.A., Jain, R.: Similarity indexing with the SS-tree. Proc. 12th IEEE International Conference on Data Engineering. New Orleans, Lousiana (1996)

5 Friedman, J. H., Bentley, J. H. , Finkel, R. A.: An algorithm for finding best matches in logarithmic expexted time, ACM Transactions on Mathematical Softeare, Vol. 3, No.3, (1977) 209-226

6 White D.A., Jain, R.: Similarity indexing: Algorithms and performance. Proceedings of the SPIE: Storage and Retrieval for Image and Video Databases IV, Vol. 2670, San Jose, CA. (1996).

7 Kurniawati, R., .Jin, J.S, Shepherd, J.A.: The SS+-tree: An Improved Index Structure for Similarity Searches in a High-Dimensional Feature Space. Private Communication.: School of Computer Science and Engineering, University of New South Wales, Sydney 2052, Australia.

Intra-Block Max-Min Algorithm for Embedding Robust Digital Watermark into Images*

F.Y. Duan[1], I. King[2], L.W. Chan[2], and L. Xu[2]

[1] Institute of Information Science
Wuyi University
Jiangmen, Guangdong, China 529020
fyduan@letterbox.wyu.edu.cn

[2] Department of Computer Science & Engineering
The Chinese University of Hong Kong
Shatin, N.T., Hong Kong, China
{king,lwchan,lxu}@cse.cuhk.edu.hk

Abstract. We present a new DCT-based block max-min algorithm for digital watermark embedding in images. Our algorithm uses intra-block bordering upon pixels relations to generate the watermarked image. The features of our method to embed the watermark are: (1) the watermark is perceptually invisible; (2) low loss of relevant information of the original image; (3) the watermark can be read only by using a secret key; (4) the watermark is robust against translation and area cropping; and (5) retrieval of embedded watermark does not need the original image. We describe the algorithm framework and its performance against translation and area cropping with experiments.

1 Introduction

Digital watermarking is becoming increasingly popular and important as users seek ways to protect their proprietary information being transmitted over the Internet. Digital watermarking is useful in the authentication of multimedia information such as text, sound, image, and video, to ensure the undisputed proof of ownership on copyrighted materials. It is one of the better ways to protect intellectual property from illicit copying [17].

The digital watermarking is not unlike image coding when an *original signal* is embedded with the *watermarking signal* with additional requirements depending on the applications. These requirements must guard against tempering of the watermarked signal. For example, the watermarking may resist geometric transformation, compression, and sampling to ensure that the watermarking signal can still be retrieved with a high degree of accuracy. Moreover, the watermark should be perceptually invisible to the viewer nor should the watermark degrade the quality of the content.

* This work is supported in part by RGC Earmarked Grant # CUHK4176/97E.

There are also other system issues to address when designing a suitable watermarking algorithm. For example, there is always the trade-off issue of the complexity of the algorithm and the computational resource required to achieve the result. In other words, although one may achieve a high degree of security with a watermarking algorithm, the computational cost to code and decode such a watermarking may be prohibitive. Hence, it is the goal in designing the watermarking algorithm to have a suitable technique which satisfies many of the requirements with good real-time performance.

2 Previous Work

There are many ways to perform information embedding in signals. Here we are interested mainly with digital images as the source signal. There are two main classes of watermarking techniques: visible [9] and perceptually invisible [1, 3, 4, 6, 7, 10, 11, 12, 13]. The perceptually invisible watermarking techniques consist of both two-dimensional spatial watermarks [1, 3, 10, 11, 12, 13] and frequency domain watermark [4, 6, 7] while IBM has developed a proprietary visible watermark to protect images that are part of the digital Vatican library project [9].

One early watermarking method obtains a checksum of the image data, then embeds the checksum into the least significant bits(LSB) of randomly chosen pixels for watermarking [14]. Others have added a maximal length linear shift register sequence to the pixel data and identified the watermark by computing the spatial cross-correlation function of the sequence and the watermarked image [10, 15, 16]. Watermarks also can modify the image's spectral or transform coefficients directly. These algorithms often modulate DCT coefficients according to a sequence known only to the owner [4, 8]. A straightforward technique is proposed to gray scale images by adding a watermark image to the original image [1, 10]. Watermarks can be image dependent. For example, [5] uses independent visual channels for watermark embedding and [2] generated watermarked images by modulating JPEG coefficients. These watermarks are designed to be invisible, or to blend in with natural camera or scanner noise.

In [7], Hsu presented a DCT-based algorithm using inter-block relation to implement the middle-band embedding. First, the author used a fast two-dimensional pseudo-random number traversing method to permute the signature image. Second, the input image was divided into blocks of size 8×8, and then each block is DCT transformed. Third, Hsu chose the middle-band coefficients for watermark embedding. Then, a 2-D sub-block mask is used to compute the residual pattern from the chosen middle-band coefficients of inter-block. After the residual pattern is obtained, for each marked pixel of the permuted signature data he then modified the DCT coefficients according to the residual mask.

Our method is different from [7] in one major aspect. Instead of using inter-block relations, our algorithm uses intra-block relations to generate the watermarked image. Specifically, our method does not require the original image to extract the watermark. Hence, this method is less cumbersome than other methods requiring the original image. From our experiments we found this method to be more robust against localized distortions since the size of the blocks is smaller and the blocks are closer together.

In Section 3, we present our new proposed watermarking algorithm. We then present our experimental results in Section 4. Discussion is presented in Section 5 and conclusion is found in Section 6.

3 Proposed Algorithm

3.1 Digital Watermark Embedding

Let OI and OW be the original grayscale image and the original binary digital watermark respectively. They are defined as,

$$OI = \{oi(i,j), 1 \leq i,j \leq N\}, oi(i,j) \in \{0, \cdots, L-1\}, \tag{1}$$

where L is the number of grayscale levels for the image.

$$OW = \{ow(i,j), 1 \leq i,j \leq N'\}, ow(i,j) \in \{0,1\}, \tag{2}$$

with $N' = N/4$ in our case.

Permutation of the Watermark Data A secret key $K(l)$ is a random permutation of the integers from 1 to N'^2. It is used to permute the watermark. The order is defined as,

$$K[(i-1) \times N' + j] - 1 \equiv i' \pmod{N'} + j' \tag{3}$$

then,

$$W'(i,j) = OW(i'+1, j'+1). \tag{4}$$

After this permutation step of the watermark data, we have a scrambled binary watermark data which can be used in the next step.

Block Transformation of the Image The image OI is divided into 8-by-8 blocks, $BOI_{m,n}$, and the two-dimensional Discrete Cosine Transform (DCT) is computed for each block using

$$D_{m,n} = DCT(BOI_{m,n}), \qquad 1 \leq m,n \leq M \tag{5}$$

$$= \{d_{m,n}(k,l), 1 \leq k,l \leq 8\} \tag{6}$$

with $M = N/8$ in our case.

Modification of DCT Coefficients The permuted watermark W' is also divided into $M \times M$ blocks $BW'_{m,n}$ with the block size 2-by-2.

When $BW'_{m,n}(i) = 1$, $1 \leq i \leq 4$, we perform an exchange of the K_i^0 with maximum grayscale value among $K_i^0, ..., K_i^5$. When $BW'_{m,n}(i) = 0$, we perform an exchange of the K_i^0 with minimum grayscale value among $K_i^0, ..., K_i^5$.

After the modification of DCT on D, we obtain the modified D' which requires the inverse to compute the watermarked image.

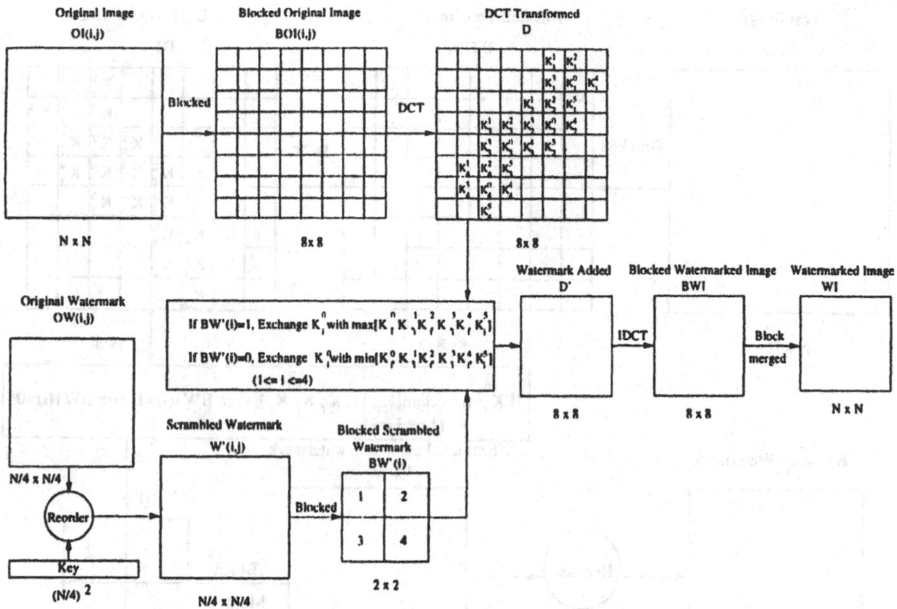

Fig. 1. The schematic of a general key-based watermarking process.

Inverse DCT We then calculate the inverse two-dimensional DCT of each block as,

$$BWI_{m,n} = IDCT(D'_{m,n}) \tag{7}$$

and then arrange the blocks back together in a single watermarked image. A schematic of the embedding watermark to image process is illustrated in Fig. 1.

3.2 Extracting Method

The extraction of watermark EW requires the test image TI and the key sequence K, without needing the original image. The schematic of extracting watermark process is shown in Fig. 2. The extraction steps are as follow:

Block Transformation The test image is divided into $M \times M$ block $BTI_{m,n}$ with the block size 8-by-8. Each block is DCT transformed.

$$D1_{m,n} = DCT(BTI_{m,n}) \tag{8}$$

Extracting the Permuted Watermark

$$BW1(i) = \begin{cases} 1 & K_i^0 > median[K_i^1, K_i^2, K_i^3, K_i^4, K_i^5], \\ 0 & \text{Otherwise.} \end{cases} \tag{9}$$

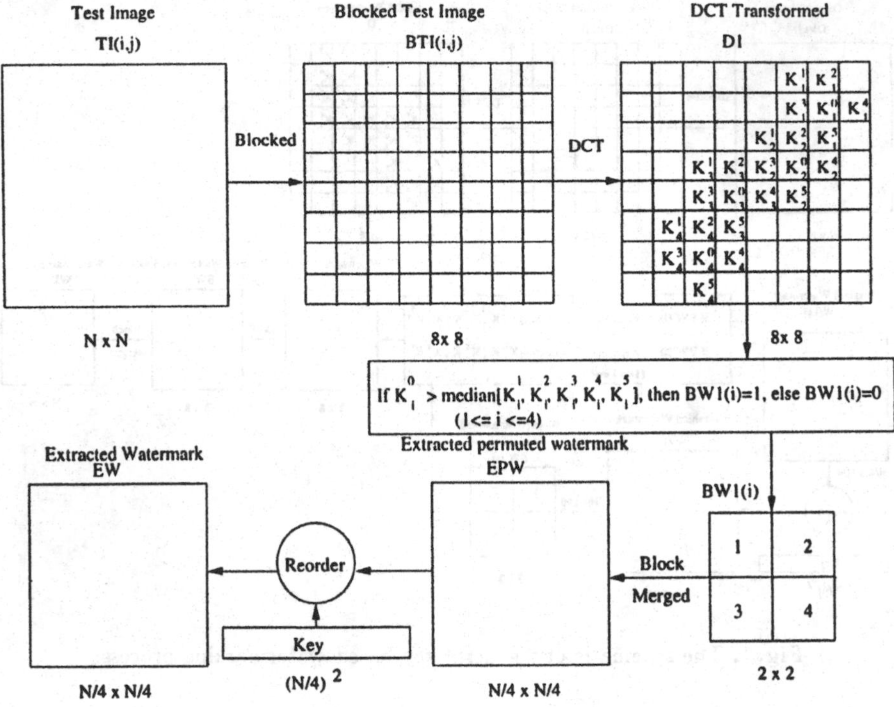

Fig. 2. The schematic of extracting watermark process.

where $1 \leq i \leq 4$, $K_i^0, K_i^1, K_i^2, K_i^3, K_i^4, K_i^5$.

We then arrange the blocks $BW1_{m,n}$ back together in a single extracted permuted watermark EPW. Finally, the extracted watermark EW is obtained by using of the follow equation.

$$EW(i'+1, j'+1) = EPW(i, j) \tag{10}$$

where the relation among i, j, i', j' meets Eq. (3). See also Fig. 2 for a more detailed schematic of the digital watermark extracting process.

4 Experiments

We perform two basic transformations: (1) translation and (2) cropping. The experiments were conducted on UltraSparc workstation with Matlab 5. The original image is a 256 level 256×256 grayscale image.

We defined the Average Intensity Error (AIE) when comparing the watermarked image with the original as,

$$\text{AIE} \equiv \frac{\sum |OI - WI|}{N \times N}. \tag{11}$$

Table 1. Values of the variables used in our experiment

Variable Name	Meaning	Variable Type	Value
OI	Original Image	matrix	256×256
WI	Watermarked Image	matrix	256×256
OW	Original Watermark	matrix	64×64
EW	Extracted Watermark	matrix	64×64
EW'	Extracted Watermark from Black Image	matrix	64×64
K	Key	vector	1×64^2
N		integer	256
N'		integer	64
M		integer	32
L		integer	8

We found that the AIE is approximately 11.9 using the watermarked image shown in Fig. 4(b). This result is quite acceptable since we cannot distinguish the original image and the watermarked image perceptually.

4.1 Translation

Using cropped watermarked image of same cropping sizes over different location of the watermarked image to extract watermark, we test our method's ability to recover watermark image under translation shift with the Relative Error Rate (RER) as the performance criterion:

$$\begin{aligned} \text{RER} &\equiv \frac{\sum |OW - EW|}{N' \times N'} \div \frac{\sum |OW - EW'|}{N' \times N'} \\ &= \frac{\sum |OW - EW|}{\sum |OW - EW'|} \times 100\%. \end{aligned} \tag{12}$$

where EW' means the extracted watermark from black image. The experimental results are shown in Fig. 3 for both horizontal and vertical translations.

Our method is translational invariant since there is no significant effect under translation with a constant size as shown in Fig. 3 when we used a cropping image of 50×230 selected horizontally and vertically. The error obtained is fairly similar which demonstrated that this method is robust against translation.

4.2 Cropping

Our experiments also showed that our method is sensitive to the size of the cropped area. The error is linear and inversely proportional to the size of the cropped areas as shown in Fig. 5. It is similar with one found in [6]. Here we varied the size of the cropped image from 100% (256×256) to 0.055% (11×11) with

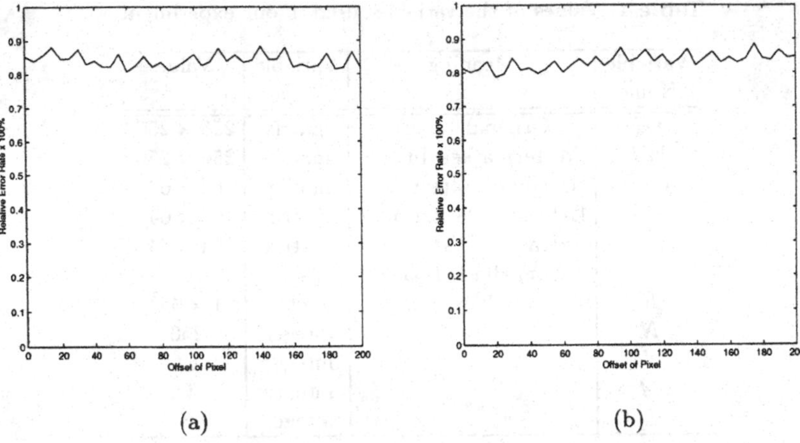

Fig. 3. We used a cropped image of 50×230 and translate it horizontally in (a) and vertically in (b).

the RER ranging from 1% to 100% respectively. Furthermore, we observed that when the RER is greater than 86.6% (the cropped image size is less than 14%) the retrieved watermark image cannot be distinguished with the original watermark perceptually. These results demonstrated that our proposed algorithm is more robust to the cropping than the ones proposed in [6, 7].

5 Discussion

The average retrieved watermark error per pixel is linear and inversely proportional to the percentage of cropping image. When the cropping image is around 14% of the original image, the retrieved watermark is illegible perceptually. Even when we used the whole watermarked original image without cropping, there is still an error arising from the DCT computation. This background error is about 0.1-0.2%. The condition of failure of the method is that the variation of the grayscale value of the original image is very small.

6 Conclusion

We present a new DCT-based block algorithm for watermark embedding in digital images. Its main advantages to [6, 7] are: (1) the watermark can be extracted without the original image and (2) its robustness against cropping and translation distortions. Instead using inter-block relations, our algorithm uses intra-block relations to generate the watermarked image. This method is more robust

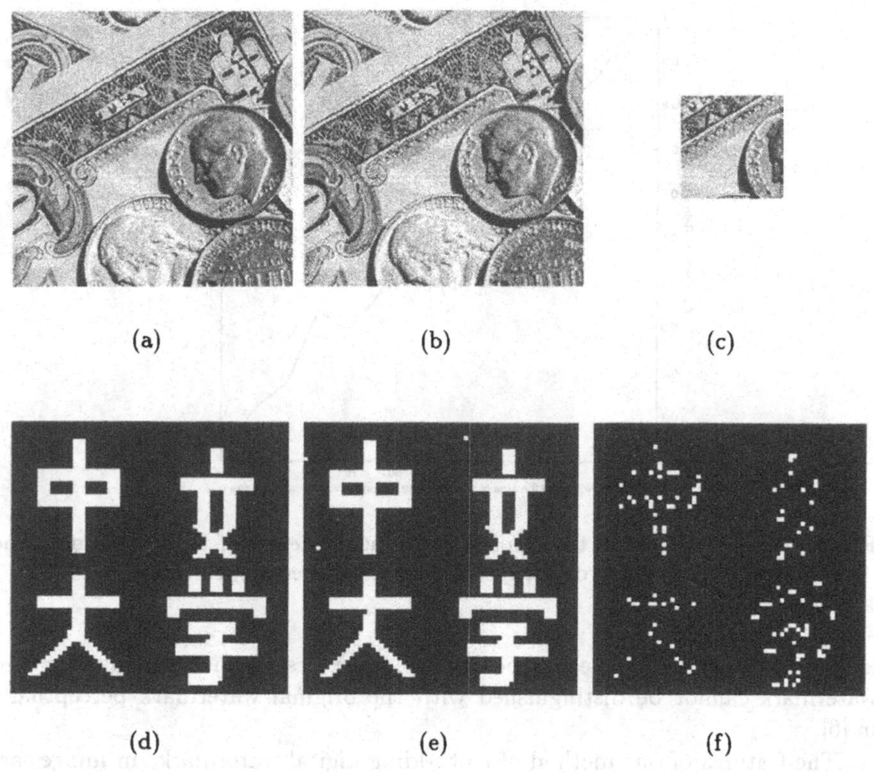

Fig. 4. (a) the original image, (b) the watermarked image, (c) the minimum size cropping watermarked image, (d) the original watermark , (e) the extracted watermark from (b), (f) a poor extracted watermark from (c) with RER of 86.6% using 14% of the image area from the watermarked image.

against localized distortions since the size of the blocks is smaller and the blocks are closer together.

In our experiments we added the watermark to the image by modifying the more perceptually significant components of the image spectrum. The difference between the original and the watermarked image is small. This result is quite acceptable since we can distinguish the original image and the watermarked image perceptually.

Our experiments showed that our method is sensitive to the size of the cropped area. The Relative Error Rate is linear and inversely proportional to the size of the cropped area. We observed that when the RER is greater than 86.6% (the cropped image size is less than 14%) the retrieved watermark cannot be distinguished with the original watermark perceptually. However, when the RER

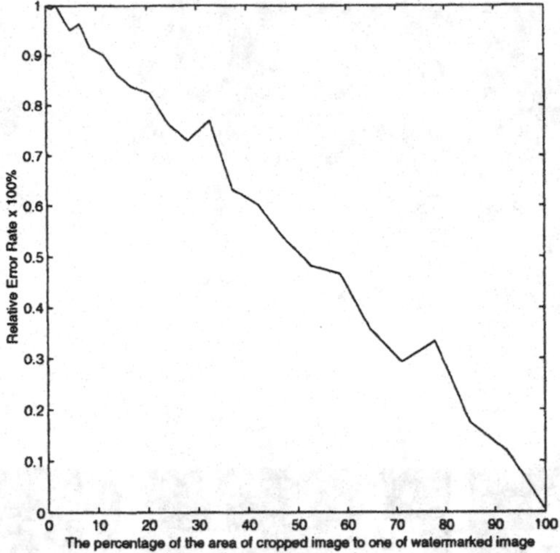

Fig. 5. We vary the size of the cropped image at the center of the image and found that the error is inversely proportional to the area of the cropped image.

is greater than 64.6% (the cropped image size is less than 37%) the retrieved watermark cannot be distinguished with the original watermark perceptually in [6].

The features of our method of embedding digital watermarks in images are as follow:

1. the watermark is perceptually invisible;
2. low loss of relevant information of original image;
3. the watermark can be read only by using a secret key;
4. the watermark is robust against translation and area cropping;
5. retrieval of embedded watermark does not need the original image.

References

1. W. Bender, D. Gruhl, N. Mormoto, and A. Lu. Techniques for data hiding. In *Proceedings of the SPIE*, San Jose CA, USA, February 1995.
2. F.M. Boland, J.J.K.O Ruanaidh, and C. Dautzenberg. Watermarking digital images for copyright protection. In *Proceedings of the International Conference on Image Processing and its Applications*, pages 321–326, Edinburgh, Scotland, July 1995.
3. G. Caronni. Assuring ownership rights for digital image. In *Proc. Reliable IT Systems*. VIS'95. Vieweg Publishing Company, 1995.
4. I.J. Cox, Joe Kilian, Tom Leighton, and Talal Shamoon. Secure spread spectrum watermarking for images audio and video. In *IEEE Int. Conf. on Image Processing*, volume 3, pages 243–246, Lausanne,Switzerland, September 1996.

5. J.F. Delaigle, C.De Vleeschouwer, and B. Macq. Digital watermarking. In *Proceedings of the IS & T/SPIE Conference on Optical Security and Counterfeit Deterrence Techniques*, volume 2659, pages 99–110, San Jose, CA, USA, Feb.1-2 1996.
6. F.Y. Duan, I. King, L.W. Chan, and L. Xu. Intra-block algorithm for digital watermarking. In *14th International Conference on Pattern Recognition (ICPR'98)*, Brisbane, Queensland., Australia, 17-20 August 1998.
7. C.T. Hsu and J.L. Wu. Hidden signatures in image. In *IEEE Int. Conf. on Image Processing*, volume 3, pages 223–226, Lausanne,Switzerland, September 1996.
8. E. Koch and J. Zhao. Towards robust and hidden image copyright labeling. In *Proceedings of the 1995 IEEE Workshop on Nonlinear Signal and Image Processing*, pages 452–455, Neos Marmaras, Greece, June 20-22 1995.
9. F. Mintzer, A. Cazes, F. Giordano, L. Lee, K. Magerlein, and F. Schiattarella. Capturing and preparing images of vatican library manuscripts for access via internet. In *Proceedings of the IS & T's 48th Annual Conference*, pages 74–77, Washington,DC, May 1995.
10. R.G.van Schyndel, A.Z. Tirkel, N.R.A. Mee, and C.F. Osborne. A digital watermark. In *Proceedings of the IEEE International Conference on Image Processing*, volume 2, pages 86–90, Austin, Texas, USA, November 1994.
11. J.R. Smith and B.O. Comiskey. Modulation and information hiding in images. In R. Anderson, editor, *Information Hiding:First Int. Workshop Proc.*, volume 1174 of Lecture Notes in Computer Science, pages 207–226. Springer-Verlag, 1996.
12. M.D. Swanson, B. Zhu, and A.H. Tewfik. Transparent robust image watermarking. In *Proceedings of the 1996 International Conference on Image Processing*, volume 3, pages 211–214, Lausanne, Switzerland, Sept. 16-19 1996.
13. L.T. Turner. Digital data security system. In *Patent IPN WO 89/08915*, 1989.
14. S. Walton. Image authentication for a slippery new age. In *Dr.Dobb's Journal*, pages 18–26, April 1995.
15. R.B. Wolfgang and E.J. Delp. A watermark for digital images. In *IEEE Int. Conf. on Image Processing*, volume 3, pages 219–222, Lausanne,Switzerland, 1996.
16. R.B. Wolfgang and E.J. Delp. A watermarking technique for digital image:further studies. In *Proceedings of the International Conference on Imaging Science,System, and Technology*, pages 279–287, Las,Vegas, June 30-Julu 3 1997.
17. J. Zhao. Look, its not there: Digital watermarking is the best way to protect intellectual property from illicit copying. In *Byte Magazine*, January 1997.

Springer and the environment

At Springer we firmly believe that an international science publisher has a special obligation to the environment, and our corporate policies consistently reflect this conviction.

We also expect our business partners – paper mills, printers, packaging manufacturers, etc. – to commit themselves to using materials and production processes that do not harm the environment. The paper in this book is made from low- or no-chlorine pulp and is acid free, in conformance with international standards for paper permanency.

Lecture Notes in Computer Science

For information about Vols. 1–1371

please contact your bookseller or Springer-Verlag

Vol. 1372: S. Vaudenay (Ed.), Fast Software Encryption. Proceedings, 1998. VIII, 297 pages. 1998.

Vol. 1373: M. Morvan, C. Meinel, D. Krob (Eds.), STACS 98. Proceedings, 1998. XV, 630 pages. 1998.

Vol. 1374: H. Bunt, R.-J. Beun, T. Borghuis (Eds.), Multimodal Human-Computer Communication. VIII, 345 pages. 1998. (Subseries LNAI).

Vol. 1375: R. D. Hersch, J. André, H. Brown (Eds.), Electronic Publishing, Artistic Imaging, and Digital Typography. Proceedings, 1998. XIII, 575 pages. 1998.

Vol. 1376: F. Parisi Presicce (Ed.), Recent Trends in Algebraic Development Techniques. Proceedings, 1997. VIII, 435 pages. 1998.

Vol. 1377: H.-J. Schek, F. Saltor, I. Ramos, G. Alonso (Eds.), Advances in Database Technology – EDBT'98. Proceedings, 1998. XII, 515 pages. 1998.

Vol. 1378: M. Nivat (Ed.), Foundations of Software Science and Computation Structures. Proceedings, 1998. X, 289 pages. 1998.

Vol. 1379: T. Nipkow (Ed.), Rewriting Techniques and Applications. Proceedings, 1998. X, 343 pages. 1998.

Vol. 1380: C.L. Lucchesi, A.V. Moura (Eds.), LATIN'98: Theoretical Informatics. Proceedings, 1998. XI, 391 pages. 1998.

Vol. 1381: C. Hankin (Ed.), Programming Languages and Systems. Proceedings, 1998. X, 283 pages. 1998.

Vol. 1382: E. Astesiano (Ed.), Fundamental Approaches to Software Engineering. Proceedings, 1998. XII, 331 pages. 1998.

Vol. 1383: K. Koskimies (Ed.), Compiler Construction. Proceedings, 1998. X, 309 pages. 1998.

Vol. 1384: B. Steffen (Ed.), Tools and Algorithms for the Construction and Analysis of Systems. Proceedings, 1998. XIII, 457 pages. 1998.

Vol. 1385: T. Margaria, B. Steffen, R. Rückert, J. Posegga (Eds.), Services and Visualization. Proceedings, 1997/1998. XII, 323 pages. 1998.

Vol. 1386: T.A. Henzinger, S. Sastry (Eds.), Hybrid Systems: Computation and Control. Proceedings, 1998. VIII, 417 pages. 1998.

Vol. 1387: C. Lee Giles, M. Gori (Eds.), Adaptive Processing of Sequences and Data Structures. Proceedings, 1997. XII, 434 pages. 1998. (Subseries LNAI).

Vol. 1388: J. Rolim (Ed.), Parallel and Distributed Processing. Proceedings, 1998. XVII, 1168 pages. 1998.

Vol. 1389: K. Tombre, A.K. Chhabra (Eds.), Graphics Recognition. Proceedings, 1997. XII, 421 pages. 1998.

Vol. 1390: C. Scheideler, Universal Routing Strategies for Interconnection Networks. XVII, 234 pages. 1998.

Vol. 1391: W. Banzhaf, R. Poli, M. Schoenauer, T.C. Fogarty (Eds.), Genetic Programming. Proceedings, 1998. X, 232 pages. 1998.

Vol. 1392: A. Barth, M. Breu, A. Endres, A. de Kemp (Eds.), Digital Libraries in Computer Science: The MeDoc Approach. VIII, 239 pages. 1998.

Vol. 1393: D. Bert (Ed.), B'98: Recent Advances in the Development and Use of the B Method. Proceedings, 1998. VIII, 313 pages. 1998.

Vol. 1394: X. Wu. R. Kotagiri, K.B. Korb (Eds.), Research and Development in Knowledge Discovery and Data Mining. Proceedings, 1998. XVI, 424 pages. 1998. (Subseries LNAI).

Vol. 1395: H. Kitano (Ed.), RoboCup-97: Robot Soccer World Cup I. XIV, 520 pages. 1998. (Subseries LNAI).

Vol. 1396: E. Okamoto, G. Davida, M. Mambo (Eds.), Information Security. Proceedings, 1997. XII, 357 pages. 1998.

Vol. 1397: H. de Swart (Ed.), Automated Reasoning with Analytic Tableaux and Related Methods. Proceedings, 1998. X, 325 pages. 1998. (Subseries LNAI).

Vol. 1398: C. Nédellec, C. Rouveirol (Eds.), Machine Learning: ECML-98. Proceedings, 1998. XII, 420 pages. 1998. (Subseries LNAI).

Vol. 1399: O. Etzion, S. Jajodia, S. Sripada (Eds.), Temporal Databases: Research and Practice. X, 429 pages. 1998.

Vol. 1400: M. Lenz, B. Bartsch-Spörl, H.-D. Burkhard, S. Wess (Eds.), Case-Based Reasoning Technology. XVIII, 405 pages. 1998. (Subseries LNAI).

Vol. 1401: P. Sloot, M. Bubak, B. Hertzberger (Eds.), High-Performance Computing and Networking. Proceedings, 1998. XX, 1309 pages. 1998.

Vol. 1402: W. Lamersdorf, M. Merz (Eds.), Trends in Distributed Systems for Electronic Commerce. Proceedings, 1998. XII, 255 pages. 1998.

Vol. 1403: K. Nyberg (Ed.), Advances in Cryptology – EUROCRYPT '98. Proceedings, 1998. X, 607 pages. 1998.

Vol. 1404: C. Freksa, C. Habel. K.F. Wender (Eds.), Spatial Cognition. VIII, 491 pages. 1998. (Subseries LNAI).

Vol. 1405: S.M. Embury, N.J. Fiddian, W.A. Gray, A.C. Jones (Eds.), Advances in Databases. Proceedings, 1998. XII, 183 pages. 1998.

Vol. 1406: H. Burkhardt, B. Neumann (Eds.), Computer Vision – ECCV'98. Vol. I. Proceedings, 1998. XVI, 927 pages. 1998.

Vol. 1407: H. Burkhardt, B. Neumann (Eds.), Computer Vision – ECCV'98. Vol. II. Proceedings, 1998. XVI, 881 pages. 1998.

Vol. 1409: T. Schaub, The Automation of Reasoning with Incomplete Information. XI, 159 pages. 1998. (Subseries LNAI).

Vol. 1411: L. Asplund (Ed.), Reliable Software Technologies – Ada-Europe. Proceedings, 1998. XI, 297 pages. 1998.

Vol. 1412: R.E. Bixby, E.A. Boyd, R.Z. Ri ´´os-Mercado (Eds.), Integer Programming and Combinatorial Optimization. Proceedings, 1998. IX, 437 pages. 1998.

Vol. 1413: B. Pernici, C. Thanos (Eds.), Advanced Information Systems Engineering. Proceedings, 1998. X, 423 pages. 1998.

Vol. 1414: M. Nielsen, W. Thomas (Eds.), Computer Science Logic. Selected Papers, 1997. VIII, 511 pages. 1998.

Vol. 1415: J. Mira, A.P. del Pobil, M.Ali (Eds.), Methodology and Tools in Knowledge-Based Systems. Vol. I. Proceedings, 1998. XXIV, 887 pages. 1998. (Subseries LNAI).

Vol. 1416: A.P. del Pobil, J. Mira, M.Ali (Eds.), Tasks and Methods in Applied Artificial Intelligence. Vol.II. Proceedings, 1998. XXIII, 943 pages. 1998. (Subseries LNAI).

Vol. 1417: S. Yalamanchili, J. Duato (Eds.), Parallel Computer Routing and Communication. Proceedings, 1997. XII, 309 pages. 1998.

Vol. 1418: R. Mercer, E. Neufeld (Eds.), Advances in Artificial Intelligence. Proceedings, 1998. XII, 467 pages. 1998. (Subseries LNAI).

Vol. 1419: G. Vigna (Ed.), Mobile Agents and Security. XII, 257 pages. 1998.

Vol. 1420: J. Desel, M. Silva (Eds.), Application and Theory of Petri Nets 1998. Proceedings, 1998. VIII, 385 pages. 1998.

Vol. 1421: C. Kirchner, H. Kirchner (Eds.), Automated Deduction – CADE-15. Proceedings, 1998. XIV, 443 pages. 1998. (Subseries LNAI).

Vol. 1422: J. Jeuring (Ed.), Mathematics of Program Construction. Proceedings, 1998. X, 383 pages. 1998.

Vol. 1423: J.P. Buhler (Ed.), Algorithmic Number Theory. Proceedings, 1998. X, 640 pages. 1998.

Vol. 1424: L. Polkowski, A. Skowron (Eds.), Rough Sets and Current Trends in Computing. Proceedings, 1998. XIII, 626 pages. 1998. (Subseries LNAI).

Vol. 1425: D. Hutchison, R. Schäfer (Eds.), Multimedia Applications, Services and Techniques – ECMAST'98. Proceedings, 1998. XVI, 532 pages. 1998.

Vol. 1427: A.J. Hu, M.Y. Vardi (Eds.), Computer Aided Verification. Proceedings, 1998. IX, 552 pages. 1998.

Vol. 1430: S. Trigila, A. Mullery, M. Campolargo, H. Vanderstraeten, M. Mampaey (Eds.), Intelligence in Services and Networks: Technology for Ubiquitous Telecom Services. Proceedings, 1998. XII, 550 pages. 1998.

Vol. 1431: H. Imai, Y. Zheng (Eds.), Public Key Cryptography. Proceedings, 1998. XI, 263 pages. 1998.

Vol. 1432: S. Arnborg, L. Ivansson (Eds.), Algorithm Theory – SWAT '98. Proceedings, 1998. IX, 347 pages. 1998.

Vol. 1433: V. Honavar, G. Slutzki (Eds.), Grammatical Inference. Proceedings, 1998. X, 271 pages. 1998. (Subseries LNAI).

Vol. 1434: J.-C. Heudin (Ed.), Virtual Worlds. Proceedings, 1998. XII, 412 pages. 1998. (Subseries LNAI).

Vol. 1435: M. Klusch, G. Weiß (Eds.), Cooperative Information Agents II. Proceedings, 1998. IX, 307 pages. 1998. (Subseries LNAI).

Vol. 1436: D. Wood, S. Yu (Eds.), Automata Implementation. Proceedings, 1997. VIII, 253 pages. 1998.

Vol. 1437: S. Albayrak, F.J. Garijo (Eds.), Intelligent Agents for Telecommunication Applications. Proceedings, 1998. XII, 251 pages. 1998. (Subseries LNAI).

Vol. 1438: C. Boyd, E. Dawson (Eds.), Information Security and Privacy. Proceedings, 1998. XI, 423 pages. 1998.

Vol. 1439: B. Magnusson (Ed.), System Configuration Management. Proceedings, 1998. X, 207 pages. 1998.

Vol. 1441: W. Wobcke, M. Pagnucco, C. Zhang (Eds.), Agents and Multi-Agent Systems. Proceedings, 1997. XII, 241 pages. 1998. (Subseries LNAI).

Vol. 1443: K.G. Larsen, S. Skyum, G. Winskel (Eds.), Automata, Languages and Programming. Proceedings, 1998. XVI, 932 pages. 1998.

Vol. 1444: K. Jansen, J. Rolim (Eds.), Approximation Algorithms for Combinatorial Optimization. Proceedings, 1998. VIII, 201 pages. 1998.

Vol. 1445: E. Jul (Ed.), ECOOP'98 – Object-Oriented Programming. Proceedings, 1998. XII, 635 pages. 1998.

Vol. 1446: D. Page (Ed.), Inductive Logic Programming. Proceedings, 1998. VIII, 301 pages. 1998. (Subseries LNAI).

Vol. 1448: M. Farach-Colton (Ed.), Combinatorial Pattern Matching. Proceedings, 1998. VIII, 251 pages. 1998.

Vol. 1449: W.-L. Hsu, M.-Y. Kao (Eds.), Computing and Combinatorics. Proceedings, 1998. XII, 372 pages. 1998.

Vol. 1451: A. Amin, D. Dori, P. Pudil, H. Freeman (Eds.), Advances in Pattern Recognition. Proceedings, 1998. XXI, 1048 pages. 1998.

Vol. 1452: B.P. Goettl, H.M. Halff, C.L. Redfield, V.J. Shute (Eds.), Intelligent Tutoring Systems. Proceedings, 1998. XIX, 629 pages. 1998.

Vol. 1453: M.-L. Mugnier, M. Chein (Eds.), Conceptual Structures: Theory, Tools and Applications. Proceedings, 1998. XIII, 439 pages. (Subseries LNAI).

Vol. 1454: I. Smith (Ed.), Artificial Intelligence in Structural Engineering. XI, 497 pages. 1998. (Subseries LNAI).

Vol. 1456: A. Drogoul, M. Tambe, T. Fukuda (Eds.), Collective Robotics. Proceedings, 1998. VII, 161 pages. 1998. (Subseries LNAI).

Vol. 1457: A. Ferreira, J. Rolim, H. Simon, S.-H. Teng (Eds.), Solving Irregularly Structured Problems in Prallel. Proceedings, 1998. X, 408 pages. 1998.

Vol. 1458: V.O. Mittal, H.A. Yanco, J. Aronis, R-. Simpson (Eds.), Assistive Technology in Artificial Intelligence. X, 273 pages. 1998. (Subseries LNAI).

Vol. 1461: G. Bilardi, G.F. Italiano, A. Pietracaprina, G. Pucci (Eds.), Algorithms – ESA'98. Proceedings, 1998. XII, 516 pages. 1998.

Vol. 1464: H.H.S. Ip, A.W.M. Smeulders (Eds.), Multimedia Information Analysis and Retrieval. Proceedings, 1998. VIII, 264 pages. 1998.